SOLID-STATE RADIATION DETECTORS

Technology and Applications

Devices, Circuits, and Systems

Series Editor
Krzysztof Iniewski
CMOS Emerging Technologies Research Inc.,
Vancouver, British Columbia, Canada

SOLID-STATE RADIATION DETECTORS

Technology and Applications

EDITED BY

SALAH AWADALLA
TAIBAH UNIVERSITY, YANBU, SAUDI ARABIA

KRZYSZTOF INIEWSKI MANAGING EDITOR
CMOS EMERGING TECHNOLOGIES RESEARCH INC.
VANCOUVER, BRITISH COLUMBIA, CANADA

CRC Press
Taylor & Francis Group
Boca Raton London New York

CRC Press is an imprint of the
Taylor & Francis Group, an **informa** business

CRC Press
Taylor & Francis Group
6000 Broken Sound Parkway NW, Suite 300
Boca Raton, FL 33487-2742

First issued in paperback 2019

© 2015 by Taylor & Francis Group, LLC
CRC Press is an imprint of Taylor & Francis Group, an Informa business

No claim to original U.S. Government works

ISBN-13: 978-1-4822-6220-9 (hbk)
ISBN-13: 978-0-367-37777-9 (pbk)

Visit the Taylor & Francis Web site at
http://www.taylorandfrancis.com

and the CRC Press Web site at
http://www.crcpress.com

Contents

Preface

This book reviews the present and emerging material-based technologies for radiation detectors. The goal is to assist institutions and research organizations in increasing the speed of developing cutting-edge radiation detector technology for medical imaging and other applications. To achieve this objective, each chapter is authored by a leading expert(s) on the respective materials and technology and is structured to address the current advantages and challenges and discuss novel research and applications of those materials and technology. The book contains 15 chapters. Chapter 1 covers modern semiconductors used for radiation monitoring. Chapters 2 through 5 discuss CdZnTe and CdTe technology for imaging applications including three-dimensional capability detectors. Chapter 6 highlights interconnect technology for current pixel detectors, while Chapters 7, 9, and 10 deal with hybrid pixel detectors and their characterizations. Chapter 8 handles the integrated analog signal processing read-out front ends for particle detectors. Chapter 11 reviews new organic materials with direct bandgap for direct energy detection, then Chapter 12 summarizes recent developments involving lanthanum halide and cerium bromide scintillators. Chapter 13 explores the potential of recent progress in the field of crystallogenesis, quantum dots, and photonics crystals toward a new concept of x- and gamma-ray detectors based on metamaterials. The last two chapters discuss the advantages and challenges of position-sensitivity photomultipliers and silicon photomultipliers for scintillation crystals, respectively.

Salah A. Awadalla

Editors

Salah Awadalla obtained his PhD from the Department of Mechanical and Materials Science and Engineering at Washington State University. Dr. Awadalla is currently a professor at the University of Taibah, Yanbu Branch, Saudi Arabia, in the Department of Mechanical Engineering. In addition, he is the director of Teaching Laboratories and the Research and Development Program. In his professional career, Dr. Awadalla has held numerous positions ranging from staff scientist to manager and director at Washington State University and Redlen Technologies. Dr. Awadalla has published more than 50 papers in the field of cadmium zinc telluride for medical and security applications and holds more than eight U.S.-granted patents in the field of cadmium zinc telluride fabrication and processing for medical imaging. Dr. Awadalla can be reached at salahawadalla@yahoo.com.

Krzysztof (Kris) Iniewski is managing R&D at Redlen Technologies, a start-up company in Vancouver, Canada. Redlen's revolutionary production process for advanced semiconductor materials enables a new generation of more accurate, all-digital, radiation-based imaging solutions. Kris is also president of CMOS Emerging Technologies Research Inc. (www.cmosetr.com), an organization of high-tech events covering communications, microsystems, optoelectronics, and sensors. In his career, Dr. Iniewski has held numerous faculty and management positions at the University of Toronto, the University of Alberta, Simon Fraser University, and PMC-Sierra Inc. He has published over 100 research papers in international journals and conferences. He holds 18 international patents granted in the United States, Canada, France, Germany, and Japan. He is a frequent invited speaker and has consulted for multiple organizations internationally. He has written and edited several books for CRC Press, Cambridge University Press, IEEE Press, Wiley, McGraw-Hill, Artech House, and Springer. His personal goal is to contribute to healthy living and sustainability through innovative engineering solutions. In his leisurely time, Kris can be found hiking, sailing, skiing, or biking in beautiful British Columbia. He can be reached at kris.iniewski@gmail.com.

Contributors

Raja Aamir
Department of Radiology
University of Otago
Christchurch, New Zealand

Nigel G. Anderson
Department of Radiology
University of Otago
Christchurch, New Zealand

Anthony P. H. Butler
MARS Bioimaging Ltd
Christchurch, New Zealand

Phil H. Butler
Physics Department
University of Canterbury

and

MARS Bioimaging Ltd
Christchurch, New Zealand

and

CERN
Geneva, Switzerland

Ezio Caroli
INAF/IASF Bologna
Bologna, Italy

Stefano del Sordo
INAF/IASF Palermo
Palermo, Italy

Sébastien Dubos
Irfu, CEA Saclay
Gif-sur-Yvette, France

Beatrice Fraboni
Department of Physics and Astronomy
University of Bologna
Bologna, Italy

Alessandro Fraleoni-Morgera
Department of Engineering and
    Architecture
University of Trieste

and

Elettra Sincrotrone Trieste SCpA
Trieste, Italy

and

CNR-NanoS3
Modena, Italy

Erik Fröjdh
CERN
Geneva, Switzerland

Olivier Gevin
Irfu, CEA Saclay
Gif-sur-Yvette, France

Eva N. Gimenez
Diamond Light Source
Didcot, United Kingdom

Son Hoang
Department of Physics
University of Houston
Houston, Texas

J. Idarraga-Munoz
Department of Physics
University of Houston
Houston, Texas

Carl Jackson
SensL Technologies Ltd
Cork, Ireland

Martin Kroupa
Department of Physics
University of Houston
Houston, Texas

Paul Lecoq
CERN
Geneva, Switzerland

Olivier Limousin
Irfu, CEA Saclay
Gif-sur-Yvette, France

Brian McGarvey
SensL Technologies Ltd
Cork, Ireland

Aline Meuris
Irfu, CEA Saclay
Gif-sur-Yvette, France

Thomas Noulis
Intel Mobile Communications GmbH
Neubiberg, Germany

Kevin O'Neill
SensL Technologies Ltd
Cork, Ireland

Raj K. Panta
Department of Radiology
University of Otago
Christchurch, New Zealand

Lawrence S. Pinsky
Department of Physics
University of Houston
Houston, Texas

Michael Prokesch
Semiconductor Technology
eV Products, Inc.
Saxonburg, Pennsylvania

Kishore Rajendran
Department of Radiology
University of Otago
Christchurch, New Zealand

Oliver J. Roberts
Space Science and Advanced Materials
    Research Group
University College Dublin
Belfield, Ireland

Michael Seimetz
Institute for Instrumentation in
    Molecular Imaging (I3M)
Polytechnic University of Valencia/CSIC
Valencia, Spain

Paul Seller
Science and Technology Facilities Council
Rutherford Appleton Laboratory
Didcot, United Kingdom

Nicholas N. Stoffle
Department of Physics
University of Houston
Houston, Texas

Matthew C. Veale
Science and Technology Facilities Council
Rutherford Appleton Laboratory
Didcot, United Kingdom

Liam Wall
SensL Technologies Ltd
Cork, Ireland

Michael F. Walsh
MARS Bioimaging Limited
Christchurch, New Zealand

1 Modern Semiconductor Pixel Detectors Used as Radiation Monitors

Martin Kroupa, Nicholas N. Stoffle, Son Hoang, J. Idarraga-Munoz, and Lawrence S. Pinsky

CONTENTS

1.1 INTRODUCTION

In our daily lives, radiation cannot be seen,* smelled, or heard; that might be why it retains an aura of mystery. The truth is, radiation is all around us, and we are exposed to it on a daily basis. Indeed, the evolution of life on Earth may have actually required radiation [2]. However, when radiation levels exceed values to which the human body is adapted, it can present significant health hazards [3]. Generally speaking, radiation measurement and monitoring on Earth is required only if some special event occurs, or in places where we deal with increased radioactivity on a

---

* Nevertheless, astronauts do see "light flashes" during their time in orbit or in space. Some have reported associated directional components as well. This phenomenon has been known since the time of the Apollo mission [1]. However, it is caused by high-energy particles crossing the eye and its surroundings. If you would like to have the same experience here on Earth, your eye would have to be irradiated by some heavy ion accelerator.

1

daily basis (e.g., nuclear power plants). There is, however, a place where radiation monitoring is a necessity—space [4]. Normally, humankind is protected from the radiation found throughout space by Earth's atmosphere and magnetic field. What would happen to us if we did not have this protection? Also, can we establish and minimize the risks associated with exposure to space radiation? These questions have to be answered if humankind is to expand its influence to other planets or to the stars beyond.

Radiation dosimetry is the science dealing with the accurate determination of the energy deposited in a material, such as living tissue, by radiation. Properly establishing the threat that radiation, ionizing or otherwise, poses to humans is a daunting task due to the complexity of the interaction between radiation and matter. Not only are there many different types of radiation, but the radiation in space has a continuous energy spectrum, ranging from energies that can be shielded by a single piece of paper to incredible energies that represent orders of magnitude higher than those we can create in the largest accelerators on Earth [5,6]. How radiation interacts with matter also depends on the material composition, and thus a detailed calculation is needed when information from a silicon-based detector is used to assess effects on human tissue.

Such complexities make dosimetry a complex, interdisciplinary field, wherein there is always a demand for new radiation monitors. Developments in microelectronics open new possibilities in this area, and development of leading-edge radiation monitors based on pixel-detector technologies is currently under way. These new monitors will be used for future crewed space missions, as well as for portable radiation monitors, which may be used both within the crewed space program and back on Earth.

1.2 RADIATION IN SPACE

Over the last several decades, since humans first ventured into space, our understanding of the ionizing radiation environment near Earth and within the solar system has expanded, but this understanding is far from complete. Several models for the radiation spectra at Earth have been developed, based on data gathered both from terrestrial measurements and from satellites in orbit around our planet as well as from interplanetary probes [7–9]. One of the biggest obstacles for all future crewed missions traveling beyond the protection of Earth's magnetic field will be the radiation environment [10]. Without the protection of Earth's magnetic field and atmosphere, radiation represents a very significant threat to astronauts' health. Moreover, the radiation in space is not constant in time. For example, solar particle events (SPEs) that erupt from the sun can significantly increase the radiation in the interplanetary environment and have the potential to seriously impact, or possibly cause the death of, insufficiently protected astronauts on interplanetary or deep-space missions. Thus, fast and precise measurement of the radiation field is one of the priorities for all future missions.

1.2.1 RADIATION NEAR EARTH

Earth's magnetic field and atmosphere provide terrestrial life with a robust shield from cosmic and solar radiation, though relatively small amounts of such radiation do reach sea level. For example, muons and neutron cascades are measurable, along with a range of other particle types, at ground-level observation stations. As altitude increases, the amount of matter for particles to interact with decreases. Hence, the dose from ionizing radiation increases with altitude, and the background radiation dose at sea level is lower than the background measured at higher elevations [11]. The surface radiation levels also decrease with latitude, the equatorial regions having lower surface radiation levels.

Earth's magnetic field can be thought of as a dipole field, which is tilted and offset relative to Earth's spin axis. Charged particles become trapped in the geomagnetic field within regions known as the Van Allen Belts. The offset and tilt of the geomagnetic field give rise to the South Atlantic Anomaly, a region of increased radiation in low Earth orbit (LEO) where the inner Van Allen Belt, containing primarily trapped protons, is nearest the surface of Earth [12,13].

At LEO altitudes (altitudes between 160 and 2000 km), the radiation field has a distinct separation in components based on geographic location. Trapped electrons populate high-latitude regions over North America and above the Southern Indian Ocean, as well as in regions near the magnetic poles. These are regions where the trapped electrons can reach LEO altitudes as they bounce between north–south magnetic mirror points. These regions are also populated by those galactic cosmic rays with sufficient energy to penetrate the magnetic field.

Closer to the planet's equator, the magnetic field is more effective at similar altitudes, resulting in only the higher-energy cosmic rays being able to penetrate to LEO altitudes. In addition, geomagnetic cusp regions allow access of solar particles to LEO altitudes. Geomagnetic cusp regions are regions at higher latitudes where geomagnetic field lines have been opened to the interplanetary magnetic field through interaction with the solar wind [14,15].

Solar phenomena, such as coronal mass ejections (CMEs) and proton events, also have an impact on radiation components in LEO. CMEs are shock fronts in the interplanetary medium composed of plasma swept up following a solar eruption. When such a shock passes Earth, it can cause disturbances in the geomagnetic field, which have the ability to cause variations in the radiation belt location and composition. The result is a widening of the areas of effect associated with the magnetic field and, in some cases, the formation of temporary belts of trapped particles [12].

SPEs are also a concern at LEO. High-energy protons and other solar products are accelerated toward Earth as a result of disturbances or eruptions in the Sun's corona [16,17]. The high-energy protons arrive at Earth within minutes to hours and can cause a dramatic increase in both the energy spectrum and in the overall proton flux [18]. If a vehicle is not well shielded by the geomagnetic field, SPEs can result in greatly increased radiation exposure relative to quiescent periods [16,19]. For deep-space crewed missions outside Earth's magnetic field, SPEs can represent potentially fatal radiation risks in the absence of sufficient shielding to protect the crew.

Finally, galactic cosmic rays (GCR) comprise a fully ionized background in the space radiation field with a component that extends well into relativistic energies and a nuclear composition that ranges from electrons and protons through iron and beyond. Thought to have been initially accelerated in supernova remnants and other interstellar media and further accelerated through various processes during propagation to our solar system, these relativistic ions are the most difficult portion of the space radiation field against which to shield [16,20]. While GCRs are a relatively small portion of the particle flux in LEO, it represents a large fraction of the biologically significant radiation exposure to spacefarers. This, combined with the difficulty in shielding against GCRs, presents a unique problem as humans transition into crewed interplanetary exploration [16,21].

In summary, any generic radiation monitor that attempts to assess the content of the radiation environment needs to be sufficiently sensitive to nuclei from protons through iron with energies from tens of megaelectronvolts to fully relativistic, as well as lower-energy electrons up to tens of megaelectronvolts. No mention has been made of neutrons, as there are essentially no primary neutrons, but there are albedo neutrons that are produced by interactions of the primary nuclei with other objects like Earth's atmosphere and planetary surfaces, and there are also secondary neutrons produced in interactions with the spacecrafts themselves.

1.3   RADIATION MONITORS

Current radiation monitoring hardware used in space exploration falls into two categories: passive and active. Passive instrumentation, such as thermoluminescent detectors, collects energy from the radiation field to which it is exposed and does not provide a signal until the detector is processed to extract the desired latent information [22]. Active instrumentation, however, provides a data stream that can be used immediately or stored for future analysis.

Passive detectors used in space dosimetry applications are well understood and are heavily relied upon in operational radiation dosimetry, especially to provide statutory records. The methods and materials have been well developed, and the material responses to radiation are understood to a high degree [23]. This translates into a reliable set of results with which to assess radiation exposures and related health risks. The downside, however, is the significant time delay between the exposure and obtaining the result, and the nonlinearity of the response to the details of the composition of the incident radiation. This is especially true for spaceborne instrumentation, which may be deployed for months before the data are extracted from the detector.

Active instrumentation based on proportional counters, such as the tissue equivalent proportional counter (TEPC), which is currently used for operational dosimetry on the International Space Station (ISS), is built on well-developed instrumentation technology [22,24]. While these instruments give real-time (or nearly so) feedback relating to the radiation exposures in a vehicle, such instruments rely on estimation of linear (or lineal) energy transfer (LET) into dose and dose equivalent values for use in assessing the dosimetric quantities of interest in the space radiation environment. The limitation is that TEPCs have a sensitive gas volume that is typically cylindrical with no tracking information. The net output is simply the total integrated charge deposited in the gas, which does directly yield the absorbed dose in an equivalent mass of tissue. However, without any measurement of the path length, only average or mean LET can be estimated with assumptions like isotropic illumination.

The benefit of TEPCs is that no correction for the material is needed as it is "tissue equivalent," unlike Si-based detectors.

Existing instrumentation based on solid-state particle telescope technology provides time-resolved energy and charge spectra of the local radiation fields. However, such instruments have large mass and power-draw requirements, which impact the overall launch and power budget available to space-based platforms.

Modern pixel-detector technology presents a novel approach to space radiation dosimetry [25]. Pixel-detector hardware can be calibrated to provide segmented-grid measurements of energy deposition across a solid-state detector for planar-type detectors, or throughout the detector volume for three-dimensional pixel-detector technologies. Such detectors are small and low-power devices (approximately 2 cm^2 and 0.5 W, respectively, for Timepix silicon detectors) in comparison to the cubic feet and tens of watts required to operate existing active space radiation detectors.

In addition to size and power advantages, the spatial and temporal resolution provided by existing and emergent pixel-detector technologies provides information previously not available to active detectors in a space environment. The spatial distribution of energy deposition is available within the pixel-detector data, and coupling the spatial information with the temporal information allows individual particle tracks to be identified and analyzed, potentially providing the most complete characterization of the radiation field.

1.4 NEW RADIATION ENVIRONMENT MONITOR BASED ON PIXEL-DETECTOR TECHNOLOGY

The great advances in microelectronics allow the creation of more sophisticated structures in the front-end electronics of modern pixel detectors. Although many institutions and companies are developing new devices, they all can be divided into two categories—charge-integrating devices or single-quantum-counting devices. These two approaches are quite different, and both have pros and cons.

1.4.1 CHARGE-INTEGRATING DEVICES

Examples of charge-integrating devices are charge-coupled devices (CCDs) (such as those found in earlier digital cameras), flat panels, and complementary metal–oxide–semiconductor (CMOS) pixel detectors. These devices integrate deposited charge within each pixel during measurement. This charge is then measured on the device output capacitance. The CCD was invented in 1969 at AT&T Bell Labs by Willard Boyle and George E. Smith [22]. The principle of the device is rather simple; a CCD is made of silicon, where each pixel contains a potential well. When the charge is collected in the pixel, it is stored in this well. When the device is read out, the charge is shifted along the row to the charge amplifier and the other read-out electronics. The output of the detector is the analog signal corresponding to the total charge integrated in the pixel during measurement.

Flat-panel detectors are based on the technology used in LCD monitors. A scintillator layer made of gadolinium oxysulfide or cesium iodide converts x-rays to visible light. This visible light is then detected by an array of pixels with photodiodes made of amorphous silicon placed on glass. The advantage of this type of detector is the possibility of building detectors with large areas. The disadvantage is limited spatial resolution resulting from the nonzero thickness of the scintillator.

Because the output of charge-integrating devices is an analog signal, such devices are not energy sensitive; that is, they cannot be used for energy measurements.

The latest CMOS pixel devices are a combination of bulk sensor material with embedded transistor structures that can store the collected charge locally. They can be addressed in sequence to supply the charge to a common read-out bus in sequence, where the analog charge is digitized externally. These devices still digitize the collected charge externally, but the charge from each pixel is provided directly to the external digitizer without being shifted through the intervening pixels.

1.4.2 Single-Quantum Counting Devices

Single-quantum counting (SQC) devices differ from charge-integrating devices in many significant ways. Charge-integrating devices collect the charge throughout the acquisition duration. This charge is processed and analyzed, giving rough information about how many quanta were detected (at least relative to other pixels). SQC devices, as the name indicates, process the signal from each quantum separately. This is a result of SQC devices having more sophisticated electronics positioned in each pixel. The added front-end electronics process signals from each event and convert them into a digital number representing the event. The number of measured quanta is exact.*

Hybrid detector technology, developed within the field of high-energy physics, is a new detector concept wherein the detector is composed of two parts—a sensor and a read-out chip. These two components are bump bonded together, and both can be optimized separately with respect to the application. Ionizing particles crossing the sensor component create electron–hole pairs. The electrons and holes are separated by an applied reverse bias voltage, allowing the charge carriers to be collected by the read-out chip. The collected charge is then processed in the front-end electronics of the appropriate pixels.

One of the advantages of the hybrid technology is that the sensor can be made from a different material than the read-out chip according to the application requirements. For example, x-ray imaging could benefit from the use of cadmium telluride (CdTe) sensors, which have much higher detection efficiency than silicon for x-rays above a few kiloelectronvolts. Silicon sensors can be made much thicker and from silicon with different properties (highly resistive silicon) as compared to common monolithic detectors. Figure 1.1 shows a schematic illustration of a hybrid pixel detector.

---

* Omitting pileups. It can be shown that because of the rather short time necessary for signal shaping and small pixel size, modern single-quantum counting devices can operate in very high fluxes without pileups producing significant errors.

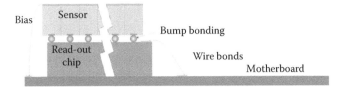

FIGURE 1.1 **(See color insert)** The hybrid detector consists of two independent parts: a sensor and a read-out chip. These two parts are joined together by bump bonding. The advantage of this technology is that the sensor material can be tailored to the specific application.

1.4.3 TIMEPIX DETECTOR

The Timepix detector is semiconductor pixel detector based on the hybrid detector technology (Figure 1.2) [26]. It was developed by the Medipix2 collaboration, consisting of more than 20 institutions around the world and was manufactured by IBM.

The sensor is equipped with a single common backside electrode, while the frontside is composed of a matrix of electrodes (256×256 square pixels with a pitch of 55 µm), giving it an active area of 2 cm^2. Each pixel on the sensor is bump bonded to the pixel pad on the read-out chip. This arrangement means that every pixel behaves as an independent detector, each with its own devoted read-out electronics. The front-end electronics within each pixel are divided into two parts: analog and digital. The analog portion collects, amplifies, and shapes the signal and is able to process signals from either electrons or holes. Typically, detectors bump bonded to either 300 or 500 µm thick n-type silicon sensors are used, resulting in the collection of holes to produce the signal. Hence, all signals from the preamplifier shown hereafter are negative. It should also be mentioned that the term time over threshold (TOT) is commonly used, independent of the sign of the charge collected.

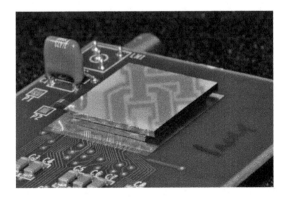

FIGURE 1.2 The detail of the Medipix/Timepix detector with a 1 mm Si sensor mounted on a standard probe card. The sensor backside is typically covered by a thin aluminum layer to provide better ohmic contact. The aluminum layer here reflects the logo of the University of Houston, which, together with the Institute of Experimental and Applied Physics in Prague and the SRAG group at NASA JSC, is responsible for pioneering pixel-detector usage for space dosimetry.

The parameters of the signal shaping can be modified by a set of digital-to-analog converters (DACs), which are used to modify the internal settings of the detector. The collected signal is compared to a set threshold level in the comparator at the end of the analog part of the pixel electronics chain. The output from each pixel, then, is a digital number, which differs depending on the mode to which the Timepix was set. The Timepix pixel can be set to one of four modes: Medipix mode, Timepix mode, one-hit mode, and TOT mode. The counter is incremented every time the signal exceeds threshold in the Medipix mode—it is essentially a counting mode. Timepix mode uses an internal clock on the chip to measure the time at which the signal crosses the threshold—this is also called time of arrival mode. The one-hit mode gives binary information indicating whether there was any signal above the threshold within the pixel (1) or not (0). The TOT mode measures how long the signal was above the threshold by counting clock pulses during this time. The counting is stopped when the signal falls again below threshold (see Figure 1.3). The time the pulse remains above the threshold corresponds to the charge deposited in the pixel. This is essentially an energy measurement mode, since application of a device-specific calibration yields the energy per pixel from the TOT measurement. In TOT mode, each pixel operates as an independent multi-channel analyzer. The energy calibration procedure must be done for each pixel, but once finished, the Timepix is capable of measuring energy deposition in each pixel.

The Timepix is thus well suited for measurements where the precise position coupled with energy or time information are required [27,28]. What all modes have in common, and what is most important from an imaging point of view, is that the threshold used in the comparator of the pixel can be always set to a value above the noise of the electronics. Together with an active dark-current suppression circuit

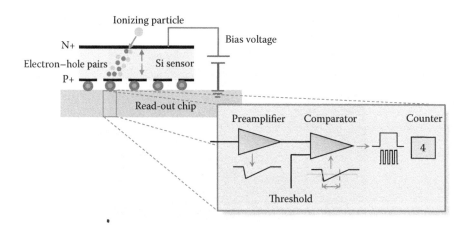

FIGURE 1.3 (**See color insert**) The schematic illustration of TOT mode. The signal from the electrode is amplified and compared in the comparator. The time the signal is over the comparator level (threshold) represents the energy of the interacting particle. The collected charge is compared in each pixel with the threshold level. If it is lower, the event is not registered and the signal is lost. As a consequence, the charge registered by all pixels is often lower than its original value, which deteriorates the energy resolution.

within each pixel, this means that, unlike a CCD, the Timepix detector data is independent of dark current; that is, the time of measurement can be virtually limitless. This also means that the Timepix detector is able to detect each particle that deposits charge above threshold—in other words, the Timepix is a single-quantum counting detector. Moreover, by analyzing Timepix data, one can obtain additional information about the measured quanta of radiation. The Timepix detector can be used to observe tracks of interacting particles, allowing it to act as a solid-state bubble chamber. Common image-processing algorithms can be used on the Timepix data to identify different types of radiation and the associated particle properties.

1.4.3.1 Energy Calibration

The TOT mode allows the measurement of the length of time the signal is above the threshold. This time corresponds to the height of the signal; that is, to the charge deposited in the pixel. As was mentioned previously, each pixel has its own analog electronics. Hence, the signal response can differ from pixel to pixel. The signal from an interacting particle can be detected in adjacent pixels. The differences between signals from different pixels results in a systematic error when the signals from several pixels are added together. In order to obtain a response that is as homogenous as possible from all pixels, some calibration procedures have to be carried out.

The energy resolution of the pixelated detector is influenced by the charge-sharing effect [29–32]. This effect originates from the fact that the charge created by the incoming particle spreads out and is registered between several adjacent pixels (depending on the place of interaction and detector parameters). In single-particle counting detectors such as the Timepix, the charge collected in each pixel is compared with a certain threshold level. If it is lower than this threshold, the event is not registered, the charge is lost, and, as a consequence, the energy resolution of these detectors is deteriorated. Although work has been done to describe the charge-sharing effect in the Timepix detector, and some models for this behavior have been proposed, the direct measurement of the charge-sharing effect is still a tricky issue. Generally speaking, the energy resolution of the Timepix detector cannot currently be compared to dedicated spectroscopic detectors, but it is good enough to obtain very precise measurement of dosimetric values.

1.4.3.2 Timepix on ISS

As previously discussed, radiation monitors currently used in space are either bulky or are passive detectors. The use of such large hardware is limited. One cannot clip tens or hundreds of kilograms of radiation-monitoring hardware to each spacesuit. Moreover, the price of bringing such massive hardware into orbit is enormous. While passive detectors are lightweight and can be easily attached to a spacesuit or put into any vehicle, they do not provide online information about the radiation field. This drawback is critical. As mentioned before, there are many events in the space radiation environment that cause significant changes in particle fluxes. The fast response and early warning capability of active monitoring hardware provides time-resolved information that is necessary to allow the appropriate reaction of the

FIGURE 1.4 Photo of an REM. The size of this device is comparable to that of a standard USB flash drive. Moreover, similarly to a flash drive, the REM is plugged in directly to the USB port of the PC or laptop. The REM is controlled by software running on the PC. The analysis of data is also performed in real time by the PC. The entire software suite is stand alone, and it provides the user with dosimetric information about the radiation environment.

crews in space. Thus, there is a pressing need for small, compact, low-power, active radiation monitors.

The radiation environment monitor (REM) [33] hardware (Figure 1.4) was designed as a demonstration project to show the possibility of creating small, lightweight, low-power radiation monitors using state-of-the-art pixel detectors. The REM is an active monitor connected to a standard PC or laptop via USB. All power and data-transfer requirements are fulfilled through the USB connection. Analysis software running on the PC provides fast and reliable information about the current radiation environment.

The analyzing software was evaluated and fine-tuned during accelerator campaigns prior to the REM hardware launch. A huge set of data, covering a broad range of particle types and energies, was measured with REM and Timepix detectors. Using accelerators has the advantage that the properties of the interacting particles are well known, and this data can serve as a verification and calibration set for current and future analysis algorithms.

Five REMs were delivered to the ISS by Russian Progress 48 in August 2012. They were deployed in October of the same year. From the day of deployment, the detectors have been collecting data, which are analyzed and give real-time information about the radiation environment aboard the ISS (Figure 1.5). Data are also transferred to Earth-based servers on a daily basis. The data are processed again after downlink to create daily, weekly, and monthly reports used by personnel at the National Aeronautics and Space Administration's Johnson Space Center (NASA JSC).

In summary, the REM hardware has proven to be a viable solution for radiation monitoring aboard the ISS. Following a successful initial deployment of the REM hardware, NASA requested two additional devices to be delivered to the ISS. One unit replaced a REM device, which failed on orbit, while the other will be used to test the advantages of employing a thicker sensor. While previous REM devices employed 300 μm silicon sensors bonded to the Timepix chip, the new unit has a 500 μm sensor. Using a thicker sensor is expected to allow better reconstruction of the angles of incident particles.

FIGURE 1.5 **(See color insert)** The picture taken aboard the ISS shows the REM deployed on a station support computer. The REM is emphasized in a gray box and shown in detail to the right, plugged into the station laptop.

1.5 ADVANTAGES OF PIXEL DETECTORS

As previously mentioned, the Timepix detectors work like solid-state bubble chambers, wherein each pixel behaves like an independent multichannel analyzer. Thus, properly calibrated detectors [34,35] allow direct measurement of the deposited dose in the sensor. Moreover, additional information about incident particles can be obtained by careful analysis of the tracks that particles create in the sensor. Indeed, for humans it is easy to distinguish different types of particles simply by looking at the track patterns (see Figure 1.6). Image-processing algorithms that analyze

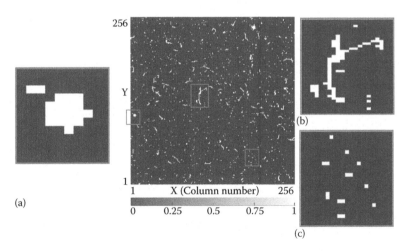

FIGURE 1.6 An example of the radiation background measured for 30 min at the University of Houston (at sea level). One can see the different types of radiation detected in the device just by looking at different patterns created in the sensor. Protons and alpha particles create small round clusters (a). Electron tracks undergo multiple scattering, creating curved tracks (b), while photons create point-like patterns (c).

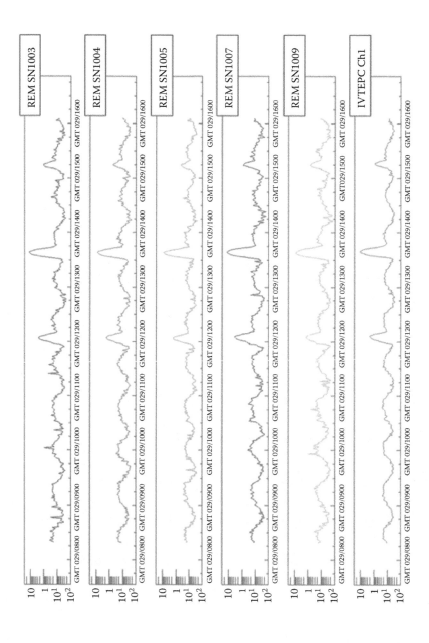

FIGURE 1.7 Comparison of absorbed dose rate measured by five REM devices compared to data obtained by IVTEPC over 8 h. The temporal changes of the radiation environment are clearly visible during the course of the day. It is clearly seen that the trends from all devices are the same. Each device is in a different part of the ISS, and small discrepancies between devices are caused by different shielding around devices.

tracks created by particle interaction within the sensor are being developed at the University of Houston.

After more than a year of measurement with REMs aboard the ISS, we can conclude that the REM hardware has exceeded expectations. The data from the REM hardware are carefully compared to other devices on board. The comparison of dose rates measured by REM units and by intravehicular tissue equivalent proportional counters (IVTEPC) is shown in Figure 1.7. It can be seen that, while much smaller, lighter, and less power demanding, the REM devices are capable of providing the same dosimetric data as IVTPEC. Moreover, the extra information obtained by analyzing the tracks of the interacting particles can be utilized to find information such as directionality, particle type, and energy estimation, which up to now were only in the domain of specialized bulky detectors. Over a year operating REM aboard the ISS has shown that pixel detectors can supply all of this information with the use of the appropriate tailor-made algorithms applied to the collected data.

1.6   FUTURE OF RADIATION MONITORS
BASED ON PIXEL DETECTORS

Following the success of the REM hardware, NASA initiated two other projects focused on developing new stand-alone radiation monitors based on Timepix technology. One of these devices, the Hybrid Environmental Radiation Assessor (HERA) device, will be the primary radiation monitor for the newly developed Orion module. It is also worth mentioning that other agencies are working on the utilization of pixel-detector technology for dosimetric application. The European Space Agency (ESA) is currently flying a prototype device on the Proba-V satellite, and the Japanese Aerospace Exploration Agency (JAXA) is preparing a device that will be operated on an upcoming satellite (both based on Timepix technology).

But that is not all; progress in microelectronics is so rapid that the Medipix3 collaboration is currently developing a new series of Timepix3 detectors. These devices will provide not only precise spatial and energy information, but also extremely precise (~3 ns) time information in each pixel. Radiation monitors based on such devices would have zero dead time and would obtain the most complete set of information about a particle that can be obtained by one detector.

Bringing these new technologies to the radiation-monitoring field is indeed very exciting. New ideas and approaches can be realized, while some strategies regarding radiation monitoring may be completely altered. One thing is certain; as we stated in the beginning of this chapter, radiation will follow humankind on its journey to the stars. It is up to us to learn how to evaluate and protect ourselves from it.

ACKNOWLEDGMENT

We would like to express our gratitude to the Medipix2 collaboration and especially to the Institute of Experimental and Applied Physics for delivering the REM hardware and software. We would also like to thank the Space Radiation Analysis Group (SRAG) of NASA JSC for their support.

REFERENCES

1. L. S. Pinsky, W. Z. Osborne, J. V. Bailey, R. E. Benson, and L. F. Thompson. Light flashes observed by astronauts on Apollo 11 through Apollo 17. *Science* 183(4128), 957–959 (1974).

2. R. Paschke, R. W. H. Chang, and D. Young. Probable role of gamma-irradiation in the origin of life. *Science* 125, 881 (1957).

3. N. B. Cook. *The Effects of Atomic Weapons*. U.S. Department of Defense (1950).

4. National Council on Radiation Protection and Measurements. Guidance on radiation received in space activities (NCRP Report No.98). NCRP: Bethesda, MD (1989).

5. J. Linsley. Evidence for a primary cosmic-ray particle with energy 10^{20} eV. *Phys. Rev. Lett.* 10, 146–148 (1963).

6. R. A. Meyers. *Encyclopedia of Physical Science and Technology* (3rd edn.), pp. 253–268. Academic Press: San Diego, CA (2003).

7. G. D. Badhwar and P. M. O'Neill. An improved model of galactic cosmic radiation for space exploration missions. *Int. J Rad. Appl. Instrum. D. Nucl. Tracks Radiat. Meas.* 20(3), 403–410 (1992).

8. G. D. Badhwar and P. M. O'Neill. Galactic cosmic radiation model and its applications. *Adv. Space Res.* 17(2), 7–17 (1996).

9. P. M. O'Neill. Badhwar–O'Neill 2010 galactic cosmic ray flux model: Revised. *IEEE Trans. Nucl. Sci.* 57(6), 3148–3153 (2010).

10. NASA Technology Roadmaps. *NASA Space Technology Roadmaps and Priorities: Restoring NASA's Technological Edge and Paving the Way for a New Era in Space.* National Research Council of the National Academies: Washington, DC (2012).

11. W. Heinrich, S. Roesler, and H. Schraube. Physics of cosmic radiation fields. *Radiat. Prot. Dosimetry* 86(4), 253–258 (1999).

12. S. Bourdarie and M. Xapsos. The near-Earth space radiation environment. *IEEE Trans. Nucl. Sci.* 55(4), 1810–1832 (2008).

13. J. A. Van Allen and L. A. Frank. Radiation around the earth to a radial distance of 107,400 km. *Nature.* 183(4659), 430–434 (1959).

14. T. I. Gombosi. *Physics of the Space Environment* (Cambridge Atmospheric and Space Science Series). Cambridge University Press: Cambridge (1998).

15. T. F. Tascione. Introduction to the Space Environment (Orbit: A Foundation Series). Krieger (2010).

16. L. W. Townsend. Implications of the space radiation environment for human exploration in deep space. *Radiat. Prot. Dosimetry* 115(1–4), 44–50 (2005).

17. S. Biswas and C. E. Fichtel. Nuclear composition and rigidity spectra of solar cosmic rays. *Astrophys. J.* 139, 941 (1964).

18. M. A. Shea and D. F. Smart. A summary of major solar proton events. *Sol Phys.* 127(2), 297–320 (1990).

19. J. R. Letaw, R. Silberberg, and C. H. Tsao. Radiation hazards on space missions outside the magnetosphere. *Adv. Space Res.* 9(10), 285–291 (1989).

20. J. A. Simpson. Elemental and isotopic composition of the galactic cosmic rays. *Annu. Rev. Nucl. Par. Sci.* 33(1), 323–382 (1983).

21. F. A. Cucinotta, S. Hu, N. A. Schwadron, K. Kozarev, L. W. Townsend, and M. H. Y. Kim. Space radiation risk limits and Earth-Moon-Mars environmental models. *Space Weather* 8(12) (2010).

22. G. F. Knoll. *Radiation Detection and Measurement.* Wiley (2011).

23. A. F. McKinlay. *Thermoluminescence Dosimetry (Medical Physics Handbooks 5).* Heyden & Son: Philadelphia (1981).

24. J. E. Turner. *Atoms, Radiation, and Radiation Protection.* Wiley: New York (2007).

25. W. S. Boyle and G. E. Smith. Charge coupled semiconductor devices. *Bell Syst. Tech. J.* 49(4), 587–593 (1970).
26. X. Llopart, R. Ballabriga, M. Campbell, L. Tlustos, and W. Wong. Timepix, a 65k programmable pixel readout chip for arrival time, energy and/or photon counting measurements. *Nucl. Instr. Methods A* 581, 485–494 (2007).
27. J. Žemlička, J. Jakůbek, M. Kroupa, and V. Tichy. Energy and position sensitive pixel detector Timepix for X-ray fluorescence imaging. *Nucl. Instrum. Methods A* 607(1), 202–204 (2009).
28. J. Žemlička, J. Jakůbek, M. Kroupa, D. Hradil, J. Hradilová, and H. Mislerová. Analysis of painted arts by energy sensitive radiographic techniques with the pixel detector Timepix. *J. Instrum.* 6(1) (2011).
29. J. Jakůbek, A. Cejnarová, T. Holý, S. Pospíšil, J. Uher, and Z. Vykydal. Pixel detectors for imaging with heavy charged particles. *Nucl. Instrum. Methods A* 591(1), 155–158 (2008).
30. J. Jakůbek. Energy sensitive x-ray radiography and charge sharing effect in pixelated detector. *Nucl. Instrum. Methods A* 607(1), 192–195 (2009).
31. J. Bouchami, A. Gutierrez, A. Houdayer, J. Idarraga, J. Jakůbek, C. Lebel, C. Leroy, J. Martin, M. Platkevič, and S. Pospíšil. Study of the charge sharing in silicon pixel detector with heavy ionizing particles interacting with a Medipix2 device. *Nucl. Instrum. Methods A* 607(1), 196–198 (2009).
32. J. Marchal. Theoretical analysis of the effect of charge-sharing on the detective quantum efficiency of single-photon counting segmented silicon detectors. *J. Instrum.* 5, P01004 (2010).
33. D. Turecek, L. Pinsky, J. Jakubek, Z. Vykydal, N. Stoffle, and S. Pospisil. Small dosimeter based on Timepix device for International Space Station. *J. Instrum.* 6, C12037 (2011).
34. M. Kroupa, J. Jakůbek, and F. Krejčí. Charge collection characterization with semiconductor pixel detector Timepix. *IEEE NSS/MIC/RTSD Conf. Proc.* R12-37 (2008).
35. J. Jakůbek. Precise energy calibration of pixel detector working in time-over-threshold mode. *Nucl. Instrum. Methods A* 633(Suppl. 1), S262–S266 (2011).

2 CdZnTe for Gamma- and X-Ray Applications

Michael Prokesch

CONTENTS

2.1 INTRODUCTION

Commercially, cadmium zinc telluride (CdZnTe; CZT) and cadmium telluride (CdTe) are currently the most relevant II–VI compound semiconductor systems for x-ray and gamma-ray radiation detection. Their main advantages over traditional semiconductor materials such as silicon and germanium are their high radiation stopping power due to the larger atomic numbers of their constituents and the low background density of free charge carriers at room temperature (RT) due to their wider bandgap [1,2]. Successful RT spectroscopy and x-ray applications utilizing

FIGURE 2.1   Typical 8" (24 kg) CZT ingot grown by the HP-EDG technique at eV Products, Inc. (2011).

CZT or CdTe detectors have been reported and have been in commercial use for many years.

2.1.1 CRYSTAL GROWTH

Various melt, solvent, and vapor phase crystal growth techniques have been employed over the years to produce CdTe and CZT single crystals [3]. Figure 2.1 shows an 8" CZT ingot grown from a melt using the high-pressure electrodynamic gradient method (HP-EDG) [4]. This growth technique is typically carried out in semiopen growth systems, which can limit the control and reproducibility of the process in a production setting.

A closed-system approach to growing radiation detector-grade CZT is the traveling heater method (THM), in which the CZT crystallizes out of a solution. This is typically done in sealed quartz ampoules at temperatures well below the CZT melting point. Figure 2.2 shows a 3" CZT ingot grown using the THM technique. Figure 2.3 displays an optical photograph of the cross section of the ingot from Figure 2.2 together with an electron backscatter diffraction (EBSD) orientation map of the adjacent slice proving that single crystallinity can be achieved over the entire ingot diameter.

FIGURE 2.2   Typical 3" CZT ingot grown by the THM technique at eV Products, Inc. (2013). Solvent zone and top (last to crystallize) slice have been taken off from the left. This ingot weighs about 6.3 kg.

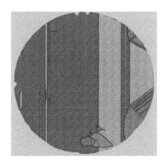

FIGURE 2.3 **(See color insert)** Cross section of the THM ingot from Figure 2.2 and electron backscatter diffraction (EBSD) orientation map of the adjacent slice revealing twin (red) and grain boundaries (black).

2.1.2 MATERIALS AND DEVICE CHALLENGES

Significant progress has been made in CZT crystal growth over the past two decades; however, a major hurdle for the broader commercial use of these materials remains the technical challenge of growing larger single crystals, which causes a relatively high cost of large monolithic volume detectors. Furthermore, charge trapping within single crystal grains can cause significant charge loss and, ultimately, poor resolution in gamma spectroscopy. In commercial semi-insulating CZT, the mobility–lifetime products of the holes are typically orders of magnitude lower than those of the electrons (10^{-6}–10^{-4}) cm^2/V versus (10^{-4}–10^{-2}) cm^2/V. This difference leads to more or less significant hole tailing in parallel-plate detectors.

This issue is addressed primarily by electron collecting device configurations and correction schemes such as semihemispherical detectors and their variants [5–7], coplanar grid [8], orthogonal and drift strip [9–11], small pixel monolithic array [12,13], and virtual Frisch-grid detector devices [14].

In practice, device design optimization can become rather involved due to the contributions of materials and processing effects originating from nonuniformities of both matrix charge transport properties and electric field distributions, macroscopic defects, electrode interface, and side and street surface properties that can impact the charge collection and noise performance of the final detector devices. In dependence on the actual fabrication process, the interface and surface properties can become more or less dependent on the crystal bulk properties and the crystallographic orientation, which can also affect the electric field distribution within the device. Furthermore, the detector always operates in connection with a particular read-out electronics chain, which opens a matrix of working points that needs to be optimized in dependence on the device design and the materials and interface/surface properties.

While uniform charge loss can be corrected by a variety of techniques, it can still cause difficulties if excessive space charge buildup distorts the internal electric field. In this case, the carrier drift velocities can become position dependent, which can lower the efficiency of, for example, depth-of-interaction-based correction schemes. High-resolution coplanar grid detectors (CPGs) [8,15,16], for example, theoretically require only uniform (and in practice, also high [17]) electron transport. Good hole transport is in first-order theory not a requirement for high-resolution CPGs. However, if the hole trapping becomes so strong that space charge buildup causes significant local variations of the electric field, the device performance can still be negatively impacted. This space charge buildup occurs upon biasing the detector and deteriorates the device performance already under low-flux conditions (dark polarization). The internal electric field distribution can be affected by the design of the contact barriers [18,19].

This mechanism of space charge buildup in the dark must not be confused with dynamic high-flux polarization, which is triggered primarily by the x-ray or infrared (IR)-generated nonequilibrium carriers and which will be discussed later.

2.1.3 HIGHER-DIMENSIONAL DEFECTS

Nonuniformities of the charge transport properties including defect-related steering effects within single crystals are typically enhanced by the incorporation of second-phase defects such as tellurium (Te) inclusions and extended dislocation networks around them. These difficulties need to be addressed by advances in the crystal growth techniques. Figure 2.4 shows the effect of an inclusion decorated twin boundary on the x–y map of charge-collection efficiency as obtained by alpha scanning [20]. Detailed studies regarding the effect of noncorrelated inclusions, dislocations, and subgrain boundaries on charge transport in CZT have been performed using high-spatial-resolution x-ray response mapping [21–23]. The fundamental role of (zero-dimensional) point defects in electrical compensation and uniform and nonuniform charge trapping will be discussed in Sections 2.2 and 2.3.

2.1.4 ANNEALING

Te inclusions are generated during crystal growth under Te-rich conditions. Within limits, their size distribution can be controlled by crystal growth parameters, such as temperature gradients at the growth interface and so on [24].

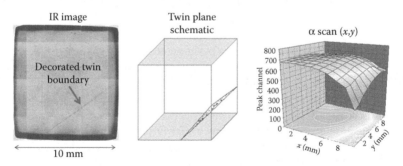

FIGURE 2.4 (See color insert) Effect of a decorated twin plan on charge-collection efficiency. (From Soldner, S.A., Narvett, A.J., Covalt, D.E., and C. Szeles, Characterization of the charge transport uniformity of high pressure grown CdZnTe crystals for large-volume nuclear detector applications, Presented at IEEE 13th International Workshop on Room-Temperature Semiconductor X- and Gamma-ray Detectors, Portland, OR, 2003.)

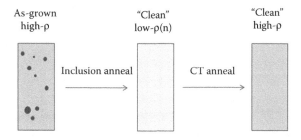

FIGURE 2.5   Annealing sequence consisting of an inclusion anneal and a charge transport (CT) anneal.

In principle, Te inclusions can be relatively easily removed by postgrowth annealing in cadmium (Cd)-rich vapor, which is practical to do on a wafer level. The resulting material appears clean in IR transmission; however, it is typically not immediately clear to what extent additional structural damage has been caused in the process as the Te removal or CdTe microgrowth process may create additional defects. Another issue is that the resulting crystals typically turn low-resistivity n-type after the inclusion anneal and a second annealing step usually has to be added to restore the semi-insulating state. This sequence is shown in Figure 2.5. The IR scans of a typical CZT sample before and after annealing are shown in Figure 2.6. After annealing, no inclusions are visible in IR microscopy, but note that this type of microscopy would not reveal any features less than about 1 μm in size nor those that are transparent to IR light.

Figure 2.7 summarizes the electrical properties and charge transport performance of an inclusion-free $(6 \times 6 \times 3)$ mm³ parallel-plate detector that was prepared from an inclusion and charge transport annealed CZT slice (two-step sequence). The semi-insulating state and radiation detector grade charge transport properties were restored and are comparable with those before annealing. The electron $\mu\tau$ as obtained from the standard "Hecht fit" [25] of the bias dependence of charge-collection efficiency (CCE) actually improved in this particular case. However, the benefit of such an annealing approach on charge transport uniformity, for example, for fine-pitch x-ray

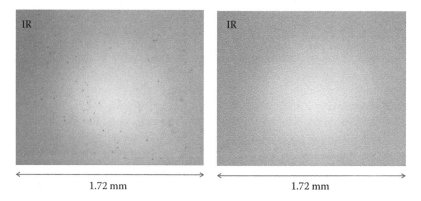

FIGURE 2.6   IR scans of the same sample before and after inclusion annealing.

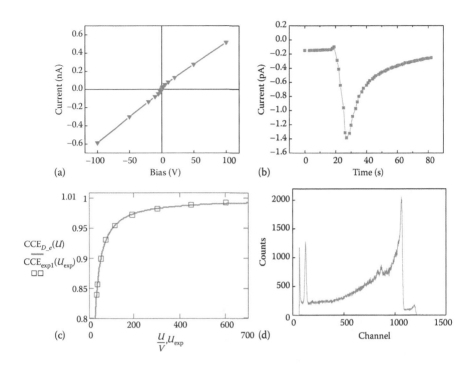

FIGURE 2.7 Electrical properties and the charge transport performance of an inclusion-free $(6 \times 6 \times 3)mm^3$ CZT sample after the two-step annealing sequence: (a) current-voltage characteristics. The resistivity taken from the linear ± 100 mV range (before barrier kicks in) is $5.7 \times 10^{10} \, \Omega$ cm at 23°C. (b) Thermoelectric effect (hot probe) measurement typical for high-resistivity, n-type CZT. (c) Electron Hecht curve revealing an electron $\mu\tau$ product of $\sim 1 \times 10^{-2} cm^2/V$. (d) ^{57}Co spectrum measured in parallel-plate configuration at a bias of 1200 V.

imaging, still needs to be proven. It may always be preferable to minimize second-phase particle incorporation during crystal growth and initial cool down, provided it is technically feasible.

2.1.5 RADIATION SENSOR DEVICES

State-of-the-art semiconductor radiation detectors for gamma- and x-ray photons typically employ a device configuration in which the detector is shaped like a rectangular parallelepiped or cylinder. The cathode is typically formed by a continuous metal film deposited on one planar face of the detector. The cathode may extend to the physical edges of the detector or, in certain configurations, partially or all the way over the sidewalls [6,7]. Certain device configurations would also employ cathode strips [9]. The anode is typically laid out as a continuous planar electrode, single pixel or small pixel array [12,13], coplanar grid [8], or strip design [9–11]. Additional electrodes such as guard rings, guard bands, steering grids, and control electrodes [26] may be added to the pattern. Multiple metal layers, diffusion barriers, or other coatings on top of the primary electrical contact may be required to ensure mechanical robustness or facilitate

FIGURE 2.8 Typical CZT detector device configurations produced at eV Products, Inc.

specific interconnect processes. Bare CZT surfaces (sidewalls and streets between pixels or grids) need to be electrically passivated to minimize leakage currents and noise. Some detector concepts require electrically insulated shielding electrodes [27].

Figure 2.8 shows examples of typical detector and electrode configurations. CPGs for high-resolution spectroscopy, linear array detectors for medium-flux x-ray, and large pitch (>1 mm) spectroscopy detectors are shown. The detector thicknesses in Figure 2.8 range from 3 to 25 mm.

The interface processing and electrode technology can significantly affect the electrical properties of the electrode barriers, which constitute for most CZT devices some form of blocking contact, that is, the high-bias resistance will be lower than expected from an ideal ohmic device of a given bulk resistivity [28]. The most common electrode structure for high-resistivity CZT is a back-to-back blocking configuration, but note that the observed barriers will in general be less pronounced than for, for example, CdTe-In diode structures [29].

Figure 2.9 shows the current–voltage (I–V) characteristics of CZT parallel-plate detectors at 23°C. The blocking–blocking (Pt–CZT–Pt) configuration in the ±800 V bias range appears linear in that scale but the zoom shows a well-pronounced, symmetric barrier kicking at about ±1 V. The CZT bulk resistivity can be obtained from guarded electrodes in the linear range below ±100 mV [28]. The I–V characteristics for an asymmetric electrode configuration are shown for comparison. Note that the dark leakage current in nonblocking polarity far exceeds the current expected from an "ohmic device," that is, additional carriers are injected at higher bias and the detector functions only in reverse polarity.

Figure 2.10 shows a ^{57}Co spectrum measured with a simple $(13 \times 13 \times 3)$ mm^3 parallel-plate detector biased at 700 V. All major features, including the Fe x-rays at 6.4 eV, are well resolved. Figure 2.11 shows a ^{137}Cs spectrum from the same detector. Note that this is not a favorable device configuration to obtain high resolution and photopeak efficiency at those high energies.

A simple yet highly efficient detector configuration to obtain higher resolution and photopeak efficiency at 662 keV is the coplanar grid design [8]. Figure 2.12 shows the spectra from a set of CZT detectors configured as CPGs. The resolution is well below 2%. The pulser peak is shown to indicate the electronic noise contribution.

Figure 2.13 shows the combined ^{137}Cs spectrum obtained with a 2×2 prototype array of $(7 \times 7 \times 20)$ mm^3 position-sensitive virtual Frisch-grid detectors with electronically segmented anodes at Brookhaven National Laboratory (BNL) yielding

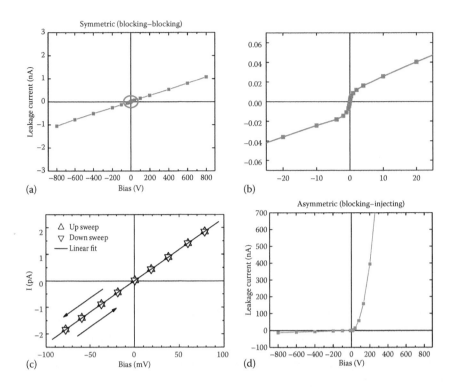

FIGURE 2.9 I–V characteristics of CZT parallel-plate detectors at 23°C: (a) Blocking–blocking configuration in ±800 V bias range. (b) Same curve zoomed into the ±20 V range to show the barrier effect. (c) Low bias range precision measurement (±80 mV) to obtain resistivity (guarded electrode). (d) Asymmetric electrode configuration.

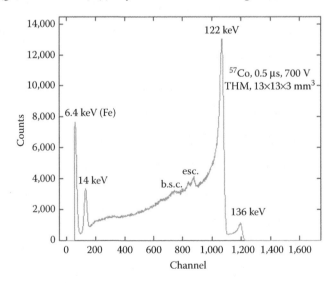

FIGURE 2.10 ^{57}Co spectrum from a $(13 \times 13 \times 3)$ mm^3 parallel-plate detector biased at 700 V.

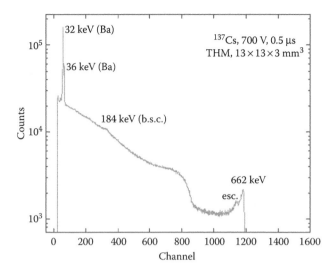

FIGURE 2.11 ^{137}Cs spectrum from the same detector biased at 700 V.

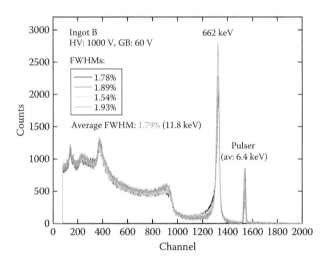

FIGURE 2.12 (See color insert) ^{137}Cs spectra taken with four different CZT detectors in CPG configuration.

about 0.8% resolution at 662 keV [30]. The four 20 mm thick CZT base detectors are bar shaped with parallel-plate electrodes as shown at the very right in Figure 2.8.

Figure 2.14 shows the results of an open x-ray beam photon-counting experiment utilizing a 500 μm pitch, 16×16 pixel monolithic array CZT detector. The count-rate response of all 256 pixels to an x-ray flux ramp and the response nonuniformity histogram are shown [31]. Note that from detector to detector, the total counts (sum over all pixels) are usually preserved within 3%. If the count-rate response is stable over time, the pixel count nonuniformities can be easily normalized/calibrated (uniformity correction).

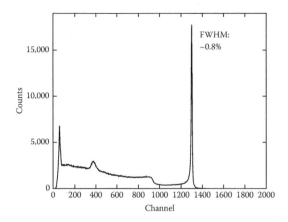

FIGURE 2.13 Combined ¹³⁷Cs spectrum from a position-sensitive virtual Frisch-grid array after drift time and lateral corrections at BNL. The 20 mm thick CZT base detectors are from eV Products. (From Bolotnikov, A., Ackley, K., Camarda, G., Cherches, C., Cui, Y., De Geronimo, G., Eger, J., et al. Using position-sensing strips to enhance the performance of virtual Frisch-grid detectors. Presented at SPIE, Hard X-Ray, Gamma-Ray, and Neutron Detector Physics XVI, San Diego, 2014.)

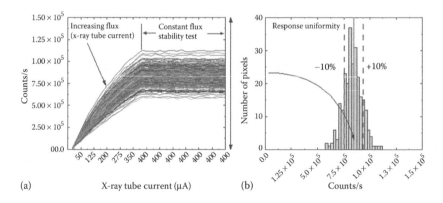

FIGURE 2.14 **(See color insert)** Test result for a typical 16 × 16 pixel CdZnTe monolithic detector array. (a) The count rate as a function of the x-ray flux (tube current), which is followed by stability and uniformity tests at a constant maximum tube current. (b) The count distribution at maximum current indicating the response uniformity of the device. (From Szeles, C., Soldner, S.A., Vydrin, S., Graves, J., and Bale, D.S., *IEEE-TNS* 54, 1350, 2007.)

2.2 BASICS OF SEMI-INSULATING CZT

2.2.1 SEMI-INSULATING STATE

The temperature and composition dependence of the CZT bandgap can be approximated by Equation 2.1:

TABLE 2.1

Cd$_{1-x}$Zn$_x$Te Parameters Used for the Calculations

RT electron drift mobility Cd$_{0.9}$Zn$_{0.1}$Te (cm^2/V·s)	1000	[32]
RT hole drift mobility Cd$_{0.9}$Zn$_{0.1}$Te (cm^2/V·s)	50	[32]
Electron effective mass CdTe (0.91×10^{-30} kg)	0.11	[33]
Hole effective mass CdTe (0.91×10^{-30} kg)	0.73	[34]
E_0 (eV)	1.606	[35]
a_1 (eV)	0.38	[28]
a_2 (eV)	0.463	[35]
a_3 (eV/K)	4.5×10^{-4}	[28]
a_4 (K)	264	[36]

Source: Prokesch, M. and Szeles, C., *J. Appl. Phys.*, 100, 14503, 2006.

$$E_g(T,x) = E_0 + a_1 x + a_2 x^2 - \frac{a_3 T^2}{a_4 + T}, \qquad (2.1)$$

where x is the zinc (Zn) mole fraction in Cd$_{1-x}$Zn$_x$Te. The other parameters are listed in Table 2.1.

The theoretical maximum resistivity of a semiconductor at a given temperature T is

$$\rho_{max}(T) = \frac{1}{2e_0 n_i(T)\sqrt{\mu_n(T)\mu_p(T)}}. \qquad (2.2)$$

This derives from $\rho = 1/e_0(\mu_n n + \mu_p p)$ for the case when the mobility–concentration products of the free electrons and holes are equal:

$$\mu_n n = \mu_p p. \qquad (2.3)$$

The elementary charge is e_0 and the intrinsic carrier concentration is n_i. Using the parameters from Table 2.1, we obtain $\rho_{max} \sim 1 \times 10^{11}$ Ω·cm for Cd$_{1-x}$Zn$_x$Te with $x = 0.1$ at 23°C. At this point, the conduction type as measured in an ideal thermoelectric effect experiment would be undefined, that is, theoretically neither n- nor p-type. Because of the higher drift mobility of the free electrons in CZT $\mu_n > \mu_p$, the Fermi level is below the intrinsic level in this case ($p > n$). The free-hole concentration $p = n_i\sqrt{\mu_n/\mu_p}$ is about 6×10^5 cm^{-3} at RT and the concentration of the free electrons, n, is a further order of magnitude lower.

This must not be confused with the following two commonly used transition points:

$$n = p, \qquad\qquad (2.4)$$

which is the point of complete compensation from a Fermi statistics standpoint. The Fermi level is now at the intrinsic level and the semiconductor is n-type as per Equation 2.3 at both sides of this transition point.

$$\mu_n^2 n = \mu_p^2 p, \qquad\qquad (2.5)$$

is the point where, for example, the Hall coefficient would change its sign. The Fermi level is even further below the intrinsic level than with Equation 2.3 and the material is p-type on both sides of this transition point. While those distinctions are of minor relevance in low-resistivity semiconductors where the Fermi level is energetically far off any of those transition points and majority carrier approximations can be used, they become important in semi-insulators when compensation models are compared with experimental data.

To actually achieve a semi-insulating state at RT, that is, free carrier concentrations of say $< 10^7\,\mathrm{cm^{-3}}$, requires effective compensation of electrically active foreign impurities, native point defects, point defect complexes, energy levels in the forbidden gap related to crystallographic discontinuities and lattice distortions, and any combinations of these. In practice, shallow donor and acceptor concentrations cannot be matched that closely by controlled (shallow) counterdoping and defects that produce deep energy levels are required to pin the Fermi level close to the middle of the bandgap. Somewhat loosely distinguishing between shallow levels that would be fully ionized when the Fermi level is close to the middle of the bandgap and deep levels that would be only partially ionized to compensate for the net difference of the concentrations of the shallow levels, the condition to obtain semi-insulating material can be generalized [37]:

$$\sum_i N_{A\_i}^{\text{shallow}} \le \sum_j N_{D\_j}^{\text{shallow}} + \sum_k N_{D\_k}^{\text{deep}}$$

$$\sum_j N_{D\_j}^{\text{shallow}} \le \sum_i N_{A\_i}^{\text{shallow}} + \sum_l N_{A\_l}^{\text{deep}}. \qquad (2.6)$$

This simply means that the total concentrations of all shallow acceptors, $N_{A\_i}^{\text{shallow}}$, must be smaller than the total concentrations of all shallow and deep donors, $N_{D\_j}^{\text{shallow}}$ and $N_{D\_k}^{\text{deep}}$, together, and at the same time, the total concentrations of the shallow donors, $N_{D\_j}^{\text{shallow}}$, has to be smaller than the total concentrations of all the acceptors, $N_{A\_i}^{\text{shallow}}$ and $N_{A\_l}^{\text{deep}}$. Condition 2.6 remains fully valid if defects with multiple charge states are individually indexed for each of their energy levels. It also provides a sufficient framework for both deep donor and deep acceptor compensation models. Note, however, that a simple comparison of the total numbers of donors and acceptors does not necessarily predict the conduction type of semi-insulating material as this will also depend on the actual deep-level ionization energies.

The calculation of the ionized defect and free carrier concentrations is based on the solution of the charge neutrality condition:

$$n + k_m \sum_m N_{A\_m}^{(-k_m)} = p + k_n \sum_n N_{D\_n}^{(+k_n)}, \tag{2.7}$$

which states that the equilibrium concentration of all negative charges (ionized acceptors N_{A_m} and free electrons n) has to equal the number of all positive charges (ionized donors N_{D_n} and free holes p); k_m and k_n indicate the actual charge states (1, 2,...). The individual concentrations of each species are given by the Fermi–Dirac statistics [38,39]. As an example, the ionized concentration, $N_{D\_j}^+$, of a single donor j with total concentration, N_{D_j}, and degeneracy, g_{D_j}, is given by the energy difference between its 0/+ transition level, E_{D_j}, and the Fermi energy, E_F:

$$\frac{N_{D\_j}^+}{N_{D\_j}}(T,x) = \frac{1}{1 + g_{D\_j} \exp\left(-\dfrac{E_{D\_j}(T,x) - E_F(T,x)}{k_B T}\right)}, \tag{2.8}$$

where:

k_B = Boltzmann constant
T = temperature
x = Zn mole fraction in $Cd_{1-x}Zn_xTe$

Similar equations apply for ionized acceptors. The degeneracy factors, however, may be different for donors and acceptors depending on the band structure of a given semiconductor [40]. Once the neutrality equation is solved for E_F, all individual ionized fractions including the free carrier concentrations can be obtained and the free electron concentration n is given by

$$\frac{n}{N_C}(T,x) = \exp\left(-\frac{E_C(T,x) - E_F(T,x)}{k_B T}\right), \tag{2.9}$$

where E_C is the conduction band minimum energy and N_C is the conduction band density of states. Note the general temperature and composition dependencies of the donor and acceptor ionization energies in Equation 2.8, which are of particular importance for the proper treatment of localized (deep) levels in compensation models [37].

Figure 2.15 shows the theoretical temperature dependence of the fundamental $Cd_{1-x}Zn_xTe$ band edge energies for a composition of $x = 0.1$ and the possible energetic positions of donor levels with respect to a host crystal independent reference level, that is, in absolute energy scale [41]. The valance band edge of pure binary CdTe at 0 K was chosen as the energetic zero point and a linear temperature correction was added:

$$E_V(T,x) = E_V(0\,K, CdTe) + \alpha_T T. \tag{2.10}$$

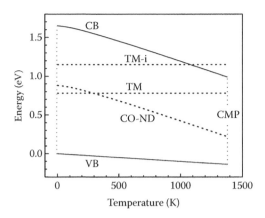

FIGURE 2.15   Temperature dependence of the $Cd_{1-x}Zn_xTe$ bandgap for $x = 0.1$ and possible energetic positions of donor levels in absolute energy scale. (From Prokesch, M. and Szeles, C., *Phys. Rev., B* 75, 245204, 2007.)

The linear temperature coefficient α_T has been taken from a theoretical work [42]. The dotted (TM) lines depict the strong temperature dependence of the ionization energies of localized, host independent donor levels that could be related to, for example, substitutional transition metal impurities [41]. The level related to TM-i would be an intermediate level in the bandgap at RT and would move into the conduction band continuum at high temperatures. The CO-ND line indicates the possible behavior of a donor state that is primarily buildup from cation orbitals, which could cause the level to somewhat follow the conduction band edge with the temperature. In this case, the donor ionization may be expected to stay roughly constant over the entire temperature range. Similar consideration can be applied to acceptor states [37].

These effects have important implications on electrical compensation models that relate high-temperature defect equilibria to spectroscopic data that are obtained at lower temperatures. They also affect the interpretation of temperature-dependent experiments in that the results will depend on assumptions regarding the actual physical nature of the defects.

Condition 2.6 immediately implies that for wide bandgap semiconductors, a semi-insulating state can theoretically always be obtained by excessive doping with deep donor impurities (if the net shallow level concentration is acceptor) or deep acceptor impurities (if the net shallow level concentration is donor), independent of the order of magnitude of the net shallow doping concentration. While this approach will theoretically always produce more or less "expensive resistors," it will not necessarily lead to semiconductor radiation detectors because the same deep levels (localized centers) that pin the Fermi level close to the middle of the bandgap would also tend to deteriorate the transport/collection of radiation or particle-generated excess charge carriers by trapping.

2.2.2 CHARGE TRAPPING AND DETRAPPING

In steady state, the statistics of trapping, detrapping, and recombination via localized centers is governed by the Shockley–Read–Hall (SRH) theory [43,44]. The basic concept is that each trap has an energy level, E_t, in the bandgap with two possible charge states. This is illustrated in Figure 2.16.

The two possible charge states of the generic trap are labeled k and $k-1$, where k is the more positive charge state in which the defect is "unoccupied." In general, the $k/k-1$ level can belong to deep donors (+/0), deep acceptors (0/–), or higher-charged transitions such as (2+/+) or (–/2–). The four basic processes are (a) electron capture by the unoccupied defect at rate $R_{c,n}$; (b) electron emission from the occupied defect at rate $R_{e,n}$; (c) hole capture by the occupied defect at rate $R_{c,p}$, which is equivalent to electron transfer from the defect to the valence band; and (d) hole emission from the unoccupied defect at rate $R_{e,p}$, with is equivalent to electron transfer from the valence band to the defect. In this framework, recombination can be understood as a sequence of electron capture (a) followed by hole capture (c), which is essentially an electron transfer from the conduction to the valence band, which eliminates one conduction band electron and one hole in the valence band.

The SRH theory derives a relation between the capture and emission rates based on the principle of detailed balance in thermal equilibrium, which requires equal rates of capture and emission separately for both electrons and holes ($R_{c,n} = R_{e,n}$ and $R_{c,p} = R_{e,p}$) with

$$R_{c,n} = v_n^{th}\sigma_n n N_t (1 - P_t),$$

$$R_{e,n} = e_n N_t P_t,$$

$$R_{c,p} = v_p^{th}\sigma_p p N_t P_t,$$

$$R_{e,p} = e_p N_t (1 - P_t), \tag{2.11}$$

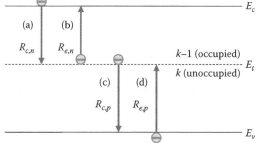

FIGURE 2.16 Schematic diagram of the four basic capture and emission paths as explained in the text.

where:

n and p = total free electron and hole concentrations
N_t = total trap density
P_t = probability of the trap being occupied, which, for a nondegenerated ground state, is

$$P_t = \frac{1}{1 + \exp\left(\dfrac{E_t - E_F}{kT}\right)}. \tag{2.12}$$

This leads to expressions for the electron and hole emission rates, $e_{n/p}$, in terms of thermal velocities, $\upsilon_{n/p}^{th}$, capture cross sections, $\sigma_{n/p}$, and band densities of states $N_{c/v}$ including the well-known SRH attempt-to-escape frequency factors $\omega_{n/p} = \upsilon_{n/p}^{th} \sigma_{n/p} N_{c/v}$:

$$e_n = \omega_n \exp\left(-\frac{E_c - E_t}{k_B T}\right)$$

$$e_p = \omega_p \exp\left(-\frac{E_t - E_v}{k_B T}\right). \tag{2.13}$$

These relations are then used to determine the net capture, emission, and recombination rates for steady-state (stationary nonequilibrium) conditions under which the net rates of electron and hole capture have to be equal to ensure constant electron and hole concentrations and constant average defect occupations [43]:

$$U = R_{c,n} - R_{e,n} = R_{c,p} - R_{e,p}. \tag{2.14}$$

With Equations 2.11 and 2.13, the average occupation probabilities, net capture, and net recombination rates can be calculated from Condition 2.14. Those quantities obviously depend on the injection level (Δn and Δp). The net recombination rate U is

$$U = \frac{N_t \upsilon_n^{th} \sigma_n \upsilon_p^{th} \sigma_p \left(np - n_i^2\right)}{\upsilon_n^{th} \sigma_n \left[n + n_i \exp\left(\dfrac{E_t - E_i}{kT}\right)\right] + \upsilon_p^{th} \sigma_p \left[p + n_i \exp\left(\dfrac{E_i - E_t}{kT}\right)\right]}, \tag{2.15}$$

where $n = n_{equ} + \Delta n$ and $p = p_{equ} + \Delta p$ denote the total carrier concentrations in the bands and E_i is the intrinsic Fermi level. Equation 2.15 necessarily implies $U=0$ for thermal equilibrium ($np = n_i^2$).

Note that while Condition 2.14 is always fulfilled in steady state, it is not straightforward to apply the SRH treatment to experiments where significant occupation transients occur, such as the analysis of emission rates in photoinduced current transient spectroscopy (PICTS) or thermoelectric effect spectroscopy (TEES) [45].

2.2.3 CAPTURE CROSS SECTIONS

Figure 2.17 shows a schematic illustrating the efficiencies of deep donors and deep acceptors with respect to electron and hole trapping in a fully compensated semiconductor. In the figure, the (+/0) donor and (0/–) acceptor levels are at the same energy in the bandgap. The shallow levels are fully ionized. The terms *weak* versus *strong* are based on simple considerations regarding the Coulomb potential of the trapping center with respect to each respective carrier type [46].

There is no well-established theory regarding the exact mechanism of the non-radiative transition of free carriers from the bands into localized levels close to the middle of the bandgap [47]. Single longitudinal optical phonon energies are about 21.3 meV in CdTe [48], which is obviously too low to facilitate the required energy transfer with the crystal lattice. One commonly considered mechanism is multiple phonon emission mitigated by significant lattice distortions [49]. This is usually conceived by plotting the electronic and elastic energies of the band extrema and defects over the lattice displacement (configuration coordinate diagrams) [50].

Without addressing the possible implications of the energy dissipation problem, the effects of modified phonon fields around defects, or experimental issues with deep-level spectroscopy that can lead to surprisingly low capture cross sections apparent for those defects [47,51,52], the following classic estimates on physical trapping cross sections may be applied to obtain estimates on upper limits for the allowed deep-level defect concentrations in radiation detector grade semi-insulating CZT. To this end, we consider only the trapping efficiencies of deep donors and deep acceptor per Figure 2.17, which compensate the net shallow level doping $N_{AS}^{tot} - N_{DS}^{tot}$

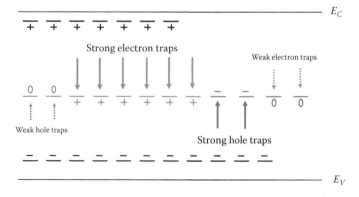

FIGURE 2.17 Schematic diagram illustrating the efficiencies of deep donor and acceptor levels with respect to electron and hole trapping in a fully compensated semiconductor.

according to Condition 2.6. To calculate the actual ionized fractions of the deep donor and acceptor states, the neutrality equation, Equation 2.7, has to be solved for E_F as described in Section 2.2.1.

The simple consideration that the attractive potential of neutral point defects should not extend much over the range of the actual lattice distortion [46] puts the effective capture radius in the order of approximately one lattice constant:

$$r_{\text{neutral}} \sim a_{\text{CdTe}}/2.$$ (2.16)

This gives a capture cross section for the neutral defect of about $\sigma_{\text{neutral}} \sim 3 \times 10^{-15}\,\text{cm}^2$ for both electron trapping on neutral acceptors and hole trapping on neutral donors. The lattice constant of CdTe is a_{CdTe}, which is about 6.48 Å at RT [53].

For charged point defects, the effective capture radius may be estimated by comparing the Coulomb energy of the singly charged defect with the thermal energy of the charge carrier [46], which gives

$$r_{\text{singly}}(T) \sim \frac{e_0^2}{6\pi\varepsilon_0\varepsilon_r k_B T},$$ (2.17)

and a capture cross section of σ_{singly} $4 \times 10^{-13}\,\text{cm}^2$ at RT for both electron trapping at positively charged donors and hole trapping at negatively charged acceptors.

In this picture, the capture cross sections for electrons and holes at the same defect would differ by two orders of magnitude with $\sigma_e > \sigma_h$ for deep donors and $\sigma_h > \sigma_e$ for deep acceptors. Looking at the simplifications of this approach, including physical insufficiencies such as the use of the bulk dielectric constant, ε_r, to calculate the attraction between point charges, the implied temperature dependence in Equation 2.17 should not be taken too seriously at this point and experimental tests and quantum-mechanical modeling sometimes seem to confirm [49,54] but often show otherwise [55].

2.2.4 DEEP-LEVEL CONCENTRATIONS

Using the rough estimates from Equations 2.16 and 2.17, the maximum allowed deep-level concentrations to maintain sufficient charge transport for radiation detector applications can be obtained by a minimum trapping time consideration. The trapping time (often referred to as "lifetime" in this context) is the average time that a free charge carrier can exist in the bands before trapping occurs. The electron lifetime components relate to the capture rates from Equation 2.11, such as

$$\tau_e^j = \frac{n}{R_{c/n}^j} = \frac{1}{v_n^{th}\sigma_n^j N_t^j},$$ (2.18)

for electron trapping at level j that has an unoccupied concentration of N_t^j (the positively charged donors or the neutral acceptors in Figure 2.17). The capture cross section of the unoccupied trap j is σ_n^j for electrons and the thermal velocity of the electrons is υ_n^{th} as the directed drift motion induced by an external bias is, in most practical applications, small compared with the random thermal motion. A similar relation can be derived for hole trapping with the only difference being that the concentrations of the occupied defects must be used. Note that the electron or hole effective masses (Table 2.1) should be used to calculate the $\upsilon_{n/p}^{th}$ as the carriers move in the respective bands before they get trapped. For the simple scenario illustrated in Figure 2.17, where each defect in a specific charge state accepts only one carrier type, the total trapping time for, for example, the electrons τ_e^{tot} can be written as

$$\frac{1}{\tau_e^{tot}} = \sum_{j=1}^{N} \frac{1}{\tau_e^j} = \frac{1}{\tau_e^{ID}} + \frac{1}{\tau_e^{NA}}, \qquad (2.19)$$

where τ_e^{ID} and τ_e^{NA} are the ionized deep donor and neutral deep acceptor–related trapping time components, respectively. Using the capture cross sections according to Equations 2.16 and 2.17 and the RT mobility data from Table 2.1, the upper limits for deep-level concentrations in radiation detector grade CZT are obtained, as shown in Table 2.2.

These data are reasonably consistent with experimental data obtained on CdTe:Cl/In and CdTe:Ge/Sn [56,57] and trapped carrier concentrations of $(10^{12}–10^{13})$ cm^{-3} were determined by space charge measurements [19,58].

Note that due to the lack of a developed theory regarding the actual capture/reemission mechanisms from localized deep centers as outlined before, fundamental misjudgments cannot be ruled out. Otherwise, the implications would be significant: If impurity concentrations beyond a few parts per billion (ppb) are detected in the radiation sensor grade CZT using macroscopic mass spectroscopy, these impurities do not generate deep levels (isolated or as complex), are almost completely compensated, or they have an active concentration in the crystal matrix that is much lower than the concentration reported by the mass spectroscopy (e.g., due to second-phase gettering or direct precipitation due to low solubility).

The relevant deep-level defects control the semiconductor's electric and charge transport properties already at concentrations well below the detection limits of standard mass spectroscopic techniques. These concentrations are also very low

---

TABLE 2.2

Maximum Deep-Level Concentrations According to Equations 2.18 and 2.19 to Obtain Mobility–Lifetime Products as Indicated

	$\mu\tau_e > 10^{-3}$ cm²/V	$\mu\tau_h > 10^{-5}$ cm²/V
Ionized deep traps	$<10^{11}$ cm^{-3} ($\sigma \sim 4 \times 10^{-13}$ cm²)	$<10^{12}$ cm^{-3} ($\sigma \sim 4 \times 10^{-13}$ cm²)
Neutral deep traps	$<10^{13}$ cm^{-3} ($\sigma \sim 3 \times 10^{-15}$ cm²)	$<10^{14}$ cm^{-3} ($\sigma \sim 3 \times 10^{-15}$ cm²)

Note: The capture cross sections are from Equations 2.16 and 2.17.

compared with any technically manageable purity of the crystals or intentionally introduced doping concentration levels.

2.2.5 SELF-COMPENSATION

Another fundamental implication comes from the requirement that the ionized deep levels have to compensate the overall net shallow level concentration offsets to achieve a semi-insulating state according to Condition 2.6. The limits per Table 2.2 simply mean that any shallow defect states need to be compensated by other shallow levels of opposite sign within about 0.1 ppb. This is an extremely close compensation requirement given the fact that intentional doping with shallow donor impurities in concentrations of several parts per million (ppm) can be technically necessary to overcompensate the residual impurity background of similar concentration.

Such close compensation can only be achieved by an energetically favorable, self-driven mechanism that compensates or deactivates a wide range of shallow level concentrations without introducing deep defects. Several specific mechanisms and their combinations have been proposed and investigated since the 1950s, mostly in context with the opposite technical problem of achieving high n- or p-type conductivity in the more ionic II–VI and III–V semiconductor systems for optoelectronic applications such as light-emitting devices [59,60].

In the latter case, self-compensation needs to be prevented or at least limited while in the radiation detector field, and it is exactly this mechanism that helps to achieve semi-insulating properties at relatively low deep-level concentrations. While successful qualitative and quantitative microscopic proof, good matching of experiment and model predictions, and full preparative control were achieved in some cases such as for ZnSe [61–63], any specifics regarding the compensation mechanism leading to semi-insulating CZT will not be detailed here given the current state of insufficient experimental evidence.

The common denominator for all possible self-driven electrical compensation mechanisms is the spontaneous formation of compensating defects, such as additional native point defects. This process is Fermi level–driven because the formation enthalpy of the charged defects with respect to the neutral ones is lowered by the energetic difference between the Fermi level and the defect level [40,64]. Under conditions close to the thermodynamic equilibrium and for the example of intentional doping with shallow donor impurities and compensation by (relatively) shallow acceptor-like native defects, the process would drive the Fermi level down to a critical value below which further generation of native defects would no longer be energetically favorable. This "optimized" Fermi level position also depends on the temperature and partial pressures in that the component partial pressures at a given temperature define the chemical potential for the generation of the neutral defects as, for example, vacancies. If the formation enthalpy and the energy levels associated with those native defects are such that the intrinsic Fermi level can be closely approached within the p-T existence region of the compound, semi-insulating material can be obtained, theoretically even without the introduction of deep levels. The spontaneous defect generation stops when the formation enthalpy of the charged native defects approaches zero (negative energies are required to drive the process). An additional practical requirement is that the corresponding defect equilibrium can be either frozen-in (quenching) or maintained by an appropriate cooling regime down to temperatures where kinetic barriers eventually stop any further uncontrolled equilibration.

The formation enthalpies and final defect states can also be affected by lattice deformations/relaxations or actual configuration changes; for example, calculations suggest that the Te antisite in the Jahn–Teller distorted configuration [65] may generate a donor level close to the middle of the bandgap in CZT [66,67]. Certain impurities can also be incorporated in multiple configurations, for example, substitutional and interstitial. If they act as donor in one configuration and as acceptor in the other configuration, they will be built into either configuration at a Fermi level–driven probability and partially compensate each other (autocompensation) [68,69]. Moreover, complex formation (e.g., A-centers) during cool-down from equilibration temperature can further alter the charge transport properties at RT.

To treat this field with appropriate respect, the reader is referred to the original works of deNobel [70] and Kroeger [71] and some excellent review articles [59,60].

2.3 DYNAMIC HIGH-FLUX POLARIZATION

2.3.1 X-RAY-GENERATED CHARGE

A primary mechanism of the energy dissipation of x-ray photons in semiconductors is multiple electron–hole pair generation by photoelectric excitation. The x-ray energy affects both the number of generated free carriers and the absorption profile.

In the case of full absorption of an incident x-ray flux, Φ, by photoelectric effect only, the pair-generation rate per detector cross section is

$$g_{\text{pair}} \sim \Phi * E_{\text{mean}}/E_{\text{pair}}. \tag{2.20}$$

The mean energy of the incident x-ray energy spectrum is E_{mean} or, in the case of thinner detectors, the mean energy of the absorbed photons. The average electron–hole pair-creation energy, E_{pair}, in semiconductors is typically about three times the bandgap energy [72], which amounts to about 4.6 eV for $\text{Cd}_{1-x}\text{Zn}_x\text{Te}$ with $x \sim 0.1$ at RT.

At higher incident x-ray energies, the probability of Compton scattering increases, in which case part of the energy is dissipated by the recoiling electron by subsequent photoelectric excitation of electron–hole pairs while the scattered photon will either dissipate its energy by photoexcitation, undergo another Compton scatter event, or escape from the detector, in which case charge is lost with respect to full energy deposition.

A general description of all possible primary and secondary processes occurring during the interaction of ionizing radiation with semiconductors is rather complex and excellent reviews can be found in the literature [73].

2.3.2 SPACE CHARGE AND DETRAPPING

Trapped charge carriers can cause electric field distortions within the detector. Applying the Poisson equation:

$$E(z) = -\frac{1}{\varepsilon_0 \varepsilon_r} \int_0^z q(r)dr, \qquad (2.21)$$

to, for example, a constant positive space charge, $q(r) = q_0$, shows that a trapped hole density of only 1.3×10^{11} cm^{-3} would be sufficient to completely shield the electric field at the anode of a 2 mm thick detector biased at 1000 V. Note that the dielectric relaxation time [74]:

$$\tau_{rel} = \varepsilon_0 \varepsilon_r \rho, \qquad (2.22)$$

is on the order of (20–30) ms in CZT at RT, which means that a semiconductor can stay charged with respect to the outside circuitry for a significant amount of time.

According to Section 2.2.2, there are basically two ways for a trapped excess charge carrier to be removed from a semiconductor: it can either be re-emitted into the respective band and (possibly after a sequence of multiple trapping/detrapping events) eventually drift out of the device or it can recombine at the localized defect site if the defect captures a charge carrier of the opposite sign. The probability of either event occurring depends on the defect properties (concentration, ionization energy, capture cross section), the temperature, and the available concentrations of nonequilibrium carriers of both types (injection level, i.e., photon or particle flux and energies). At low photon fluxes, the net recombination rates at the deep levels are very low and the dominant process at low photon fluxes is detrapping.

2.3.3 POLARIZATION IN PHOTON COUNTING

The phenomenon, which can be called "ballistic deficit assisted high-flux polarization" under intense x-ray irradiation is caused by the space charge accumulation of photogenerated holes close to the cathode, which leads to electric field pinching [75,76]. Electron charge clouds deposited on the cathode side must travel through the low-field region and therefore slow down. As the flux increases, the transit time can increase dramatically. Due to the ballistic deficit in the counting electronics, x-ray deposition events are recorded at lower energies. The entire spectrum shifts and more and more counts vanish below the counting threshold. This can lead to a sudden decrease in the photon count rate with increasing x-ray flux if a critical flux is exceeded. Independent experimental confirmation regarding the field deformation can be found in, for example, the work by Prekas et al. [77]. Note that in direct current (DC) mode, the steady-state photocurrent will always monotonically increase with flux, even in the high-flux polarization regime. Figure 2.18 shows the concept of electric field pinching due to space charge buildup [75].

2.3.4 DYNAMIC LATERAL LENSING

Another effect related to the internal charging of defects is dynamic lateral polarization, which can cause flux-dependent charge steering in high-flux applications [78]. It is caused by the same mechanism that leads to photon-counting polarization and is observed as lateral charge lensing [79].

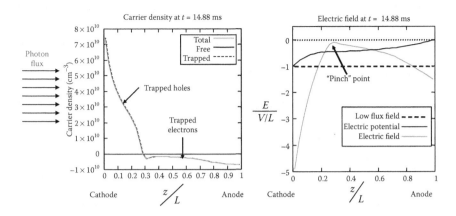

FIGURE 2.18 Illustration of electric field pinching due to space charge buildup. (From Bale, D.S. and Szeles, C., *Phys. Rev., B* 77, 35205, 2008.)

Figure 2.19 shows the pixel counts from a 500 μm pitch monolithic CZT detector array that polarizes [78]. The open x-ray beam was collimated to a 4 mm diameter area. The thin red circle marks the irradiated area. All pixels respond in a typical way to low flux. With increasing flux, a nonpolarizing device would show increasing intensity in the same area. In the polarizing device, however, counts are steered inward. The outer irradiated pixels progressively lose counts as the flux increases and eventually the whole device shuts off and only a few counts are seen in a donut-shaped pattern [78].

2.3.5 X-Ray Photocurrents

Without recombination losses, the photocurrent density, j_{photo}, should approach a bias-independent value on the order of

$$j_{photo} \sim e_0 g_{pair}. \tag{2.23}$$

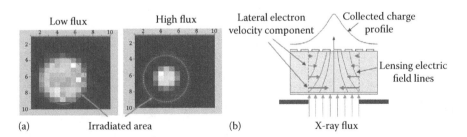

FIGURE 2.19 (a) Dynamic lateral lensing in a polarizing pixel array detector. (b) Pictorial with collimated flux on a polarizing pixelated detector showing the lensing electric field lines and the lateral component of the electron velocity. Top: the collected charge profile is shown for this situation. (From (a,b) Soldner, S.A., Bale, D.S., and Szeles, C., *Trans. Nucl. Sci.,* 54, 1723, 2007; (b) Bale, D.S., Soldner, S.A., and Szeles, C., *Appl. Phys. Lett.*, 92, 082101, 2008.)

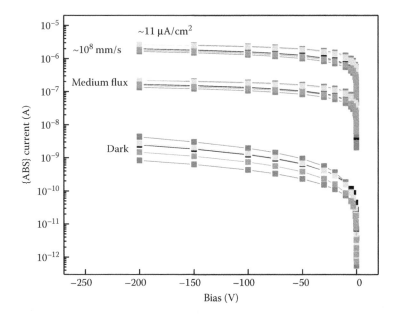

FIGURE 2.20 **(See color insert)** Bias dependence of dark and x-ray-induced steady-state currents from parallel-plate detectors of different CZT ingots (different colors).

If currents beyond the value per Equation 2.23 are measured, additional carrier injection via the electrodes must have occurred (photo-assisted breakdown).

Figure 2.20 shows dark and steady-state x-ray photo I–V curves from parallel-plate detectors of different CZT ingots. The two flux levels are high enough that the measured total currents at any given bias are much higher than the DCs and approximate the x-ray-generated steady-state photocurrents. The nominal incident x-ray flux densities coincide reasonably well with the numbers estimated from the photocurrents per Equation 2.23.

To understand the minor differences between samples from different ingots, the actual photocurrent transients must be looked at, as more or less severe recombination losses can lead to a decay of the photocurrent $j_{photo}(t)$ from its initial ($t=0$) value and Figure 2.20 resembles only the steady state.

The polarization dynamics will manifest themselves in the photocurrent transients. After a step-like flux change, it will take a finite time for the defect occupations and recombination-relevant transition rates to settle, that is, the photocurrent stabilizes at the rate with which the carrier system attains dynamic equilibrium with the photon field after flux changes [80]. In the case of sudden exposure, this is seen as a decay of the initial photocurrent toward a lower steady-state value (DC polarization) [80,81]. A similar time delay will be seen if the flux is reduced as the dynamics are simply reversed ("after-glow" if the detector was polarized in the high-flux state).

Figure 2.21 shows typical RT photocurrent transients of CZT detectors with different defect structures. While the electron mobility–lifetime products are similar for both detectors ($\sim 5 \times 10^{-3}$ cm²/V), the hole transport is significantly better for detector (a).

The photocurrent transients are controlled by the same dynamics that lead to electronically assisted count-rate polarization but there is an important difference

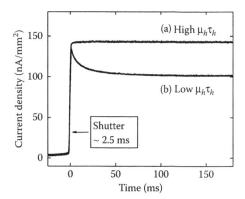

FIGURE 2.21 X-ray photocurrent temporal response from 2 mm thick CdZnTe detectors at 300 V bias: (a) high $\mu\tau_h$ material and (b) low $\mu\tau_h$ material. (From Prokesch, M., Bale, D.S., and Szeles, C., *Trans. Nucl. Sci.*, 57, 2397, 2010.)

between the two modes: The measured DC photocurrent is independent of the temporal trapping of the photogenerated charge as long as detrapping is more likely than actual charge loss due to recombination. There is no ballistic deficit in current integrating measurements so that a simple slowdown of the transient charge carriers cannot reduce the photocurrent. However, trapping can still indirectly affect the photocurrent as it deteriorates the electric field by space charge buildup (in the same way as described earlier for photon counting), at which point local charge accumulation together with high-generation rates can increase the recombination losses.

At higher bias, recombination is the only mechanism that can reduce the DC photocurrent from the expected levels per Equation 2.23 and DC polarization is therefore expected to occur at much higher fluxes than the (electronically assisted) count-rate polarization. The steady-state photocurrent will always increase monotonically with the flux even if the detector is significantly polarized as increased recombination balances the electric field distortion by removing space charge. The degree of polarization at given operating conditions (flux, bias, temperature, etc.) depends critically on the defect structure of a specific CZT sample.

Independent of the particular defect structure, CZT x-ray detectors can be "prepolarized" by subbandgap IR illumination in which case the photocurrent response to sudden x-ray exposure no longer shows any delay [80]. Figure 2.22 shows the temporal photocurrent response of a low $\mu\tau_h$ detector to a sudden high-flux x-ray exposure: curve (a) has been measured under standard conditions in the dark and curve (b) was obtained when the same detector was "conditioned" by constant IR illumination at a wavelength of 880 nm (~1.41 eV), which excites the entire detector volume. Note, however, that at this point, the internal electric field is massively distorted and the transit time for individual electron clouds traveling through the whole thickness of the device is dramatically increased.

2.3.6 INFRARED PHOTOCURRENTS

For materials characterization experiments aimed at studying DC polarization or preselecting material for high-flux applications, IR illumination can be alternatively

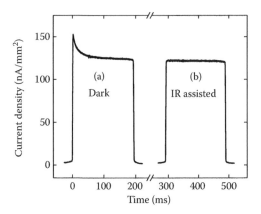

FIGURE 2.22 X-ray photocurrent temporal response to high-flux steps from a 2 mm thick, low $\mu\tau_h$ CZT detector at 600 V bias: (a) without IR and (b) with 880 nm IR illumination. The curves are corrected for the IR-induced photocurrent baseline. The curved start and end point are caused by the specifics of the x-ray shutter construction. (From Prokesch, M., Bale, D.S., and Szeles, C., *Trans. Nucl. Sci.*, 57, 2397, 2010.)

used instead of x-ray radiation to probe the high-flux response of the detector. This allows for more simple, less expensive, and more efficient setups, especially in terms of fast-flux switching.

Figure 2.23 compares the x-ray and IR-induced photocurrent responses of a CZT detector. Three different x-ray energies have been used, which affects the absorption profile. The steady-state values of all x-ray and IR-induced photocurrents were adjusted to the same value by regulating the x-ray flux and the IR diode current, respectively. The initial decay is more pronounced in the case of the IR measurement

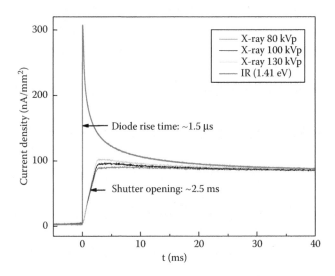

FIGURE 2.23 (**See color insert**) X-ray vs. IR-induced photocurrent transients.

for two reasons: In the case of IR illumination, there is no convolution of the fast initial decay with the relatively slow x-ray shutter motion. In fact, the photocurrent follows the nominal diode rise upon switching instantaneously (90% in 1.5 μs). The other difference is in the absorption characteristics. While the mean absorption depth for x-rays increases with the energy (largest decay for 130 kVp in Figure 2.23), IR illumination with energies below the bandgap energy generates electron–hole pairs almost uniformly throughout the entire detector volume.

Another fundamental difference between x-ray and IR excitation is the generation path of the electron–hole pairs. At subbandgap photoexcitation, the electron–hole pair generation can only occur via defect states in the forbidden gap. This has the important implication that the bulk generation efficiency in a certain exposure regime depends on the specifics of the defect structure of the semiconductor, that is, already the initial photocurrent ($t = 0$ value before polarization onset) contains information about the defect structure. The photocurrent evolution thus represents a convolution of charge generation and flux-dependent charge loss due to recombination (polarization dynamics) and both the initial photocurrent and the steady-state current depend on the defect structure. This is not the case for x-ray irradiation where the generation rates are simply defined by Equation 2.20. However, in the limit of full charge collection and no higher-order effects, similar generation rates, g_{pair}, will produce similar photocurrents.

Figure 2.24 shows the IR flux dependence of the photocurrent transients at constant bias. A notable manifestation of the convoluted dynamics is the appearance of temporal quasi-stable photocurrent plateaus upon flux onset. The time period during which the photocurrent stays stable at its initial, nonpolarized value increases as the flux decreases (delayed polarization).

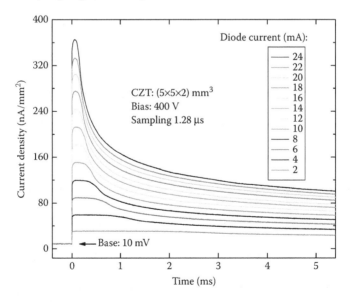

FIGURE 2.24 (**See color insert**) Example of the IR flux dependence of the photocurrent transients at constant bias.

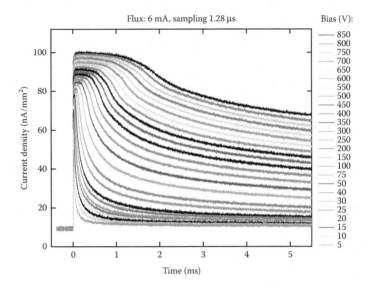

Figure 2.25 shows the bias dependence of the photocurrent transients at constant IR flux. At higher bias, the peak photocurrents are essentially bias independent if plateaus develop, that is, if polarization does not kick in instantaneously upon flux onset. A quasi-stable plateau response at higher bias is reached after ~80 μs and the plateaus extend in time as the bias increases (delayed polarization).

The flux and bias dependencies of the nonpolarized initial peak (plateau) and the polarized steady-state currents are summarized in Figure 2.26. As expected, the peak photocurrents are essentially bias independent (if the bias is not too low). At any given bias, both the peak and the steady-state photocurrents increase nearly linearly with the flux so that the ratio between the peak and the steady-state currents becomes essentially flux independent. This implies that at any flux, about the

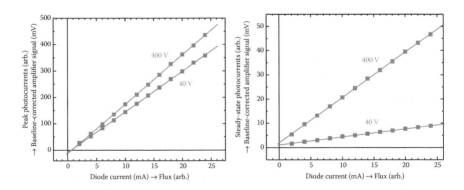

FIGURE 2.26 Peak and steady-state photocurrents from Figure 2.24 compared with the corresponding numbers at one-tenth of the original bias voltage.

same relative fraction of the photogenerated charge is lost due to recombination. This can be understood considering that in these experiments, even the lowest IR flux still generates a charge equivalent of several 10^7 x-ray photons/(mm^2 s) at 120 kVp. Eventually, the steady-state photocurrents are naturally strongly bias dependent in polarizing material as the degree of polarization is reduced as the bias increases.

2.4 SUMMARY

This chapter provided a brief introduction to a variety of materials and device aspects of CZT radiation sensors for gamma- and x-ray applications. The basics of semi-insulating semiconductor detector crystals were explained in terms of electrical compensation and charge trapping statistics. The fundamental implications of practically observed charge transport properties in detector grade CZT on the allowed deep-level defect concentrations were discussed in detail and the concept of electrical self-compensation was outlined. The final section focused on high-flux polarization phenomena in both photon counting and photocurrent modes.

REFERENCES

1. K. R. Zanio, Cadmium telluride, in: *Semiconductors and Semimetals*, Vol. 13, New York: Academic Press (1978).
2. T. E. Schlesinger, J. E. Toney, H. Yoon, E. Y. Lee, B. A. Brunett, L. Franks, and R. B. James, Cadmium zinc telluride and its use as a nuclear radiation detector material, *Mater. Sci. Eng.* R 32 (2001) 103.
3. R. Triboulet and P. Siffert, CdTe and CdZnTe growth, in: *CdTe and Related Compounds: Physics, Defects, Hetero- and Nano-structures, Crystal Growth, Surfaces and Applications*, 1st edn, Oxford: Elsevier (2009).
4. C. Szeles, S.E. Cameron, S. Soldner, J.-O. Ndap, and M.D. Reed, Development of the high-pressure electro-dynamic gradient crystal-growth technology for semi-insulating CdZnTe growth for radiation detector applications, *J. Electr. Mater.* 33 (2004) 742.
5. H. L. Malm, D. Litchinsky, and C. Canali, Single carrier charge collection in semiconductor nuclear detectors, *Rev. Phys. Appl.* 12 (1977) 303.
6. K. Parnham, J. B. Glick, C. Szeles, and K. Lynn, Performance improvement of CdZnTe detectors using modified two-terminal electrode geometry, *J. Cryst. Growth* 214–215 (2000) 1152.
7. D. Bale and C. Szeles, Design of high-performance CdZnTe quasi-hemispherical gamma-ray CAPture plus detectors. *Proc. SPIE* 6319 (2006) 63190B.
8. P. N. Luke, Single-polarity charge sensing in ionization detectors using coplanar electrodes, *Appl. Phys. Lett.* 65 (1994) 2884.
9. F. P. Doty, J. F. Butler, P. L. Hink, and J. R. Macri, Performance of submillimeter CdZnTe strip detectors, *Proceedings of IEEE Nuclear Science Symposium and Medical Imaging Conference*, San Francisco (1995) 80.
10. M. L. McConnell, J. R. Macri, J. M. Ryan, K. Larson, L.-A. Hamel, G. Bernard, C. Pomerleau, O. Tousignant, J.-C. Leroux, and V. T. Jordanov, Three-dimensional imaging and detection efficiency performance of orthogonal coplanar CZT strip detectors. *Proc. SPIE* 4141 (2000) 157.
11. M. A. J. van Pamelen and C. Budtz-Jørgensen, Novel electrode geometry to improve the performance of CdZnTe detectors, *Nucl. Instrum. Methods Phys. Res. A* 403 (1997) 390.
12. H. H. Barrett, J. D. Eskin, and H. B. Barber, Charge transport in arrays of semiconductor gamma-ray detectors, *Phys. Rev. Lett.* 75 (1995) 156.

13. A. Shor, Y. Eisen, and I. Mardor, Optimum spectroscopic performance from CZT γ- and x-ray detectors with pad and strip segmentation, *Nucl. Instrum. Methods Phys. Res. A* 428 (1999) 182.

14. D. S. McGregor, Z. He, H. A. Seifert, D. K. Wehe, and R. A. Rojeski, Single charge carrier type sensing with a parallel strip pseudo-Frisch-grid CdZnTe semiconductor radiation detector, *Appl. Phys. Lett.* 72 (1998) 792.

15. P. N. Luke, Unipolar charge sensing with coplanar electrodes: Application to semiconductor detectors, *IEEE Trans. Nucl. Sci.* 207–213 (1995) 42.

16. P. N. Luke, M. Amman, T. H. Prettyman, P. A. Russo, and D. A. Close, Electrode design for coplanar-grid detectors, *IEEE Trans. Nucl. Sci.* 44 (1997) 713.

17. S. A. Soldner, private communication (2002).

18. H. L. Malm and M. Martini, Polarization phenomena in CdTe nuclear radiation detector, *IEEE Trans. Nucl. Sci.* 21 (1974) 322.

19. A. Cola and I. Farella, The polarization mechanism in CdTe Schottky detectors, *Appl. Phys. Lett.* 94 (2009) 102113.

20. S. A. Soldner, A. J. Narvett, D. E. Covalt, and C. Szeles, Characterization of the charge transport uniformity of high pressure grown CdZnTe crystals for large-volume nuclear detector applications, Presented at the IEEE 13th International Workshop on Room-Temperature Semiconductor X- and Gamma-ray Detectors, Portland, OR (2003).

21. G. A. Carini, A. E. Bolotnikov, G. S. Camarda, G. W. Wright, R. B. James, and L. Li, Effect of Te precipitates on the performance of CdZnTe detectors, *Appl. Phys. Lett.* 88 (2006) 143515.

22. G. A. Carini, A. E. Bolotnikov, G. S. Camarda, and R. B. James, High-resolution X-ray mapping of CdZnTe detectors, *Nucl. Instrum. Methods Phys. Res. A* 579 (2007) 120.

23. A. Hossain, A. E. Bolotnikov, G. S. Camarda, Y. Cui, R. Gul, K. Kim, B. Raghothamachar, G. Yang, and R. B. James, Effects of dislocations and sub-grain boundaries on X-ray response maps of CdZnTe radiation detectors, *MRS Proceedings*, 1341 (2011).

24. P. Rudolph, Non-stoichiometry related defects at the melt growth of semiconductor compound crystals–A review, *Cryst. Res. Technol.* 38 (2003) 542.

25. K. Hecht, Zum Mechanismus des lichtelektrischen Primärstromes in isolierenden Kristallen, *Z. Phys.* 77 (1932) 235.

26. M. Mayer, L. A. Hamel, O. Tousignant, J. R. Macri, J. M. Ryan, M. L. McConnell, V. T. Jordanov, J. F. Butler, and C. L. Lingren, Signal formation in a CdZnTe imaging detector with coplanar pixel and control electrodes, *Nucl. Instrum. Methods Phys. Res. A* 428 (1999) 190.

27. A. E. Bolotnikov, N. M. Abdul-Jabbar, S. Babalola, G. S. Camarda, Y. Cui, A. Hossain, E. Jackson, H. Jackson, J. R. James, A. L. Luryi, and R. B. James, Optimization of virtual Frisch-grid CdZnTe detector designs for imaging and spectroscopy of gamma rays, *Proc. SPIE* 6706 (2007) 670603.

28. M. Prokesch and C. Szeles, Accurate measurement of electrical bulk resistivity and surface leakage of CdZnTe radiation detector crystals, *J. Appl. Phys.* 100 (2006) 14503.

29. T. Takahashi, T. Mitani, Y. Kobayashi, M. Kouda, G. Sato, S. Watanabe, K. Nakazawa, Y. Okada, M. Funaki, R. Ohno, and K. Mori, High resolution Schottky CdTe diodes, *Trans. Nucl. Sci.* 20 (2001) 100.

30. A. Bolotnikov, K. Ackley, G. Camarda, C. Cherches, Y. Cui, G. De Geronimo, J. Eger, et al., Using position-sensing strips to enhance the performance of virtual Frisch-grid detectors. Presented at SPIE, Hard X-Ray, Gamma-Ray, and Neutron Detector Physics XVI, San Diego (2014).

31. C. Szeles, S. A. Soldner, S. Vydrin, J. Graves, and D. S. Bale, Ultra high flux 2D CdZnTe monolithic detector arrays for x-ray imaging applications, *Trans. Nucl. Sci.* 54 (2007) 1350.

32. Y. Eisen and A. Shor, CdTe and CdZnTe materials for room temperature x-ray and gamma ray detectors, *J. Cryst. Growth* 184/185 (1998) 1302.

33. D. T. Marple, Effective electron mass in CdTe, *Phys. Rev.* 129 (1963) 2466.

34. D. Kranzer, Hall and drift mobility of polar p-type semiconductors. II. Application to ZnTe, CdTe, and ZnSe, *J. Phys. C: Solid State Phys.* 6 (1973) 2967.

35. J. L. Reno and E. D. Jones, Determination of the dependence of the band-gap energy on composition for $Cd_{1-x} Zn_x Te$, *Phys. Rev.* B45 (1992) 1440.

36. E. López-Cruz, J. González-Hernández, D. D. Allred, and W. P. Allred, Photoconductive characterization of $Zn_x Cd_{1-x} Te$ (0≤x≤0.25) single crystal alloys, *J. Vac. Sci. Technol.* A8 (1990) 1934.

37. M. Prokesch and C. Szeles, Effect of temperature- and composition-dependent deep level energies on electrical compensation: Experiment and model of the $Cd_{1-x} Zn_x Te$ system, *Phys. Rev. B* 75 (2007) 245204.

38. E. Fermi, Zur Quantelung des einatomigen idealen Gases, *Z. Phys.* 36 (1926) 902.

39. P. A. M. Dirac, On the theory of quantum mechanics, *Proc. Roy. Soc.* A112 (1926) 661.

40. J. A. van Vechten, A simple man's view of the thermochemistry of semiconductors, in: *Handbook on Semiconductors III*, Amsterdam: North-Holland (1980) 1.

41. J. M. Langer and H. Heinrich, Deep-level impurities: A possible guide to prediction of band-edge discontinuities in semiconductor heterojunctions, *Phys. Rev. Lett.* 55 (1985) 1414.

42. S. Krishnamurthy, A.-B. Chen, A. Sher, and M. van Schilfgaarde, Temperature dependence of band gaps in HgCdTe and other semiconductors, *J. Electron. Mater.* 24 (1995) 1121.

43. W. Shockley and W. T. Read, Statistics of the recombinations of holes and electrons, *Phys. Rev.* 87 (1952) 835.

44. R. N. Hall, Electron-hole recombination in germanium, *Phys. Rev.* 87 (1952) 387.

45. C. Szeles, private communication (2004).

46. S. W. S. McKeever, *Thermoluminescence of Solids*, Cambridge Solid State Science Series, New York: Cambridge University Press (1985).

47. R. Grill, private communication (2014).

48. R. E. Halsted, M. R. Lorenz, and B. J. Segall, Band edge emission properties of CdTe, *J. Phys. Chem. Solids* 22 (1961) 109.

49. C. H. Henry and D. V. Lang, Nonradiative capture and recombination by multiphonon emission in GaAs and GaP, *Phys. Rev. B* 15 (1977) 989.

50. M. W. Rowell, Coursework for applied physics 273, Stanford University, Stanford (2007).

51. S. K. Estreicher, T. M. Gibbons, By. Kang, and M. B. Bebek, Phonons and defects in semiconductors and nanostructures: Phonon trapping, phonon scattering, and heat flow at heterojunctions, *J. Appl. Phys.* 115 (2014) 012012.

52. R. Grill, J. Franc, H. Elhadidy, E. Belas, Š. Uxa, M. Bugár, P. Moravec, and P. Höschl, Theory of deep level spectroscopy in semi-insulating CdTe, *Trans. Nucl. Sci.* 59 (2012) 2383.

53. R. O. Bell, Review of optical applications of CdTe, *Rev. Phys. Appl.* 12 (1977) 391.

54. J. H. Zheng, H. S. Tan, and S. C. Ng, Theory of non-radiative capture of carriers by multiphonon processes for deep centres in semiconductors, *J. Phys. Condens Matter* 6 (1994) 1695.

55. T. Takebe, J. Saraie, and H. Matsunami, Detailed characterization of deep centers in CdTe: Photoionization and thermal ionization properties, *J. Appl. Phys.* 53 (1982) 457.

56. H. Elhadidy, J. Franc, P. Moravec, P. Höschl, and M. Fiederle, Deep level defects in CdTe materials studied by thermoelectric effect spectroscopy and photo-induced current transient spectroscopy, *Semicond. Sci. Technol.* 22 (2007) 537.

57. K. Suzuki, T. Sawada, and K. Imai, Effect of DC bias field on the time-of-flight current waveforms of CdTe and CdZnTe detectors, *Trans. Nucl. Sci.* 58 (2011) 1958.

58. J. Franc, R. Grill, J. Kubát, P. Hlídek, E. Belas, and P. Höschl, Influence of space charge on lux-ampere characteristics of high-resistivity CdTe, *J. Electron. Mater.* 35 (2006) 988.

59. G. F. Neumark, Defects in wide band gap II–VI crystals, *Mater. Sci. Eng. R* 21 (1997) 1.

60. U. V. Desnica, Doping limits in II–VI compounds: Challenges, problems and solutions, *Prog. Cryst Growth Charact.* 36 (1998) 291.

61. M. Prokesch, K. Irmscher, J. Gebauer, and R. Krause-Rehberg, Reversible conductivity control and quantitative identification of compensating defects in ZnSe bulk crystals, *J. Cryst. Growth* 214/215 (2000) 988.

62. K. Irmscher and M. Prokesch, Spectroscopic evidence and control of compensating native defects in doped ZnSe, *Mater. Sci. Eng.* B80 (2001) 168.

63. J. Gebauer, R. Krause-Rehberg, M. Prokesch, and K. Irmscher, Identification and quantitative evaluation of compensating Zn-vacancy-donor complexes in ZnSe by positron annihilation, *Phys. Rev. B* 66 (2002) 115206.

64. T. Y. Tan, H.-M. You, and U. M. Gösele, Thermal equilibrium concentrations and effects of negatively charged Ga vacancies in n-type GaAs, *Appl. Phys.* A56 (1993) 249.

65. H. A. Jahn and E. Teller, Stability of polyatomic molecules in degenerate electronic states. I. Orbital degeneracy, *Proc. Roy. Soc.* A161 (1937) 220.

66. S. H. Wei and S. B. Zhang, Chemical trends of defect formation and doping limit in II–VI semiconductors: The case of CdTe, *Phys. Rev.* B66 (2002) 155211.

67. A. Shepidchenko, S. Mirbt, B. Sanyal, A. Håkansson, and M. Klintenberg, Tailoring of defect levels by deformations: Te-antisite in CdTe, *J. Phys. Cond. Mat.* 41 (2013) 415801.

68. J. L. Merz, K. Nassau, and J. W. Shiever, Pair spectra and the shallow acceptors in ZnSe, *Phys Rev. B* 8 (1973) 1444.

69. C. G. van de Walle, D. B. Laks, G. F. Neumark, and S. T. Pantelides, First-principles calculations of solubilities and doping limits: Li, Na, and N in ZnSe, *Phys. Rev. B* 47 (1993) 9425.

70. D. de Nobel, Phase equilibria and semiconducting properties of cadmium telluride, *Philips Res. Rep.* 14 (1959) 430.

71. F. A. Kröger and H. J. Fink, Relations between the concentrations of imperfections in crystalline solids, in: *Solid State Physics*, Vol. 3, New York: Academic Press (1956) 307.

72. R. C. Alig and S. Bloom, Electron-hole-pair creation energies in semiconductors, *Phys. Rev. Lett.* 35 (1975) 1522.

73. G. F. Knoll, *Radiation Detection and Measurements*, New York: John Wiley & Sons (1979).

74. I. Ben-Yaacov and U. K. Mishra, *Unipolar Space Charge Limited Transport*, Berkeley: University of Southern California (2004) 1.

75. D. S. Bale and C. Szeles, Nature of polarization in wide-bandgap semiconductor detectors under high-flux irradiation: Application to semi-insulating $Cd_{1-x}Zn_xTe$, *Phys. Rev. B* 77 (2008) 35205.

76. D. S. Bale and C. Szeles, Multiple-scale analysis of charge transport in semiconductor radiation detectors: Application to semi-insulating CdZnTe, *J. Electron. Mater.* 38 (2009) 126.

77. G. Prekas, P. J. Sellin, P. Veeramani, A. W. Davies, A. Lohstroh, M. E. Özsan, and M. C. Veale, Investigation of the internal electric field distribution under in situ x-ray irradiation and under low temperature conditions by the means of the Pockels effect, *J. Phys.* D43 (2010) 085102.

78. S. A. Soldner, D. S. Bale, and C. Szeles, Dynamic lateral polarization in CdZnTe under high flux x-ray irradiation, *Trans. Nucl. Sci.* 54 (2007) 1723.

79. D. S. Bale, S. A. Soldner, and C. Szeles, A mechanism for dynamic lateral polarization in CdZnTe under high flux x-ray irradiation, *Appl. Phys. Lett.* 92 (2008) 82101.

80. M. Prokesch, D. S. Bale, and C. Szeles, Fast high-flux response of CdZnTe x-ray detectors by optical manipulation of deep level defect occupations, *Trans. Nucl. Sci.* 57 (2010) 2397.

81. Y. Du, J. LeBlanc, G. E. Possin, B. D. Yanoff, and S. Bogdanovich, Temporal response of CZT detectors under intense irradiation, *Trans. Nucl. Sci.* 50 (2003) 1031.

3 CdTe and CdZnTe Small Pixel Imaging Detectors

Matthew C. Veale

CONTENTS

3.1 INTRODUCTION

Semiconductor-based x-ray imaging detectors have been under development for many years. Ideally, these detectors should have high spatial resolution and excellent energy resolution and provide stable operation for extended periods. Substantial time and resources have led to the development of silicon-based detectors capable of meeting these requirements for x-ray energies <20 keV [1,2]. At higher x-ray energies, the poor mass attenuation coefficient of Si leads to a drastic reduction in detector efficiency. Alternative technologies, such as hyperpure germanium detectors (HPGe) [3], are capable of providing excellent energy resolution at these higher energies but are difficult to finely segment and require large cryogenic cooling systems.

The properties of compound semiconductors such as gallium arsenide (GaAs), mercuric iodide (HgI_2), thallium bromide (TlBr), cadmium telluride (CdTe), and cadmium zinc telluride (CdZnTe) are desirable for the production of high-energy x-ray imaging detectors [4]. The wide bandgap of these compounds means that they have high resistivity (>10^9 Ω·cm) and are capable of room-temperature operation, removing the need for cooling systems.

Despite the many advantageous properties of these compound semiconductors, many are still under development and face significant technological challenges before they are ready for mass distribution. GaAs-based detectors are known to contain high concentrations of traps that limit spectroscopic resolution and stability [5], while HgI_2- and TlBr-based detectors, while having very high resistivity (>10^{12} Ω·cm), still suffer from relatively poor spectroscopic performance, and their toxicity and structural stability represent challenges for some applications [6,7]. Of all the potential compound semiconductors, CdTe and CdZnTe have been the focus of most investigation.

3.2 CdTe/CdZnTe-BASED DETECTORS

3.2.1 Material Properties

The binary II–VI compound semiconductor CdTe was first investigated as an x-ray and γ-ray detector in the 1960s and produced encouraging results [8,9]. Typically formed from equal parts of Cd and Te, atomic numbers 48 and 52, respectively, the volume density is more than twice that of Si (5.85 g cm^{-3} compared with 2.33 g cm^{-3}). The high density of the material also means that the photoelectric cross section is much larger than that of Si for x-ray and γ-ray energies. The bandgap of CdTe is also much larger compared with other semiconductor materials, such as high-purity germanium (HPGe), with bandgaps of 1.57 eV and 0.67 eV, respectively. The large bandgap means that the material has inherently high resistivity, on the order of 10^9 Ω·cm, which allows detectors to run at room temperature without the need for large, expensive cryogenic cooling systems. The charge transport properties of electrons in CdTe are very good, with mobility (μ_e) on the order of 10^3 cm^2 V^{-1} s^{-1} and lifetimes (τ_e) of the order 10^{-6} s. In contrast, the transport properties of holes are much poorer than those of electrons. Typical hole drift mobilities are on the order of 10^1–10^2 cm^2 V^{-1} s^{-1}, at least an order of magnitude lower than those of electrons [10].

The properties of CdTe material may be improved by the addition of zinc during the growth of the material. The ternary alloy $Cd_{1-x}Zn_xTe$ consists of Zn atoms randomly substituted throughout the crystal lattice for Cd atoms. The addition of the Zn has two main effects: it raises the bandgap energy (for a 10% Zn fraction the bandgap is 1.572 eV compared with 1.50 eV for CdTe) as well as helping to reduce the dislocation density in the crystal. The increased bandgap decreases the number of thermal-generated carriers at room temperature, resulting in a higher-resistivity material and a reduction in detector leakage current (dark current): this allows thicker detectors to be produced. The lower dislocation density also improves the charge-transport properties by removing sources of trapping in the material. The charge transport of electrons in CdZnTe is also good, with electron mobilities on the order of $10^3 \, \text{cm}^2 \, \text{V}^{-1} \, \text{s}^{-1}$ and with lifetimes of around $10^{-6} \, \text{s}$. As with CdTe, hole transport is considerably poorer than electron transport in the alloy. Hole mobilities have been found to lie in the range $10^1-10^2 \, \text{cm}^2 \, \text{V}^{-1} \, \text{s}^{-1}$, and shorter lifetimes have also been reported, on the order of $10^{-5}-10^{-6} \, \text{s}$. The poor hole transport is due to trapping phenomena in the material: possible sources of trapping include impurities and crystal defects introduced during growth [11]. Table 3.1 summarizes the key properties of CdTe and $Cd_{0.9}Zn_{0.1}Te$.

3.2.2 SMALL PIXEL DETECTORS

Due to the poor charge transport properties of holes in CdTe and CdZnTe material, different electrode geometries have been developed to screen out the hole contribution to the detector signal. These single-charge sensing geometries include coplanar grids, Frisch-collar detectors and small pixel detectors [12]. Of these three geometries, only small pixel detectors are suitable for x-ray imaging with high spatial resolution, and these are the focus of the rest of this chapter.

Small pixel detectors make use of the Shockley–Ramo theorem, which describes the process of charge induction on the detector electrodes during a radiation interaction in the detector volume [13]. The magnitude of the charge induced on a given electrode due to charge drifting in the detector volume is given by Equation 3.1:

$$Q = -q\varphi_W(x) \tag{3.1}$$

where:

Q = charge induced on an electrode
q = charge moving within the detector
φ_W = weighting potential

Physically, the weighting potential represents the electrostatic coupling between the moving charges and the induced charge on the electrode. The weighting potential

---

TABLE 3.1

Comparison of Some of the Key Properties of CdTe and CdZnTe Materials

Material	CdTe	$Cd_{0.9}Zn_{0.1}Te$
Density (g cm^{-3})	5.78	5.85
Bandgap (eV)	1.57	1.51
Resistivity (W cm)	1×10^9	$1 \times 10^{10}-1 \times 10^{11}$
$\mu_e\tau_e$ (cm^2 V^{-1})	1×10^{-3}	1×10^{-2}
$\mu_h\tau_h$ (cm^2 V^{-1})	$<1 \times 10^{-4}$	$<1 \times 10^{-4}$
Pair creation value (eV ehp^{-1})	5	4.42
FWHM$_{@60 \, keV}$ (keV)	0.8	1.2
FWHM$_{@662 \, keV}$ (keV)	6.0	6.5

is not related to the physical potential, which describes the carrier velocity and trajectory, but is instead defined as "the potential at position x when the selected electrode is at unit bias, 1V, and all others are at zero potential" [13] and is dependent only on the carrier motion and electrode geometry.

In a small pixel detector, when the electrode size relative to the detector thickness becomes small, the weighting potential becomes constrained close to the pixel. In this instance only carriers that drift close to the pixel, where the magnitude of the weighting potential is large, will induce a significant charge on the electrode. Figure 3.1 shows examples of the weighting potentials produced using the Sentaurus Technology Computed Aided Design (TCAD) package [14] for a 2 mm thick CdTe detector for pixel pitches in the range 1.0–0.1 mm. As the pixel pitch is reduced, the weighting potential becomes progressively more concentrated at the anode.

For radiation interactions occurring close to the cathode, far from the anode pixels, the contribution to the detector signal from holes will be negligible, and the spectroscopic performance of the detector will be good. If the energy range over

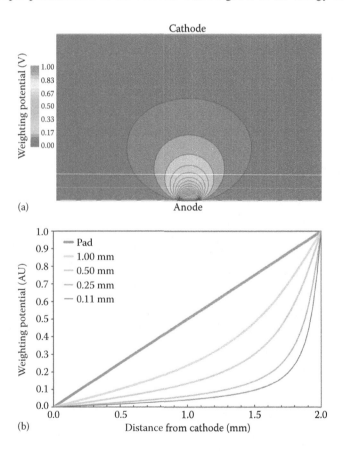

FIGURE 3.1 (**See color insert**) (a) A simulation of the weighting potential of a 250 μm pitch pixel in a 2 mm thick CdTe detector produced using Sentaurus TCAD. (b) A cross section of the weighting potential for varying pixel pitches in the range 0.11–1.00 mm.

which the detector will operate is known, the pixel pitch and thickness can be tuned to suit the application.

3.2.3 CHARGE-SHARING EFFECTS IN SMALL PIXEL DETECTORS

When processing data produced by small pixel detectors, the effects of charge sharing must be considered. On interacting within the detector, ionizing radiation produces a charge cloud of electron–hole pairs (ehp) that then drift under the applied potential. During this drift, the diameter of the charge cloud will increase due to the processes of carrier diffusion and charge repulsion. Whenever there is a carrier concentration gradient within the detector, Fick's law of carrier diffusion states that the charge will diffuse in the direction of the highest rate of decrease of the current density. For dense charge clouds, created by high-energy interactions, the electrostatic repulsion of carriers in the cloud may also lead to additional broadening of the charge cloud. If the drift time is long enough, the size of the charge cloud may become comparable to the pixel pitch, and when this occurs significant amounts of charge may be shared between pixels [15].

If the initial size of the charge cloud, and the effects of carrier repulsion, is assumed to be negligible, a reasonable assumption for interactions of energy <100 keV, an analytical solution of the diffusion equation gives Equation 3.2 [16]:

$$\sigma^2 = 2Dt_{drift} \tag{3.2}$$

where:

σ = width of the charge cloud
t_{drift} = time the charge has drifted
D = diffusion coefficient as described by Einstein's relationship:

$$D = \frac{\mu k_B T}{q} \tag{3.3}$$

where:

μ = mobility of the carrier
k_B = Boltzmann constant
T = temperature of the semiconductor
q = electrical charge

At room temperature, the diffusion coefficient for electrons in CdTe has a value on the order of 25 $cm^2 s^{-1}$, while holes have a value of 0.25 $cm^2 s^{-1}$ [17]. It is worth noting the magnitude of the diffusion coefficient, which is 40 times larger than the carrier mobility.

Figure 3.2a shows how the width of the charge cloud, as described by Equation 3.2, evolves during the drift of the carriers. For a 2 mm thick CdTe detector with an applied bias of −300 V, drift times will be on the order of 100 ns, by which time the charge cloud will have a diameter on the order of 60 μm. Figure 3.2b shows a

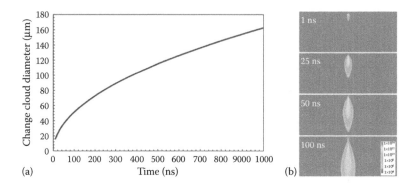

FIGURE 3.2 **(See color insert)** (a) The variation of charge-cloud diameter with drift time for a CdTe detector under –300 V bias. (b) A TCAD simulation of the dispersion of the electron cloud created by a 60 keV interaction in a 2 mm thick, 5 mm wide CdTe detector under the same bias voltage. The color scale depicts the electron density within the detector.

2-D Sentaurus TCAD simulation of the evolution of the electron cloud created by a 60 keV photon interacting in a 2 mm thick CdTe detector under a bias voltage of −300 V. During the drift, the charge cloud forms a distinct "teardrop" shape. The increase in the lateral width of the charge cloud with time is due to the diffusion and repulsion processes described above, while the length of the cloud is due to the application of the electric field, which causes the carriers to drift away from the original point of interaction. If the lateral width of the charge cloud represents a significant proportion of the pixel size, the percentage of charge-sharing events will be high, and this is an important consideration when designing a small pixel geometry.

Figure 3.3 a demonstrates the effect of charge sharing on the response of a small pixel detector. A 1 mm thick CdTe detector bonded to the HEXITEC ASIC was mapped using a collimated 10×10 μm beam of 20 keV x-rays. The response of the detector was mapped in steps of 25 μm. At each step, the spectra in each pixel were examined. When the beam is in the interpixel region, the number of counts increases despite the beam flux remaining constant, due to the charge being shared between multiple pixels. The number of pixels involved in the interaction, and the proportion of energy they acquire, is highly dependent on the beam position and the width of the charge cloud. Figure 3.3b shows how, for a beam position between four pixels, the interaction energy is distributed between the pixels. In this instance, the beam is positioned closer to pixel 1, and so this collects the majority of the 20 keV interaction energy. Without correction, these shared events lead to degradation of the spectroscopic performance of the detector and require correction [18].

A number of methods exist for correcting charge-sharing effects, and these can be broadly grouped into two categories: correction in hardware and postprocessing in software. A recent example of hardware correction is the Medipix3 application-specific integrated circuit (ASIC), in which the detector may be read out in single-pixel mode or charge summing mode, whereby interpixel communication is used to correct the effects of charge sharing [19]. In charge summing mode, the outputs of each group of four adjacent pixels on the ASIC are added together at a summing

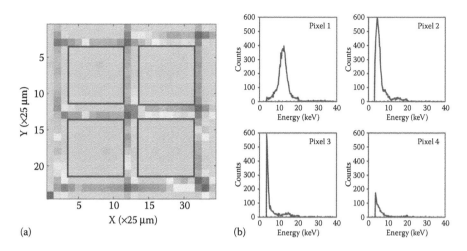

FIGURE 3.3 **(See color insert)** (a) A map of the measured x-ray intensity on a small pixel CdTe detector measured in steps of 25 μm. Data were taken with a 10 μm × 10 μm beam of 20 keV x-rays at the Diamond Light Source, UK. (b) The x-ray spectra measured in the four pixels when the beam is positioned at X12, Y13. (Adapted from Veale, M.C., Bell, S.J., Seller, P., Wilson, M.D. and Kachkanov, V., *J. Instrumentation*, 7, P07017, 2012.)

node, ensuring that at least one summing node will detect the full photon energy. This method allows rapid correction of data but is susceptible to pixel-to-pixel threshold variations, which can lead to errors in the summing process. The accuracy of the charge-sharing correction is also limited by the flux on the detector. At high flux rates, the probability of two interactions occurring in two neighboring pixels becomes high and can lead to individual events being erroneously summed together as a charge-sharing event.

In a fully spectroscopic ASIC, such as high-energy x-ray imaging technology (HEXITEC), the precise charge induced on each pixel is recorded for each frame of data, and this allows the user to process the data with multiple charge-sharing corrections to optimize efficiency and energy resolution. Figure 3.4 demonstrates the two correction algorithms that are used. The first technique is charge-sharing discrimination. In this instance, each frame of data is inspected for events that occur in groups of two or more pixels. Once these events are identified, they are simply discarded from the data set. Figure 3.4b clearly shows that the use of a discrimination algorithm leads to a large improvement in the spectroscopic performance of the detector compared with the raw data (Figure 3.4a), but this is at the expense of counting efficiency, which is decreased by 25%. In the case of the charge-sharing addition algorithm, each pixel is carefully calibrated in terms of energy prior to the acquisition, allowing the energy in each pixel to be accurately determined. When a charge-sharing event occurs between pixels, the energy deposited in each pixel is calculated, and the total energy is calculated from the sum of the individual responses. The addition technique recovers the majority of the shared events and improves the energy resolution of the detector relative to the raw data. Compared with the discriminated data, the charge-sharing addition leads to a broadening of the

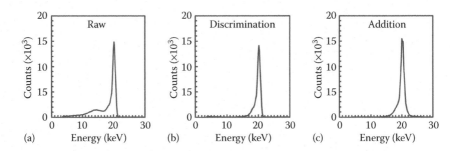

FIGURE 3.4 A comparison of different charge-sharing correction techniques in a single pixel of a small pixel CdTe detector. (a) The raw spectroscopic data measured by the detector. (b) The discriminated spectroscopic data, from which all charge-sharing events are removed. (c) The spectrum produced after charge-sharing addition, where shared events are combined to recreate the original interaction.

photopeak due to errors in calibration and the addition of electronic noise from each pixel. The accuracy of both algorithms is also limited by the noise edge of the detector, which determines the minimum detectable fraction of energy for each pixel. For the data shown in Figure 3.4, the position of the low-energy noise edge is at 4 keV, and so events involving fractions of charge <4 keV will remain uncorrected. Like the hardware corrections described earlier, these methods are also susceptible to high x-ray fluxes in which the probability of separate interactions occurring in neighboring pixels becomes high, resulting in real events being discarded or an overestimation of interaction energies. Whichever technique is used, charge-sharing correction is essential for any small pixel detector.

3.3 HIGH-ENERGY X-RAY IMAGING DETECTOR

3.3.1 HEXITEC PROJECT

In early 2000, the Science & Technology Facilities Council (STFC) identified a need within the United Kingdom's major science facilities, such as the fourth-generation Diamond synchrotron light source, for a new generation of x-ray imaging detector systems in support of cutting-edge science programs. An early grant from the UK Department of Trade and Industry funded the development of the energy-resolving detector (ERD) for x-ray fluorescence applications. The ERD was intended for the readout of solid-state pixel detectors, such as CdZnTe, and was sensitive to x-rays in the range 1 to 100 keV. The detector consisted of two different ASICs. The first was the MAC04 ASIC, which contained an array of 16×16 charge-sensitive preamplifiers on a 300 μm pitch that was flip-chip-bonded directly to CdZnTe detectors from eV Products (Saxonburg, PA) [20]. Each module also contained two SHC04 ASICs that provided signal shaping ($\tau = 2$ μs) and peak height sampling to produce pulse height spectra: a fully assembled ERD module can be seen in Figure 3.5a.

ERD modules were tested extensively at STFC and by the NASA Marshall Space Flight Center, the latter for use in balloon-borne and space-based high-energy x-ray astronomy projects [21]. The energy resolution of the detectors was good, with a

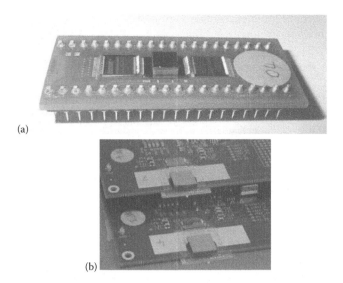

FIGURE 3.5   (a) An ERD module bump bonded to a $5 \times 5 \times 2$ mm^3 eV Products CdZnTe detector. (b) Two $20 \times 20 \times 5$ mm^3 eV Products CdZnTe detectors mounted on NUCAM test cards. (Adapted from Wilson, M.D., Seller, P., Hansson, C., Cernik, R., Marchal, J., Xin, Z-J., Perumal, V., Veale, M.C. and Sellin, P., X-ray performance of pixilated CdZnTe detectors, *IEEE Nuclear Science Symposium Conference Records,* 239–245, 2008.)

typical full width at half maximum (FWHM) of 2 keV measured for the 59.54 keV ^{241}Am photopeak with 2 mm thick CdZnTe detectors. One of the major issues with the ERD performance was the presence of spatial variations in the detector response; this was attributed to nonuniformities in the bulk CdZnTe material.

In parallel to the ERD project, an EU Framework Programme 5 grant was awarded to develop the NUCAM (Nuclear Camera) ASIC. The ASIC consisted of 128 low-noise channels, each containing a preamplifier, a CR-RC shaping circuit, and a time-over-threshold circuit [22]. While the ERD only provided pulse height spectroscopy, the NUCAM chip recorded both the charge deposited by a radiation interaction and the duration of the interaction. The time-over-threshold circuit differentiated the signal from the preamplifier, and, if the current was above the programmable (two-level) derivative threshold, a counter was enabled until the differentiated signal returned below threshold. As the hole mobility is two orders of magnitude lower than that of electrons, signals with significant contribution from hole transport consist of a fast electron-induced component and a slower hole component. Using the timing information provided by the ASIC, the time-over-threshold was correlated with the total signal height, giving an estimate of the depth of interaction that could be used to estimate the contribution from each carrier type.

The NUCAM ASIC [22] is currently being used by groups at the Institut de Ciències de l'Espai (CSIC-IEEC) and the University of Liverpool in Compton cameras for imaging of γ-rays for high-energy astrophysics [23] and medical imaging [24], respectively; a prototype system can be seen in Figure 3.5b. Measurements with single 2 mm thick Ohmic CdTe detectors from Acrorad have demonstrated an average energy resolution of 9.2 keV at 356 keV. Work is now under way in both groups to produce stacks of detectors for incorporation in camera systems.

Based on the experience gained during the ERD and NUCAM projects, the UK-based HEXITEC consortium was formed with the aim of developing a spectroscopic CdTe/CdZnTe x-ray imaging detector [25] suitable for use in a broad range of application areas, including the fields of materials science, security screening, medical imaging, and space science. The project has been led by the University of Manchester and includes the STFC, the University of Surrey, Durham University, and University of London, Birkbeck. In 2010, the consortium expanded to include the Royal Surrey County Hospital and University College London.

Each of the members of the consortium led work packages tailored to their specific expertise. Durham University led the development of the multitube physical vapor transport (MT-PVT) growth method for CdTe and CdZnTe material [26], while the University of Surrey is responsible for characterization of the material to be used in the project [27]. The Rutherford Appleton Laboratory (RAL) is responsible for the development of the readout ASICs used in the project as well as the data acquisition system. The remaining members of the collaboration are developing different imaging modalities and data-processing techniques that make use of the additional spectroscopic data provided by the detector.

3.3.2 HEXITEC DETECTOR SYSTEM

3.3.2.1 HEXITEC ASIC

The HEXITEC ASIC was developed for imaging applications in which important additional information regarding the object's properties is contained within the energy spectrum of the absorbed, transmitted, and emitted x-rays. To make use of this additional information requires a detector with excellent spectroscopic performance. The ASIC design was optimized to provide very good hard x-ray spectroscopy but at the expense of readout speed. It consists of 80×80 pixels on a pitch of 250 μm and has dimensions of 20.2 mm $\times$ 21.7 mm. Each pixel contains a 52 μm bond pad, which is used for gold stud and silver epoxy bonding to detectors. Figure 3.6 shows an image of the chip as well as a block diagram of the electronics contained within each pixel of the ASIC. The induced charge formed on the electrode of each pixel is read out using a charge preamplifier feedback circuit, which can compensate for a detector leakage current of up to 250 pA per pixel. The output of each pixel is filtered by a 2 μs peaking circuit that contains a CR-RC shaper in series with a second-order low-pass filter. A peak-track-and-hold circuit maintains the maximum voltage output of the shaper until the detector is read out using a rolling-row readout scheme. The readout is operated with a 20 MHz master clock with separate outputs for four blocks of 20×80 pixels; with this configuration, a full frame of 80×80 pixels is read out at a rate of 10 kHz. The ASIC is only sensitive to electron readout and has a noise performance of better than 800 eV per channel [28].

As described earlier in this chapter, the ability to identify and correct charge-sharing events is important for detector performance. In order to be able to correct charge-sharing events without significant degradation of the spectroscopic

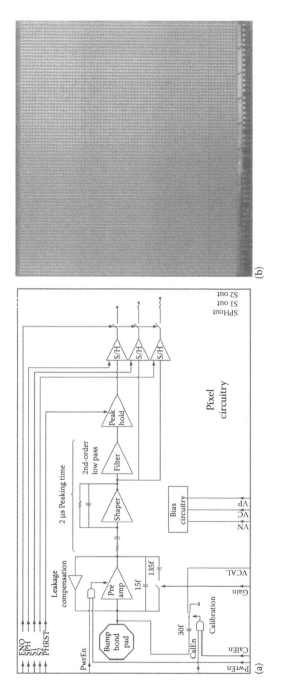

FIGURE 3.6 **(See color insert)** (a) A block diagram of the HEXITEC ASIC pixel electronics. (b) The 80×80 HEXITEC ASIC: the bottom edge of the ASIC is dedicated to I/O connections. (Adapted from Jones, L.L., Seller, P., Wilson, M.D. and Hardie, A., *Nucl. Inst. Meth. Phys. Res. A*, 604, 34–37, 2009.)

performance of the detector, only a limited number of pixels can detect an event in a single frame. It has been estimated that 10 million photons per second per detector ($\sim 2.5 \times 10^6$ photons s^{-1} cm^{-2}) is the maximum rate that can be detected with the ability to properly correct for charge sharing.

Since the first 20×20 pixel device was fabricated in 2008, the ASIC has undergone a number of iterations. The first version of the full 80×80 pixel ASIC had high- and low-gain modes, covering energy ranges of 4–200 keV and 40–1200 keV, respectively, while in the second version the low-gain range was adjusted to 12–600 keV to provide better energy resolution for some of the target applications. Until recently, all versions of the ASIC have been limited to tiling in $2 \times n$ arrays due to a 1.5 mm wide area along one edge that contains the I/O pads for wire bonding. Recently, a new version of the ASIC has been produced in which the I/O pads have been redistributed using through-silicon-via (TSV) technology onto the rear of the ASIC [29]. This allows detectors to be tiled in $n \times n$ arrays with minimum (~ 150 µm) dead space between individual modules [30].

3.3.2.2   Detector Hybridization

Materials such as CdTe and CdZnTe are sensitive to the temperature used during bonding processes. Above temperatures of $\sim 150°$C (300°F), structural changes may occur within the CdTe/CdZnTe material, and impurities and structural defects may become mobile within the crystal lattice; this may lead to degradation of detector performance. For this reason, conventional solder-reflow flip-chip assembly using eutectic tin–lead solder is unsuitable for detector hybridization [31].

At present, two low-temperature flip-chip assembly techniques are widely used for the assembly of CdTe and CdZnTe detectors. The technique of choice for assembling small pixel detectors with pixel pitches <150 µm, such as the Medipix family of detectors, is cold-weld indium bump bonding. In this process, indium bumps are deposited on both the sensor and the ASIC using standard photolithographic techniques and an indium evaporator. Following the deposition, the photoresist is removed, leaving behind bumps that are then brought together in a flip-chip bonder, and a small pressure is applied to form the cold weld. While this technique is very successful, it is an expensive process and may be unsuitable for larger pixel devices. The second low-temperature process, adhesive bonding, is much less expensive and suitable for larger pixel pitches.

Since the beginning of the HEXITEC project, a variety of different detector materials, including Si, GaAs(Cr), CdZnTe, and CdTe, have been hybridized using the gold stud and silver epoxy adhesive bonding technique. Silver-loaded epoxy is deposited on the sensor pixel pads using a stencil-printing technique, producing glue dots with a diameter of ~ 150 µm. In parallel to the epoxy deposition, gold studs are ultrasonically bonded to the ASIC bond pads to provide mechanical stability. The two components are then flip-chip-bonded together using a Suss FC150 flip-chip bonder and are cured at 80°C for 12 h. Finally, the assembled modules are wire bonded either to a standalone PCB or on an aluminum carrier that allows the detectors to be mounted in larger detector arrays; see Figures 3.7 and 3.8. Testing of detectors hybridized using this technique has consistently produced bonding yields of >99.9%.

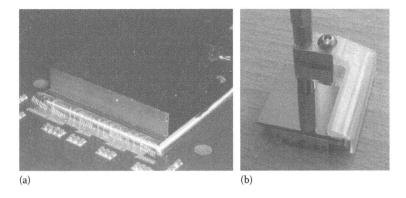

(a)   (b)

FIGURE 3.7   (a) A CdZnTe detector flip-chip-bonded to the HEXITEC ASIC and wire-bonded to a PCB. (b) A CdTe module mounted on an aluminum carrier for mounting in a multimodule system.

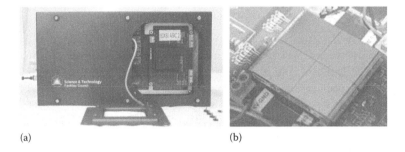

(a)   (b)

FIGURE 3.8   (a) The HEXITEC data acquisition system. (b) A 2×2 version of the HEXITEC system containing four individual modules mounted on aluminum carriers.

3.3.2.3   HEXITEC Data Acquisition System

The assembled detector modules "plug and play" directly into the HEXITEC data acquisition system (DAQ) (see Figure 3.8a). The DAQ contains all the electronics to control the detector system and acquire data. The voltage magnitudes produced by the HEXITEC modules are digitized with a 14-bit analog-to-digital converter (ADC) in the DAQ system and are transmitted to a dedicated PC over a base camera link connection. Prior to data collection, pixel offset voltages are measured and are applied as a correction to the ASIC output in hardware. The collected data is then processed in bespoke software, written in C++, that applies the desired charge-sharing corrections and arranges the detector output into peak height histograms for each pixel. A linear calibration per pixel using known lines from a known x-ray spectrum can then be used to convert the peak height spectra into an energy spectrum per pixel. A detail discussion of the calibration process for a small pixel spectroscopic detector can be found in a recent paper by Scuffham et al. [32].

While CdTe/CdZnTe systems are described as room-temperature detectors, systems still require temperature stabilization to achieve high-resolution performance. Each ASIC consumes ~1.5 W of power, and this produces a significant amount of heat. If this heat load is not managed, the ASIC temperature, measured using an onboard diode, can exceed 320 K. As described earlier, each channel of the ASIC can compensate up to a maximum of 250 pA of leakage current; beyond this, the spectroscopic performance of the detector is adversely affected. The power consumption of the ASIC means that relatively high temperatures (~60°C) can be reached if the temperature of the detector limits the bias voltages that can be used. To overcome these issues, the DAQ system contains a Peltier cooling system, run with a proportional-integral-derivative (PID) controller, which allows the temperature of the detector to be controlled in the range 275–310 K with an accuracy of ±0.1 K. The stability of the detector temperature is important, as changes of as little as 1 K can shift the pixel voltage offsets, causing the recorded peak heights to shift and degrading the spectroscopic performance of the detector. The atmospheric conditions within the DAQ are also controlled and monitored during acquisitions to ensure that no element of the detector system falls below the dew point; this prevents condensation forming, which could cause serious damage to the system.

3.3.3   CdTe Detector Performance

At present, the detectors of choice for use with the HEXITEC system are Schottky diode CdTe detectors fabricated by Acrorad Ltd. (Japan). Detectors have dimensions of $20.45 \times 20.35 \times 1.00$ mm^3 and consist of an array of 80×80 pixels on a 250 μm pitch, 50 μm spacing, with a 100 μm wide guard band running around three of the detector edges; the remaining edge has a 200 μm thick guard band. The anode pixel metal/semiconductor interface is produced from sputtered aluminum, producing a Schottky barrier. Additional layers of titanium and gold are deposited on top of the aluminum to aid adhesion; a final AlN passivation layer is deposited, containing openings for bonding of the detectors. The pad cathode that covers the second face of the CdTe is formed through electroless deposition of Pt. The full electrode structure of the detector is AlN/Au/Ti/Al/CdTe/Pt. Detectors are fabricated on a wafer scale and are diced into individual devices at the end of processing using a diamond saw. The edges of the detectors are left as cut and do not receive any additional processing [33].

The choice of a Schottky diode detector was made due to the relatively low resistivity of CdTe (~1×10^9 Ω·cm), which limits the stable operating voltage to −200 V in an ohmic arrangement; this has an adverse effect on the detector performance. The use of Schottky contacts allows higher operating voltages, typically −500 V, to be used, which leads to large improvements in the spectroscopic performance of the detector. Figure 3.9 shows a typical reverse bias current–voltage (I–V) curve measured for an Acrorad detector. The total leakage current of each module is measured as well as the leakage current collected through the guard band; this allows the edge leakage currents to be separated from the bulk pixels. The current density measured at the crystal edge is consistently an order of magnitude higher than in the bulk pixels, which is due to damage introduced into the edge during the dicing of detectors from wafers. At high voltages (>450 V), the total leakage current is dominated by the edge of the crystal and becomes unstable. The operating voltage of each CdTe detector is carefully chosen to ensure that the current drawn by the detector is stable over the expected acquisition time [34].

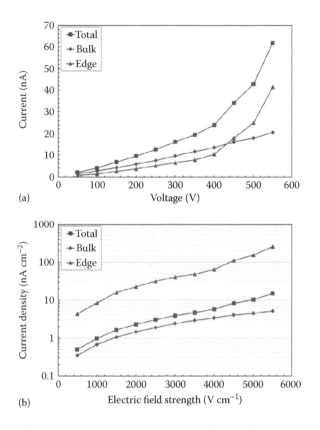

FIGURE 3.9 (a) A typical reverse bias I–V curve for a HEXITEC $20.45 \times 20.35 \times 1.00$ mm^3 Al-anode Schottky CdTe detector from Acrorad. The leakage current contributions from the crystal edges and bulk crystal are shown. (b) The same data but displayed as current density vs. field strength. The current density measured at the crystal edges (guard) is an order of magnitude higher than in the bulk.

Figure 3.10 a shows the spectroscopic performance of a typical single pixel of an Acrorad detector. The spectrum is from an ^{241}Am γ-ray source with the principal emission at 60 keV; the data has undergone a charge-sharing discrimination correction to optimize spectroscopic performance. The distribution of the measured FWHM for all 6400 pixels of a single detector is also shown in Figure 3.10b. Typically, >90% of the pixels have a FWHM of better than 1 keV measured at 60 keV, with an average value of 0.78 ± 0.19 keV. This level of spectroscopic performance is one of the best published to date for a CdTe in an imaging geometry [18].

Despite the excellent energy resolution of small pixel CdTe detectors fabricated with Schottky contacts, the effects of detector polarization require serious consideration. Spectroscopic performance is intimately related to the strength of the applied electric field, as this determines the charge-collection efficiency of the detector. The polarization phenomenon refers to instabilities in the electric field that evolve over time and lead to a progressive degradation of the energy resolution and counting efficiency. During the application of the bias voltage, holes are swept from deep

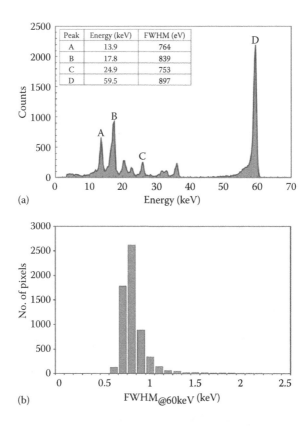

Peak	Energy (keV)	FWHM (eV)
A	13.9	764
B	17.8	839
C	24.9	753
D	59.5	897

(a)

(b)

FIGURE 3.10 (a) A typical single-pixel ^{241}Am γ-ray spectrum taken with an Acrorad Schottky CdTe detector bonded to the HEXITEC ASIC. (b) A histogram showing the typical distribution of FWHM for an entire 80×80 array.

acceptor levels and out of the detector volume. The remaining deep acceptors are now negatively ionized and result in a buildup of space charge that strengthens the electric field at the Schottky contact, lowering the barrier height and leading to an increase in leakage current with time. An associated decrease of the electric field at the cathode is also observed [35]. Polarization also occurs under extreme conditions, such as at low temperature (<270 K) or under high x-ray flux (>1 × 10^7 photons mm^{-2} s^{-1}) [36], but, as the HEXITEC detector does not operate under these conditions, they are not considered an issue.

Figure 3.11 a shows data recorded for a continuous exposure of a HEXITEC CdTe detector with a ^{241}Am source at −500 V and 280 K. Spectra are shown for the first and 30th hours of the acquisition. A comparison of the spectra shows a decrease in the number of counts detected across the entire energy range on the order of 25%. Additional tailing is also observed in the photopeaks, and this is particularly pronounced at lower energies. The reduction in the number of counts is consistent with an increase in the proportion of events that experience charge sharing. Figure 3.11b shows how the proportion of shared events varied over the 30 h acquisition; an increase of ~15% was observed. This increase is consistent with a broadening of

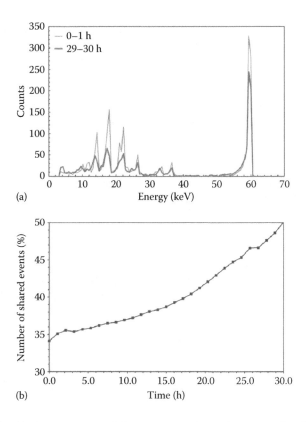

FIGURE 3.11 (a) A comparison of the spectroscopic performance of a Schottky CdTe detector at the beginning and end of a 30 h continuous acquisition. (b) The variation in the number of charge-sharing events over the same time period.

the charge cloud due to increased drift times that are a result of the reduction in the electric field close to the cathode due to polarization [37].

In the HEXITEC system, the effects of polarization are avoided through the use of a bias refresh scheme. Periodically, the operating bias is switched to 0 V and the detector left to settle for a long enough time that the space charge that has built up can recombine. In the majority of the CdTe detectors tested, typical bias refresh periods are between 60 and 120 s, where the bias is set to 0 V for 5–10 s. Data acquisition is paused prior to the bias refresh to avoid the introduction of noise into the spectroscopy; with these settings, no polarization effects are observed [38].

While the use of a bias refresh scheme successfully prevents polarization within the detector, it does introduce a significant dead time into measurements. For many of the target applications of the technology, this is not a significant problem, but in some instances, such as time-resolved imaging, it may be an issue. For HEXITEC, due to the individual ASIC pixel leakage-current constraints, CdTe detectors are limited to a thickness of 1.00 mm. For x-ray energies >100 keV, the mean free path of the photons is larger than the thickness of the detector, and this leads to progressively worse efficiency for higher energies. For these reasons, small pixel detectors fabricated from high-resistivity CdZnTe material are an attractive alternative.

3.3.4 CdZnTe Detector Performance

In recent years, access to high-quality spectroscopic-grade CdZnTe material, suitable for the fabrication of small pixel detectors, has improved [39]. Despite this, to date, there is still a lack of suppliers offering fabrication services for small pixel detectors. To address this need, a research program has been established at the RAL to develop a reliable process for the fabrication of small pixel CdZnTe detectors.

$Cd_{0.9}Zn_{0.1}Te$ spectroscopic-grade material grown by Redlen Technologies is lapped and polished using alumina slurry before chemomechanical polishing with bromine–methanol solution to the desired detector thickness, typically 2–3 mm. A standard photolithographic liftoff process is then used to pattern the detector anode before deposition of gold contacts using an electroless gold ($HAuCl_4 \cdot 3H_2O$) process. The gold contacts that are formed are typically 40–100 nm thick and form a complex metal–semiconductor interface. Due to a lack of material large enough to accommodate a full 80×80 pixel array, detectors are fabricated with an array of 74×74 pixels on the same 250 µm pitch. The pixel array is surrounded by a 200 µm guard band, giving detector dimensions of $19.3 \times 19.3 \times 2.0$–$3.0$ mm^3.

An example of the I–V characteristics of the detectors fabricated at the RAL is shown in Figure 3.12a. The I–V curve displays a "quasi-Ohmic" response, which is a characteristic of a device with back-to-back Schottky barriers. A focused ion beam was used to mill a trench through a gold contact, exposing the metal–semiconductor interface. An SEM image of this section can be seen in Figure 3.12b. The SEM image suggests that the interface between the gold and bulk CdZnTe material is complex, containing a number of layers of different stoichiometry, topography, and conductivity. The significance of these different layers is not yet fully known and is currently under investigation at the RAL [40].

Figure 3.13a shows an example of a single pixel spectrum, after charge-sharing discrimination, measured with a 2 mm thick small pixel CdZnTe detector fabricated at the RAL. The detectors are typically operated at 290 K and with an electric field strength of 2500 V cm^{-1}. For the spectrum shown in Figure 3.13a, a FWHM of 1.5 keV is measured for the 60 keV ^{241}Am photopeak. The distribution of FWHM across an entire CdZnTe detector is shown in Figure 3.13b. Of the 5746 pixels, 90% have a FWHM <2 keV, with an average value of 1.72 ± 0.95 keV. While this result is one of the best presented for CdZnTe material, the average value of the FWHM, as well as variation in values, is larger than that observed in CdTe detectors. If the performance of the CdZnTe detectors is to be improved further, the source of these variations requires investigation.

Microbeams of x-rays produced at synchrotrons are a powerful tool for the characterization of detectors, as shown in Figure 3.3. A 10 µm × 10 µm beam of 20 keV x-rays produced at the Diamond Light Source Synchrotron (UK) was used to map the detection efficiency of a CdZnTe fabricated at the RAL in 25 µm steps. Figure 3.14 compares the measured x-ray intensity map for two different regions of a CdZnTe detector. The first region represents an area of the detector where previous

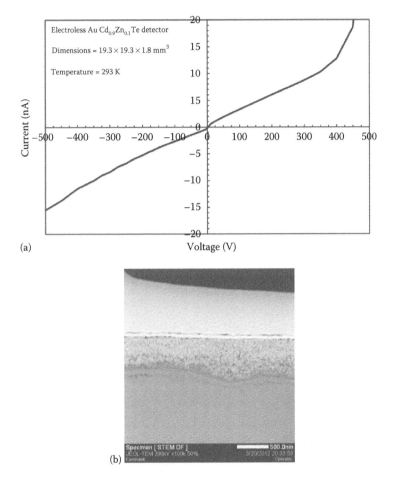

(a)

(b)

FIGURE 3.12 (a) The current–voltage characteristics of a $Cd_{0.9}Zn_{0.1}Te$ detector fabricated at the Rutherford Appleton Laboratory with gold contacts deposited via electroless deposition. (b) An SEM image of a focused ion beam (FIB) section of a contact deposited via electroless deposition. Areas of high contrast indicate conducting regions, while low-contrast regions are insulating. Multiple layers are visible in the image, indicating a complex metal–semiconductor interface.

measurements with sealed sources had shown good energy resolution, while in the second area the performance of pixels had been poor. In the well-performing region, the pixels appear as expected, in a mostly uniform array of square pixels. The map in the poorly performing region shows gross nonuniformities, with major deformations clearly visible in the pixel shape. A full analysis of these data (see Veale et al. [41]) suggests that these nonuniformities are due to the presence of potential barriers at the metal–semiconductor interface that lead to a buildup of space charge within the detector. The presence of this space charge distorts the field within the detector, causing variations in the effective field strength. If the metal–semiconductor interface shows spatial variations, as suggested by Figure 3.14b, changes in the

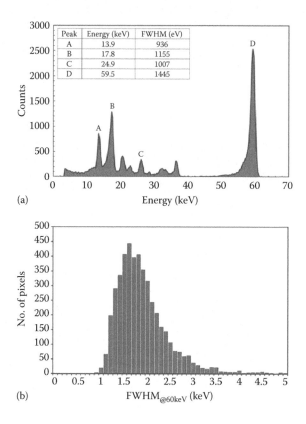

FIGURE 3.13 (a) A typical single-pixel ^{241}Am γ-ray spectrum taken with a detector fabricated from Redlen Technologies CdZnTe at the Rutherford Appleton Laboratory. (b) A histogram displaying the distribution of FWHM at 60 keV for the entire 74×74 pixel module.

concentration of space charge can lead to focusing, and defocusing, of the electric field, modifying the volume over which the pixel collects charge [42].

While small pixel detectors fabricated from CdZnTe show significant promise for the development of hard x-ray imaging systems, the overall uniformity of sensors needs to improve further to compete with CdTe detectors. While research into CdZnTe detector fabrication continues at the RAL, applications using the HEXITEC technology are being developed using CdTe as the detector material of first choice.

3.4   IMAGING APPLICATIONS

The ability to measure not only the number, but also the energy, of x-rays or γ-rays absorbed in a detector provides information that is useful across a wide range of applications. In the following section, a number of case studies are presented that demonstrate the use of spectroscopic imaging with the HEXITEC detector in different application areas.

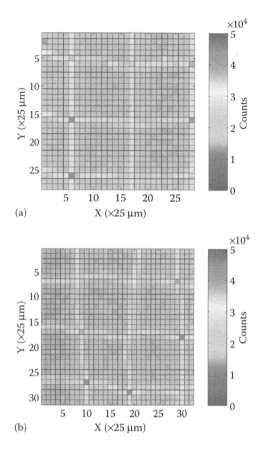

FIGURE 3.14 **(See color insert)** Maps of the measured x-ray intensity before charge-sharing correction in a good (a) and bad area (b) of a CdZnTe detector. The intensity maps were produced using a 10 μm × 10 μm beam of 20 keV x-rays in steps of 25 μm. (Adapted from Veale, M.C., Bell, S.J., Duarte, D.D., Schneider, A., Seller, P., Wilson, M.D., Kachkanov, V. and Sawhney, K.J.S., *Nucl. Inst. Meth. Phys. Res. A*, 729, 265–272, 2013.)

3.4.1 HOMELAND SECURITY: EXPLOSIVE DETECTION

O'Flynn et al. at University College London (UK) are currently developing the HEXITEC spectroscopic imaging technology for homeland security applications [43–45]. The detection of illicit materials, such as concealed drugs and explosives, is of vital importance to national security, most notably at airports and border controls.

In traditional detection systems, the attenuation of x-rays through the sample is used to detect the presence of organic materials. The drawback of this technique is that the transmission fingerprints of many of the materials of interest appear similar and are difficult to differentiate. An alternative method of detection is, instead of looking at the transmitted x-rays, to measure the x-rays diffracted by the sample. The diffraction process is governed by Bragg's law:

$$n\lambda = 2d \sin \frac{\theta}{2} \qquad\qquad (3.4)$$

where:

    λ = incident x-ray wavelength
    d = interatomic spacing in the diffracting material
    θ = angle through which the x-ray photons are scattered

The diffracted x-rays provide a unique fingerprint for different materials and allow identification of substances that may look similar under an x-ray transmission experiment—especially those with low effective atomic numbers.

Traditionally, x-ray diffraction measurements have been made in either energy-dispersive (EDXRD) or angular-dispersive (ADXRD) modes. In the EDXRD technique, the scattering angle is fixed and a polychromatic beam illuminates the sample, while in ADXRD a monochromatic x-ray beam illuminates the sample, and the diffracted intensity is measured while the angle of the beam is changed. Each of these techniques has its own advantages and disadvantages. ADXRD measurements are fast and produce data with very good Bragg peak resolution, but the technique requires mechanical movement of the sample. In EDXRD, the sample remains stationary, but the measured signal is blurred due to the angular range covered by the detector and requires accurate collimation to define the scattering geometry. The use of a spectroscopic imaging detector, such as HEXITEC, allows the EDXRD and ADXRD techniques to be combined into one single technique. Each pixel of the HEXITEC detector subtends a fixed angle in the range 0.6°–15.5° that can be accurately calibrated using known samples. If the angular data are combined with the energy-dispersive spectra collected by each pixel, the diffraction signature of different materials can be accurately determined.

Figure 3.15 shows diffraction data taken from a caffeine sample using a polychromatic x-ray source operated at a voltage of 60 kV. The caffeine sample is polycrystalline and contains many small randomly orientated particles. Diffraction from such a sample produces distinct "diffraction rings." Each pixel of the detector records every incident photon in the energy range 4–200 keV, and this allows multiple diffraction patterns to be captured at once; Figure 3.15a–f shows the different patterns recorded at 20, 30, and 40 keV [44]. The caffeine sample has known diffraction peaks and is used to calibrate the experimental arrangement and detector response in terms of angle. Once the detector and setup are accurately calibrated, the diffraction signatures of illicit materials can be studied.

An explosives simulant was produced, containing polycrystalline hexamine and pentaerythritol, which are key components of many plastic explosives such as RDX, C-4, and PETN. The simulant sample was embedded in a 24 mm thick plastic binding material, and diffraction data were collected. Figure 3.16 shows the distinctive patterns collected from the simulant sample at different energy windows. Unlike the caffeine sample, a number of focused high-intensity regions are detected, as well as diffuse diffraction rings. These regions of high intensity are due to the large grain

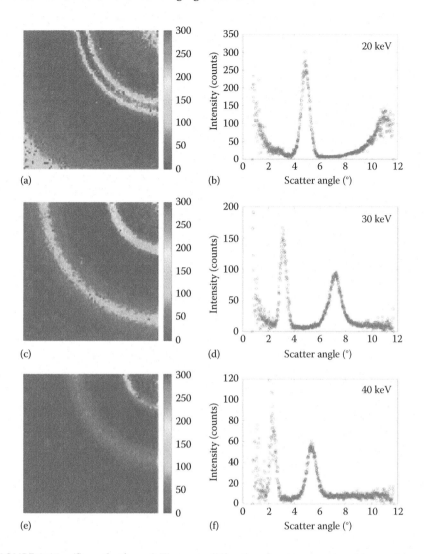

FIGURE 3.15 **(See color insert)** The x-ray diffraction signatures produced by a polycrystalline caffeine sample. The diffraction patterns and angular resolved signatures observed for photons at (a)–(b) 20 keV, (c)–(d) 30 keV, and (e)–(f) 40 keV. (From O'Flynn, D., Reid, C.B., Christodoulou, C., Wilson, M.D., Veale, M.C., Seller, P. and Speller, R.D., *Proc. SPIE*, 8357, 83570X-4, 2012.)

sizes of hexamine crystals relative to the x-ray beam size (in this instance 1 mm²) resulting in the production of characteristic Bragg peaks [43].

The unique signature produced by the simulant shows components of both hexamine and pentaerythritol, allowing both materials to be identified. Work is now under way to trial the technology in more realistic situations, for example concealing simulants in baggage, to demonstrate how diffraction data can be used to increase the detection sensitivity for these important materials.

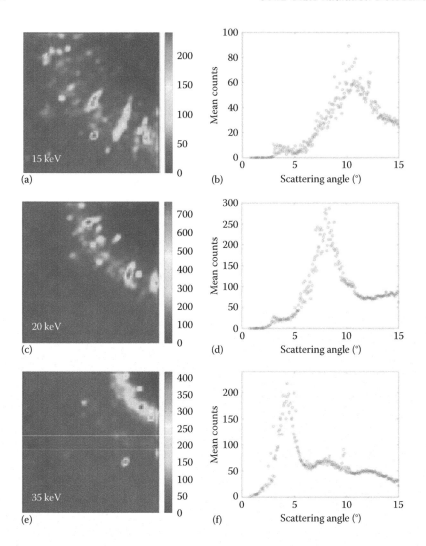

FIGURE 3.16 (See color insert) The x-ray diffraction signatures produced by an explosives simulant containing polycrystalline hexamine and pentaerythritol embedded in a plastic binding material. The diffraction patterns observed for photons at (a) 20 keV, (b) 30 keV, and (c) 40 keV. The angular resolved diffraction signatures detected at each energy are also shown in (d–f). (From O'Flynn, D., Desai, H., Reid, C. B., Christodoulou, C., Wilson, M. D., Veale, M. C., Seller, P., Hills, D., Wong, B., and Speller, R. D., *J. Crime Sci.*, 2, 4, 2013.)

3.4.2 MATERIALS SCIENCE: AEROSPACE MATERIALS CHARACTERIZATION

High-energy x-rays have the ability to penetrate deeply into materials, allowing the examination of dense objects such as welds in steel and geological core sections and the internal observation of chemical reactions inside machinery. Different experimental techniques such as x-ray fluorescence imaging (XRF) and x-ray diffraction imaging are powerful experimental techniques that provide qualitative information about the elemental composition and internal stresses and strains within a specimen. The use of these techniques requires x-ray detectors that are sensitive over a broad range of energies, such as the HEXITEC detector. In this example, spectroscopic imaging techniques developed by Jacques et al. [45,46] at the University of Manchester for the HEXITEC detector are used to characterize an important modern joining technique, friction stir welding (FSW) [47,48]. The use of spectroscopic imaging provides important information on not only the reorganization of the crystal lattice within the central weld but also the redistribution of impurities around the weld.

The FSW technique is a modern advanced joining process that uses frictional forces to fuse two surfaces together without melting either surface, as is the case in traditional fusion welding. A rapidly rotating tool is used to join the two surfaces, which are butted and clamped together. The friction generated by the rotating tool as it moves across the butted surfaces causes the two materials to soften (without melting) and mechanically mix [49]. The weld that is formed has good mechanical properties with a reduced number of defects compared with fusion welds. However, insufficient welding temperatures, due to low tool rotational speeds or high transverse speed, can lead to defects in the FSW joint. Understanding the microstructural properties of these welded joints, and the effect of key parameters such as the shape of the welding tool and the speed of rotation, is critical to improving weld strength and toughness [47].

A polychromatic x-ray beam of 50 kV x-rays produced at the Diamond Light Source Synchrotron was used to illuminate a section cut from an FSW joint with dimensions of $6 \times 6 \times 40$ mm^3. The joint was produced from an aluminum alloy (AA7050-T6) designed for light weight, strength, toughness, and resistance to stress-corrosion cracking, and commonly used in aircraft and other aerospace structures. The HEXITEC detector was mounted off-axis with a pinhole aperture positioned at low angle to the direct beam (typically $1°–5°$) to measure the x-rays scattered by the sample. The projected scattering angle varies across the detector array, and an angular calibration is used to replot the spectra onto a common axis of reciprocal d-spacing. The results of these imaging measurements can be seen in Figure 3.17.

The spectra measured by the HEXITEC detector display a number of characteristic Bragg peaks corresponding to the interplanar spacing of the crystalline material (see Figure 3.17). The diffraction peaks observed from the aluminum FSW joint were typically from the (111), (200), and (220) planes, demonstrating a strong preferred orientation. The microstructure of the center of the weld shows a preferred alignment to the (111) plane, which is parallel to the aluminum sheet surface, consistently with previous findings [50]. The fine grain structure observed in the central weld is due to the dynamic recrystallization of the material from the extreme thermomechanical action of the rotating tool. Outside the central weld, the intensity of the (111) peak drops to zero and the orientation is typically (200) and (220), indicative of the as-rolled aluminum plate.

Inspection of the individual diffraction patterns outside the central weld shows a small broad peak at a reciprocal d-spacing of 0.46 nm^{-1}; the intensity of this peak is shown in Figure 3.17. Regions of increased intensity are observed just outside the central weld, while inside the intensity is greatly reduced. The type of aluminum

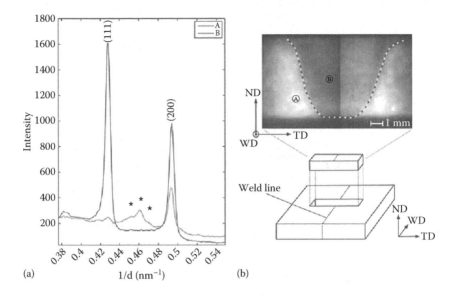

FIGURE 3.17 (a) The diffraction patterns observed from outside (A) and inside (B) the central weld. Stars indicate the theoretical positions for hexagonal $MgZn_2$ diffraction peaks (112), (201), and (004). (b) The scattering intensity at an equivalent reciprocal d-spacing of 0.46 nm^{-1}; the central weld region is highlighted by a black/white dotted line. (Adapted from Egan, C.K., Jacques, S.D.M., Connolley, T., Wilson, M.D., Veale, M.C., Seller, P. and Cernik, R.J., *Proc. Royal Soc. A: Math, Phys. & Eng. Sci.*, 470 (2165), 20130629, 2014.)

alloy studied here is known to contain minority phase impurities, including $MgZn_2$ precipitates. The theoretical positions of the (112), (201), and (004) diffraction peaks from hexagonal $MgZn_2$ are also shown in Figure 3.17; these correspond well to the peaks observed at 0.46 nm^{-1}. The FSW technique is known to dissolve precipitates inside both the central weld and the thermomechanically affected zone [51]. The increased intensity observed outside the central weld is due to a redistribution of these impurities during the FSW process.

This example demonstrates how the fully spectroscopic data collected with the HEXITEC detector can be used to simultaneously determine chemical and structural information about a sample. The techniques being developed provide a powerful new analytical tool that will further the understanding of chemical and physical structures within materials. Potentially exciting applications include the *in situ* analysis of materials exposed to hostile regimes (e.g., ceramic matrix composites or superalloys at high temperature), *in operando* chemical systems (e.g., fuel cells, batteries, or heterogeneous catalysts), or mechanical engineering components (e.g., stress-corrosion cracking or strain).

3.4.3 MEDICAL IMAGING: MULTI-ISOTOPE SPECT IMAGING

Single photon emission computed tomography (SPECT) is a functional medical imaging modality that plays a key role in the diagnosis of a wide range of clinical

conditions. Unlike structural imaging modalities, such as CT and magnetic resonance imaging (MRI), SPECT imaging allows the specific function of different
organs to be studied. The patient is injected with radiopharmaceuticals that incorpo-
rate a radioactive tracer atom, such as ^{99m}Tc or ^{123}I, as part of a larger pharmaceuti-
cally active molecule. These tracers become localized in the target organs, and the
decay of the radionuclides is subsequently detected with a γ-ray camera. Recently,
interest has grown in SPECT imaging in the fields of cardiology and neurology.
Increases in the prevalence of heart and degenerative brain diseases in the world
population have increased the demand for these functional imaging modalities [52].

At present, most SPECT imaging is carried out using a single radiotracer, but
often this cannot provide the full clinical picture. SPECT imaging using multiple
radiotracers offers the prospect of imaging several functional processes simultaneously, providing images that are intrinsically spatially and temporally coregistered.
Multi-isotope imaging has the potential to reduce the dose to the patient and reduce
the number of outpatient visits to the clinic, shortening the time to diagnosis. To
implement multi-isotope imaging requires detectors with excellent spectroscopic
performance that can distinguish between the closely spaced emissions from different radioisotope tracers. For example, a number of common radiopharmaceuticals
are produced with ^{99m}Tc and ^{123}I, which have γ-ray emissions at 141 keV and 159 keV,
respectively.

The current generation of medical γ cameras use scintillator-based detectors to
measure the radioactive decay of the radiopharmaceuticals. While scintillator detectors are a well-developed technology, are widely available, and have good counting
efficiency, they have poor spectroscopic performance and struggle to distinguish
between different radiopharmaceuticals, requiring significant correction of images.
Direct-conversion semiconductor detectors, such as CdTe, offer much better spectroscopic performance but have only recently become more widely available.

Multi-isotope imaging is currently being developed by Scuffham et al. [53] at
the University of Surrey and the NHS Royal Surrey Hospital using the HEXITEC
technology. Figure 3.18 demonstrates the typical spectroscopic performance of the
detector system for a number of γ-ray emissions from common radiopharmaceuticals: ^{67}Ga (93 keV), ^{99m}Tc (141 keV), and ^{123}I (159 keV). The spectrum was recorded
with a 1 mm thick, 80×80 pixel, Schottky CdTe detector operated at 20°C; the spectroscopy has been corrected for charge sharing using a discrimination algorithm.
The average FWHM measured by the detector at the different energies are given in
Table 3.2. The measured energy resolutions are an order of magnitude better than
typical values measured with current clinical γ cameras based on scintillators.

The multi-isotope SPECT imaging performance of the HEXITEC system was
compared with that of a GE Infinia γ camera (GE Healthcare, United States) [54].
The main chamber of an anthropomorphic brain phantom (see Figure 3.19a) was
filled with approximately 1200 MBq of ^{99m}Tc, while the striatal compartments,
whose function in a real brain is to help coordinate motivation with body movement,
were filled with 580 MBq of ^{123}I. The phantom was imaged in vertex view using
both detectors and images created for energy windows (±5%) around the ^{99m}Tc and
^{123}I γ-ray emissions. Figure 3.19b compares the SPECT images produced by both
cameras for each of the energy windows.

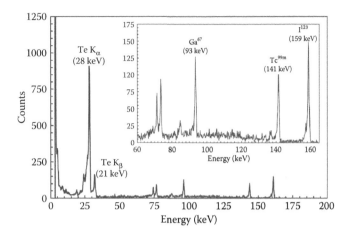

FIGURE 3.18 An example of the spectroscopic performance of a HEXITEC 1 mm thick CdTe detector at high energies (>100 keV). The spectrum is for a single pixel after charge-sharing discrimination and contains γ-ray peaks from the radioisotopes ^{67}Ga (93 keV), ^{99m}Tc (141 keV), and ^{123}I (159 keV).

TABLE 3.2

Spectroscopic Performance of HEXITEC Detector for Common γ-Ray Emissions from Radiopharmaceuticals

Radioisotope	Energy (keV)	FWHM (keV)	FWHM (%)
^{67}Ga	93.3	1.0 ± 0.4	1.1 ± 0.4
^{99m}Tc	140.5	1.1 ± 0.4	0.8 ± 0.3
^{123}I	159.0	1.2 ± 0.4	0.8 ± 0.3

The images demonstrate that the HEXITEC detector outperforms the conventional γ camera in terms of both the image contrast and the amount of cross talk observed between the two energy windows. The use of a high-resolution spectroscopic detector allows the emissions from the two different radiopharmaceuticals to be clearly distinguished, and there is no reason why additional tracers, such as ^{67}Ga, could not also be simultaneously imaged. The next stage of this project is to produce tiled modules of detectors to create a system with a field of view appropriate for human imaging. It is hoped that future generations of the instrument will also make use of thicker CdZnTe detectors, which will optimize detection efficiency at the highest energies.

3.5   SPECTROSCOPIC IMAGING: AN OUTLOOK

As shown previously in this chapter, the parallel developments of high-quality CdTe and CdZnTe material and the production of highly pixelated, low-noise electronics,

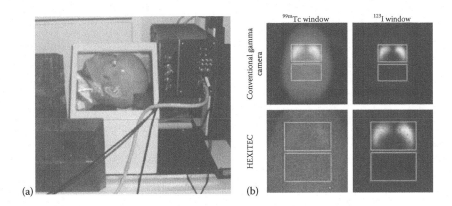

FIGURE 3.19 **(See color insert)** (a) Dual isotope SPECT imaging of an anthropomorphic brain phantom using the HEXITEC detector. The majority of the phantom is filled with ^{99m}Tc, while the striatal compartments are filled with ^{123}I. (b) A comparison of the imaging performance of a traditional GE Infinia γ camera and the HEXITEC detector system. Images are shown for energy windows around the principal emissions of the ^{99m}Tc and ^{123}I radioisotopes. The use of the HEXITEC detector greatly reduces the cross talk between the two images.

such as the HEXITEC ASIC, have made spectroscopic imaging detectors a reality. While this technology has been demonstrated in the laboratory, there are still significant challenges that must be overcome to produce industrially relevant instruments for real-world applications. A significant challenge that still remains is the production of large-area, highly pixelated detectors.

CdTe and CdZnTe material is now routinely grown on a 3" (75 mm) wafer scale, with major manufacturers now working toward 4" processes (100 mm). The main roadblock to large-area detectors is no longer the size of the available detector crystals, but is, instead, due to the limitations in the maximum ASIC size. The typical sizes of reticles (photomasks) used during the fabrication of ASICs by foundries limit the maximum chip areas to the order of 6 cm^2. Based on these limits, any large-area detector has to be fabricated from multiple modules in a tiled array requiring detectors that can be mounted together with the minimum amount of inactive area between tiles. To achieve this necessitates the development of so-called "four-side-buttable" technology.

Standard ASIC geometries require areas around the edge of each chip for I/O connections so that power can be supplied to the chip and signals delivered to and from data acquisition systems. To reduce the size of these areas, or to remove them completely, requires the development of new technologies such as through-silicon-via (TSV) processing. TSV processing is used to redistribute connections from the top surface of the ASIC to the reverse, as described earlier in Chapter 6. With the I/O connections redistributed on the rear of the ASIC, detectors can be tiled in $N \times N$ arrays, with only cost and data volumes limiting the size of detectors. To date, only a small number of trials of TSV technology for imaging detectors have been carried out, but, as this technology matures and becomes adopted commercially, it will allow new detector geometries to be developed.

While the use of TSV technology allows the production of four-side-buttable ASICs, there is currently a lack of suitable detectors for large arrays. Typically, detectors are produced with guard-ring structures that limit the effect of the crystal edges, which tend to act as paths for excess leakage current due to the presence of high densities of crystalline defects. These guard bands represent a significant dead area between individual modules and ideally would be completely removed. While a number of different groups are currently investigating so-called edgeless (or active edge) silicon detectors, little has been done for either CdTe or CdZnTe detectors. The production of active-edge sensors will require the development of edge-processing techniques and a thorough understanding of phenomena occurring at crystal edges.

Perhaps the greatest technological challenge faced by the x-ray and γ-ray imaging community is the development of detector systems for high-flux applications. The development of medical imaging equipment, next-generation synchrotron light sources, and x-ray free electron lasers (XFEL) is making ever-increasing demands on detectors. To operate at these high rates, both electronics and materials issues must be overcome. Using current ASIC technology, detectors tend to be optimized for either high-resolution spectroscopic performance, like HEXITEC, or for high-rate operation, like Medipix. To produce detectors that are capable of both spectroscopic performance and high-rate operation will require the development of new ASIC technology, such as TSV interconnects, which will allow greater functionality to be incorporated onto each chip.

In the case of CdTe and CdZnTe materials, at very high rates they are susceptible to polarization effects whereby the buildup of trapped charge, space charge, causes the electric field to collapse within the detector. It is hoped that further improvement of the crystalline quality, purity, and fabrication processes may allow higher-rate operation, but, for extreme count rates like those at XFEL, this may ultimately require the development of a new generation of detector materials.

ACKNOWLEDGMENTS

The author would like to thank Steven Bell and Diana Duarte for their hard work on materials characterization and detector simulation, which has contributed to this chapter. Thanks are also given to the many students, postdocs, and researchers whose hard work over the last decade has led to the success of the HEXITEC project.

REFERENCES

1. Ch. Broennimann, E. F. Eikenberry, B. Henrich, R. Horisberger, G. Huelsen, E. Pohl, B. Schmitt, et al. The PILATUS 1M detector, *Journal of Synchrotron Radiation*, 13, 120–130 (2006).
2. P. Pangaud, S. Basolo, N. Boudet, J.-F. Berarb, B. Chantepie, J.-C. Clemens, P. Delpierre, et al. XPAD3-S: A fast hybrid pixel readout chip for x-ray synchrotron facilities, *Nuclear Instruments and Methods in Physics Research. Section A*, 591, 159–162 (2008).
3. P. Sangsingkeow, K. D. Berry, E. J. Dumas, T. W. Raudorf and T. A. Underwood. Advances in germanium detector technology, *Nuclear Instruments and Methods in Physics Research. Section A*, 505, 183–186 (2003).

4. A. Owens and A. Peacock. Compound semiconductor radiation detectors, *Nuclear Instruments and Methods in Physics Research. Section A*, 531, 18–37 (2004).

5. M. Rogalla, J. W. Chen, R. Geppert, M. Kienzle, R. Irsigler, J. Ludwig, K. Runge, et al. Characterisation of semi-insulating GaAs for detector application, *Nuclear Instruments and Methods in Physics Research. Section A*, 380, 14–17 (1996).

6. E. Ariesanti, A. Kargar and D. S. McGregor. Fabrication and spectroscopy results of mercuric iodide Frisch collar detectors, *Nuclear Instruments and Methods in Physics Research. Section A*, 624, 656–661 (2010).

7. H. Kim, A. Kargar, L. Cirignanog, A. Churilov, G. Ciampi, W. Higgins, F. Olschner and K. Shah. Recent progress in thallium bromide gamma-ray spectrometer development, *IEEE Transactions on Nuclear Science*, 59, 243–248 (2012).

8. D. A. Jenny and R. H. Bube. Semiconducting cadmium telluride, *Physical Review*, 96, 1190–1191 (1954).

9. R. O. Bell, N. Hemmat and F. Wald. Cadmium telluride, grown from tellurium solution, as a material for nuclear radiation detectors, *Physica Status Solidi A: Applications and Materials Science*, 1, 375–387 (1970).

10. T. Takahashi and S. Watanabe. Recent progress in CdTe and CdZnTe detectors, *IEEE Transactions on Nuclear Science*, 48, 950–959 (2001).

11. T. E. Schlesinger, J. E. Toney, H. Yoon, E. Y. Lee, B. A. Brunett, L. Franks and R. B. James. Cadmium zinc telluride and its use as a nuclear radiation detector material, *Materials Science and Engineering Reports*, 32, 103–189 (2001).

12. Q. Zhang, C. Zhang, Y. Lu, K. Yang and Q. Ren. Progress in the development of CdZnTe unipolar detectors for different anode geometries and data corrections, *Sensors*, 13, 2447–2474 (2013).

13. Z. He. Review of the Shockley-Ramo theorem and its application in semiconductor gamma-ray detectors, *Nuclear Instruments and Methods in Physics Research. Section A*, 463, 250–267 (2001).

14. M. D. Wilson, P. Seller, M. C. Veale and P. J. Sellin. Investigation of the small pixel effect in CdZnTe detectors, *IEEE Nuclear Science Symposium Conference Record*, 2, 1255–1259 (2007).

15. A. Meuris, O. Limousin and C. Blondel. Charge sharing in CdTe pixilated detectors, *Nuclear Instruments and Methods in Physics Research. Section A*, 610, 294–297 (2009).

16. K. Iniewski, H. Chen, G. Bindley, I. Kuvvetli and C. Budtz-Jorgensen. Modelling charge-sharing effects in pixelated CZT detectors, *IEEE Nuclear Science Symposium Conference Record*, 6, 4608–4611 (2007).

17. A. E. Bolotnikov, S. E. Boggs, W. R. Cook, F. A. Harrison and S. M. Schindler. Use of a pulsed laser to study properties of CdZnTe pixel detectors, *SPIE Proceedings*, 3769, 52–58 (1999).

18. M. C. Veale, S. J. Bell, P. Seller, M. D. Wilson and V. Kachkanov. X-ray micro-beam characterisation of a small pixel spectroscopic CdTe detector, *Journal of Instrumentation*, 7, P07017 (2012).

19. D. Pennicard, R. Ballabriga, X. Llopart, M. Campbell and H. Graafsma. Simulations of charge summing and threshold dispersion effects in Medipix3, *Nuclear Instruments and Methods in Physics Research. Section A*, 636, 74–81 (2011).

20. C. C. T. Hansson, M. D. Wilson, P. Seller and R. J. Cernik. Performance limitations of the pixelated ERD detector with respect to imaging using the rapid tomographic energy dispersive diffraction imaging system, *Nuclear Instruments and Methods in Physics Research. Section A*, 604, 119–122 (2009).

21. J. A. Gaskin, D. P. Sharma, B. D. Ramsey, S. Mitchell and P. Seller. Characterization of a pixelated cadmium-zinc-telluride detector for astrophysical application, *Proceedings of SPIE*, 5165, 45 (2004).

22. P. Seller, A. L. Hardie, L. L. Jones, A. J. Boston and S. V. Rigby. Nucam: A 128 chan-
    nel integrated circuit with a pulse-height and rise-time measurement on each chan-
    nel including on-chip 12bit ADC for high-Z X-ray detectors, *IEEE Nuclear Science
    Symposium Conference Record*, 6, 3786–3789 (2006).
23. J. M. Alvarez, J. L. Galvez, M. Hernanz, J. Isern, M. Llopis, M. Lozano, G. Pellegrini
    and M. Chmeissani. Imaging detector development for nuclear astrophysics using pix-
    elated CdTe, *Nuclear Instruments and Methods in Physics Research. Section A*, 623,
    434–436 (2010).
24. D. S. Judson, A. J. Boston, P. J. Coleman-Smith, D. M. Cullen, A. Hardie, L. J. Harkness,
    L. L. Jones, et al. Compton imaging with the PorGamRays spectrometer, *Nuclear
    Instruments and Methods in Physics Research. Section A*, 652, 587–590 (2011).
25. M. D. Wilson, P. Seller, C. Hansson, R. Cernik, J. Marchal, Z-J. Xin, V. Perumal, M. C.
    Veale and P. Sellin. X-ray performance of pixilated CdZnTe detectors, *IEEE Nuclear
    Science Symposium Conference Record*, 239–245 (2008).
26. A. Choubey, J. Toman, A. W. Brinkman, J. T. Mullins, B. J. Cantewell, D. P. Halliday
    and A. Basu. Heteroepitaxial growth and properties of crystals of CdTe on GaAs sub-
    strates, *SPIE Proceedings*, 6706, 67060Z (2007).
27. A. Choubey, P. Veeramani, A. T. G. Pym, J. T. Mullins, P. J. Sellin, A. W. Brinkman,
    I. Radley, A. Basu and B. K. Tanner. Growth by the multi-tube physical vapour trans-
    port method and characterisation of bulk (Cd,Zn)Te, *Journal of Crystal Growth*, 352,
    120–123 (2012).
28. L. L. Jones, P. Seller, M. D. Wilson and A. Hardie. HEXITEC ASIC: A pixelated readout
    chip for CZT detectors, *Nuclear Instruments and Methods in Physics Research. Section
    A*, 604, 34–37 (2009).
29. P. Seller, S. J. Bell, M. C. Veale and M. D. Wilson. Through silicon via redistribution
    of I/O pads for 4-side butt-able imaging detectors, *IEEE Nuclear Science Symposium
    Conference Record*, 4142–2146 (2012).
30. M. D. Wilson, S. J. Bell, R. J. Cernik, C. Christodoulou, C. K. Edgan, D. O'Flynn, S.
    Jacques, et al. Multiple module pixellated CdTe spectroscopic X-ray detector, *IEEE
    Transactions on Nuclear Science*, 60, 1197–1200 (2013).
31. J. E. Clayton, C. M. H. Chen, W. R. Cook and F. A. Harrison. Assembly technique for a
    fine-pitch, low-noise interface; joining a CdZnTe pixel-array detector and custom VLSI
    chip with Au stud bumps and conductive epoxy, *IEEE Nuclear Science Symposium
    Conference Record*, 5, 3513–3517 (2003).
32. J. Scuffham, M. C. Veale, M. D. Wilson and P. Seller. Algorithms for spectral calibration
    of energy-resolving small-pixel detectors, *Journal of Instrumentation*, 8, P10024 (2013).
33. M. Funaki, T. Ozaki, K. Satoh and R. Ohno. Growth and characterization of CdTe single
    crystals for radiation detectors, *Nuclear Instruments and Methods in Physics Research.
    Section A*, 436, 120–136 (1999).
34. D. D. Duarte, S. J. Bell, J. Lipp, A. Schneider, P. Seller, M. C. Veale, M. D. Wilson, et al.
    Edge effects in a small pixel CdTe for X-ray imaging, *Journal of Instrumentation*, 8,
    P10018 (2013).
35. H. Toyama, A. Higa, M. Yamazato and T. Maehama. Quantitative analysis of polariza-
    tion phenomena in CdTe radiation detectors, *Japanese Journal of Applied Physics*, 45,
    8842–8847 (2006).
36. G. Prekas, P. J. Sellin, P. Veeramani, A. W. Davies, A. Lohstroh, M. E. Ozsan and M. C.
    Veale. Investigation of the internal electric field distribution under in situ X-ray irradia-
    tion and under low temperature conditions by the means of the Pockels effect, *Journal
    of Physics D: Applied Physics*, 43, 085102 (2010).
37. M. C. Veale, J. Kalliopuska, H. Pohjonen, H. Andersson, S. Nenonen, P. Seller and
    M. D. Wilson. Characterization of M-π-n CdTe pixel detectors coupled to HEXITEC
    readout chip, *Journal of Instrumentation*, 7, C01035 (2012).

38. P. Seller, S. Bell, R. J. Cernik, C. Christodoulou, C. K. Egan, J. A. Gaskin, S. Jacques, et al. Pixellated Cd(Zn)Te high-energy X-ray instrument, *Journal of Instrumentation*, 6, C12009 (2011).

39. H. Chen, S. A. Awadalla, J. Mackenzie, R. Redden, G. Bindley, A. E. Bolotnikov, G. S. Camarda, G. Carini and R. B. James. Characterization of travelling heater method (THM) grown $Cd_{0.9}Zn_{0.1}Te$ crystals, *IEEE Transactions on Nuclear Science*, 54, 811–816 (2007).

40. S. J. Bell, M. A. Baker, H. Chen, P. Marthandam, V. Perumal, A. Schneider, P. Seller, M. C. Veale and M. D. Wilson. A multi-technique characterization of electroless gold contacts on single crystal CdZnTe radiation detectors, *Journal of Physics D: Applied Physics*, 46, 455502 (2013).

41. M. C. Veale, S. J. Bell, D. D. Duarte, A. Schneider, P. Seller, M. D. Wilson, V. Kachkanov and K. J. S. Sawhney. Synchrotron characterisation of non-uniformities in a small pixel cadmium zinc telluride imaging detector, *Nuclear Instruments and Methods in Physics Research. Section A*, 729, 265–272 (2013).

42. A. E. Bolotnikov, G. S. Camarda, Y. Cui, A. Hossain, G. Yang, H. W. Yao and R. B. James. Internal electric-field-lines distribution in CdZnTe detectors measured using X-ray mapping, *IEEE Transactions on Nuclear Science*, 56, 791–794 (2009).

43. D. O'Flynn, H. Desai, C. B. Reid, C. Christodoulou, M. D. Wilson, M. C. Veale, P. Seller, D. Hills, B. Wong and R. D. Speller. Identification of simulants for explosives using pixelated X-ray diffraction, *Journal of Crime Science*, 2, 4 (2013).

44. D. O'Flynn, C. B. Reid, C. Christodoulou, M. D. Wilson, M. C. Veale, P. Seller and R. D. Speller. Pixelated diffraction signatures for explosives detection. *Proceedings of SPIE*, 8357, 83570X (2012).

45. D. O'Flynn, C. B. Reid, C. Christodoulou, M. D. Wilson, M. C. Veale, P. Seller, D. Hills, H. Desai, B. Wong and R. Speller. Explosive detection using pixelated X-ray diffraction (PixD), *Journal of Instrumentation*, 8, P03007 (2013).

46. S. D. M. Jacques, C. K. Egan, M. D. Wilson, M. C. Veale, P. Seller and R. J. Cernik. A laboratory system for element specific hyperspectral X-ray imaging, *Analyst*, 138, 755–759 (2013).

47. R. Nandan, T. DebRoy and H. K. D. Bhadeshia. Recent advances in friction-stir welding: process, weldment structure and properties, *Progress in Material Science*, 53, 980–1023 (2008).

48. C. K. Egan, S. D. M. Jacques, T. Connolley, M. D. Wilson, M. C. Veale, P. Seller and R. J. Cernik. Dark-field hyperspectral X-ray imaging, *Proceedings of the Royal Society A: Mathematical, Physical and Engineering Sciences*, 470 (2165), 20130629 (2014).

49. Womfalcs7. Pipe - friction stir welding, YouTube, http://youtu.be/niVsJPFlg1Y. (Accessed October 31, 2014.)

50. M. Ahmed, B. Wynne, W. Rainforth and P. Threadgill. Through-thickness crystallographic texture of stationary shoulder friction stir welded aluminium, *Scripta Materialia*, 64, 45–48 (2011).

51. M. Mahoney, C. Rhodes, J. Flintoff, W. Bingel and R. Spurling. Properties of friction-stir-welded 7075 T651 aluminium, *Metallurgical and Materials Transactions A*, 29, 1955–1964 (1998).

52. W. L. Duvall, L. B. Croft, T. Godiwala, E. Ginsberg, T. George and M. J. Henzlova. Reduced isotope dose with rapid SPECT MPI imaging: Initial experience with a CZT SPECT camera, *Journal of Nuclear Cardiology*, 17, 1009–1014 (2010).

53. J. W. Scuffham, M. D. Wilson, P. Seller, M. C. Veale, P. J. Sellin, S. D. M. Jacques and R. J. Cernik. A CdTe detector for hyperspectral SPECT imaging, *Journal of Instrumentation*, 7, P08027 (2012).

54. J. W. Scuffham, M. D. Wilson, S. Pani, D. D. Duarte, M. C. Veale, S. Bell, P. Seller, P. J. Sellin and R. J. Cernik. Evaluation of a new small-pixel CdTe spectroscopic detector in dual-tracer SPECT brain imaging, *IEEE Nuclear Science Symposium Conference Record*, 3115–3118 (2012).

4 CdTe/CZT Spectrometers with 3-D Imaging Capabilities

Ezio Caroli and Stefano del Sordo

CONTENTS

4.1 INTRODUCTION

Semiconductor detector technology has dramatically changed the broad field of x-ray and γ-ray spectroscopy and imaging. Semiconductor detectors, originally developed for particle physics applications, are now widely used for x/γ-ray spectroscopy and imaging in a wide variety of fields, including, for example, x-ray fluorescence, γ-ray monitoring and localization, noninvasive inspection and analysis, astronomy, and diagnostic medicine. The success of semiconductor detectors is due to several unique characteristics, such as excellent energy resolution, high detection efficiency, and the possibility of development of compact and highly segmented detection systems. Among semiconductor devices, silicon (Si) detectors are the key detectors in the soft x-ray band (<15 keV). Si-PIN diode detectors [1] and silicon drift detectors (SDDs) [2], operated with moderate cooling by means of small Peltier cells, show excellent spectroscopic performance and good detection efficiency below 15 keV. On the other hand, germanium (Ge) detectors are unsurpassed for high-resolution spectroscopy in the hard x-ray energy band (>15 keV) and will continue to be the first choice for laboratory-based high-performance spectrometers [3].

However, in the last decades, there has been an increasing demand for the development of room-temperature detectors with compact structure having the portability and convenience of a scintillator but with a significant improvement in energy resolution and spatial resolution. To fulfil these requirements, numerous high-Z and wide bandgap compound semiconductors have been exploited [4,5]. In fact, as demonstrated by the impressive increase in the scientific literature and technological development around the world, cadmium telluride (CdTe) and cadmium zinc telluride (CdZnTe or simply CZT)-based devices are today dominating the room-temperature semiconductor applications scenario, being widely used for the development of x/γ-ray instrumentation [6–8].

As already mentioned, for purely spectroscopic applications in the hard x-ray and γ-ray regime, germanium detectors have retained their supremacy, while applications that require imaging capabilities with high spatial resolution, possibly coupled with good spectroscopic performance (at room temperature), define the field in which the potential and advantages of CdTe/CZT-technology sensors can be fully exploited. In fact, the possibility of quite easily segmenting the charge-collecting electrodes in strips or arrays, as well as assembling a mosaic of even small sensitive units (i.e., crystals), makes it possible to obtain devices with intrinsically excellent bidimensional spatial resolution (down to tens of microns) and, according to the type of readout electronics, also to measure the energy released by the interaction of photons with the material (see [9,10] and other chapters in this book).

In this chapter, we will focus on a particular type of detector based on sensitive elements of CZT/CdTe, namely, spectrometers with spatial resolution in three dimensions. These, in fact, represent the new frontiers for applications in different fields that require increasing performance, such as high-energy astrophysics, environmental radiation monitoring, medical diagnostics with positron emission tomography (PET), and inspections for homeland security. The advantages offered by the possibility of reconstructing both the position of interaction of the incident photons in three dimensions (3-D) and the energy deposited by each interaction are of fundamental importance for applications that require high detection efficiency even at high energies (>100 keV), that is, in the Compton-scattering regime, as well as a wide-field localization of the direction of incidence and a uniform spectroscopic response throughout the sensitive volume. In fact, the possibility of reconstructing the photon interaction position in 3-D will allow correction for signal variations due to charge trapping and material nonuniformity, and will therefore allow the sensitive volume of each detector unit to be increased without degrading the spectroscopic performance.

4.2   X- AND γ-RAY SPECTROSCOPY WITH CZT/CdTe SENSORS

The typical operation of semiconductor detectors is based on collection of the charges created by photon interactions through the application of an external electric field. The choice of the proper semiconductor material for a radiation detector is mainly influenced by the energy range of interest. Among the various interaction mechanisms of x-rays and γ-rays with matter, three effects play an important role in radiation measurements: photoelectric absorption, Compton scattering, and pair production. In photoelectric absorption, the photon transfers all its energy to an atomic electron, while a photon interacting through the Compton process transfers only a fraction of its energy to an outer electron, producing a hot electron and a degraded photon; in pair production, a photon with energy above a threshold energy of 1.02 MeV interacts within the Coulomb field of the nucleus, producing an electron and positron pair. Neglecting the escape of characteristic x-rays from the detector volume (the so-called fluorescent lines), only the photoelectric effect results in the total absorption of the incident energy and thus gives useful information about the photon energy. The interaction cross sections are highly dependent on the atomic number. In photoelectric absorption, it varies as $Z^{4,5}$, Z for Compton scattering, and Z^2 for pair production.

4.2.1 CDTe/CZT USED AS X- AND γ-RAY SPECTROMETER

An optimum spectroscopic detector must favor photoelectric interactions, and so semiconductor materials with a high atomic number are preferred. Figure 4.1a shows the linear attenuation coefficients, calculated by using tabulated interaction cross-section values [11], for photoelectric absorption and Compton scattering of Si, CdTe, HgI$_2$, NaI, and BGO; NaI and BGO are solid scintillator materials typically used in radiation measurement. As shown in Figure 4.1a, photoelectric absorption is the main process up to about 200 keV for CdTe. The efficiency for CdTe detectors versus detector thickness and at various typical photon energies is reported in Figure 4.1b. A 10 mm thick CdTe detector ensures good photoelectric efficiency at 140 keV (>90%), while a 1 mm thick CdTe detector is characterized by a photoelectric efficiency of 100% at 40 keV.

Semiconductor detectors for x- and γ-ray spectroscopy behave as solid-state ionization chambers operated in pulse mode. The simplest configuration is a planar detector, which is a slab of a semiconductor material with metal electrodes on the opposite faces of the semiconductor (Figure 4.2). Photon interactions produce electron–hole pairs in the semiconductor volume through the discussed interactions. The interaction is a two-step process, whereby the electrons created in the photoelectric or Compton process lose their energy through electron–hole ionization. The most important feature of photoelectric absorption is that the number of electron–hole pairs is proportional to the photon energy. If E is the incident photon energy, the number of electron–hole pairs N is equal to E/w, where w is the average

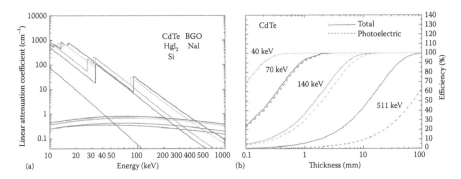

FIGURE 4.1 (See color insert) (a) Linear attenuation coefficients for photoelectric absorption and Compton scattering of CdTe, Si, HgI$_2$, NaI, and BGO. (b) Efficiency of CdTe detectors as a function of detector thickness at various photon energies.

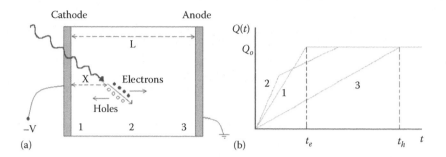

FIGURE 4.2 Planar configuration of a semiconductor detector: (a) Electron–hole pairs, generated by radiation, are swept toward the appropriate electrode by the electric field; (b) the time dependence of the induced charge for three different interaction sites in the detector (positions 1, 2, and 3). The fast-rising part is due to the electron component, while the slower component is due to the holes.

pair-creation energy. The generated charge cloud is $Q_0 = eE/w$. The electrons and holes move toward the opposite electrodes, anode and cathode for electrons and holes, respectively (Figure 4.2).

The movement of the electrons and holes causes variation ΔQ of induced charge on the electrodes. It is possible to calculate the induced charge ΔQ by the Shockley–Ramo theorem [12,13], which makes use of the concept of a weighting potential, defined as the potential that would exist in the detector with the collecting electrode held at unit potential, while all other electrodes are held at zero potential. According to the Shockley–Ramo theorem, the charge induced by a carrier q, moving from x_i to x_f, is given by Equation 4.1:

$$\Delta Q = -q\left[\varphi(x_f) - \varphi(x_i)\right] \tag{4.1}$$

where $\varphi(x)$ is the weighting potential at position x. It is possible to calculate the weighting potential by analytically solving the Laplace equation inside a detector [14]. In a semiconductor, the total induced charge is given by the sum of the induced charges due to both electrons and holes.

Charge trapping and recombination are typical effects in compound semiconductors and may prevent full charge collection. For a planar detector, having a uniform electric field, neglecting charge detrapping, the charge-collection efficiency (CCE), that is, the induced charge normalized to the generated charge, can be evaluated by the Hecht equation [15] and derived models [16] and is strongly dependent on the photon interaction position. This dependence, coupled with the random distribution of photon interaction points inside the sensitive volume, increases the fluctuations on the induced charge and produces peak broadening in the energy spectrum as well as the characteristic low tail asymmetry in the full energy peak shape observed in the planar CdTe/CZT sensor.

The charge transport properties of a semiconductor, expressed by the hole and electron mobility lifetime products ($\mu_h \tau_h$ and $\mu_e \tau_e$), are key parameters in the development of radiation detectors. Poor mobility lifetime products result in short mean drift length λ and therefore small λ/L ratios, which limit the maximum thickness and energy range of the detectors. Compound semiconductors, generally, are characterized by poor charge transport properties due to charge trapping. Trapping centers are mainly caused by structural defects (e.g., vacancies), impurities, and irregularities (e.g., dislocations, inclusions). In compound (CdTe and CZT) semiconductors, the $\mu_e \tau_e$ is typically of the order of 10^{-5}–10^{-3} cm^2 V^{-1}, while $\mu_h \tau_h$ is usually much worse, with values around 10^{-6}–10^{-4} cm^2 V^{-1}. Therefore, the corresponding mean drift lengths of electrons and holes are 0.2–20 mm and 0.02–2 mm, respectively, for typical applied electric fields of 2000 V cm^{-1} [17].

The charge-collection efficiency is a crucial property of a radiation detector, affecting spectroscopic performance, and in particular energy resolution. High charge-collection efficiency ensures good energy resolution, which also depends on the statistics of the charge generation and the noise of the readout electronics. Therefore, the energy resolution (FWHM) of a radiation detector is mainly influenced by three contributing factors:

$$\Delta E = \sqrt{(2.355)^2 (F \cdot E \cdot w) + \Delta E_{\text{el}} + \Delta E_{\text{coll}}} \qquad (4.2)$$

The first contribution is the Fano noise due to the statistics of the charge-carrier generation. In semiconductors, the Fano factor F is much smaller than unity (0.06–0.14) [18]. The second contribution is the electronic noise, which is generally measured directly using a precision pulser, while the third term takes into account the contribution of the charge-collection process. Different semiempirical relations have been proposed for the charge-collection contribution evaluation of different detectors [19].

Figure 4.3 shows a typical spectroscopic system based on a semiconductor detector. The detector signals are amplified by a charge-sensitive preamplifier (CSA) and then shaped by a linear amplifier (shaping amplifier). Energy spectra are obtained by a multichannel analyzer (MCA), which samples and records the shaped signals.

As pointed out in the foregoing discussions, poor hole-transport properties of CdTe and CdZnTe materials are a critical issue in the development of x- and γ-ray detectors. Hole trapping reduces the charge-collection efficiency of the detectors and produces asymmetry and a long tail in the photopeaks in the measured spectra (hole tailing). Several methods have been used in order to minimize this effect.

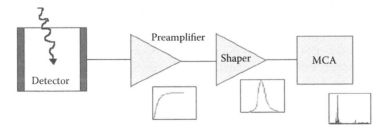

FIGURE 4.3 Block diagram of a standard spectroscopic detection system for x- and γ-rays.

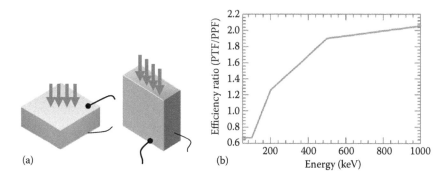

(a)   (b)

FIGURE 4.4 (a) Usual irradiation configuration in which photons impinge (arrows) the detector through the cathode surface (PPF: planar parallel field) and the PTF (planar transverse field) in which the photons impinge on the sensor orthogonally with respect to the charge collecting field. (b) Ratio between PTF and PPF efficiency calculated in [21] for impinging photon energies from 50 to 1000 keV, assuming PTF thickness equal to 10 mm and distance between electrodes (i.e., the PPF absorption thickness) 2.5 mm.

Some techniques concern the particular irradiation configuration of the detectors (see Figure 4.4a). Planar parallel field (PPF) is the classical configuration used in overall planar detectors, in which the detectors are irradiated through the cathode electrode, thus minimizing the hole-trapping probability. In an alternative configuration, denoted planar transverse field (PTF) [20], the irradiation direction is orthogonal (transverse) to the electric field. In such a configuration, different detector thicknesses can be chosen, in order to fit the detection efficiency required, without modifying the interelectrode distance and thus the charge-collection properties of the detectors. This technique is particularly useful to develop detectors with high detection efficiency in the γ-ray energy range. Figure 4.4b shows a plot of the ratio between the efficiency achievable by a CdTe spectrometer with lateral sides of 10 mm and a distance between electrodes of 2.5 mm [21]. This plot shows that the PTF irradiation configuration becomes favorable in terms of detection efficiency above 200 keV.

4.2.2 Spectroscopic Performance Improvement Techniques

Different methods have been proposed to compensate for the trapping effects in CdTe/CZT semiconductor detectors in order to improve their spectroscopic resolution toward its theoretical limit and to increase their full energy efficiency. The most frequently used methods rely on the possibility of avoiding the contribution of holes to the formation of the charge signal, thus using the CZT/CdTe detector as a single-charge sensing device. In this configuration, only electrons are collected. Because of their mobility characteristics, the effect of trapping is limited and can be further compensated for using information derivable from simple signal manipulation.

Several techniques are used in the development of detectors based on the collection of electrons (single-charge carrier sensitive), which have better transport properties

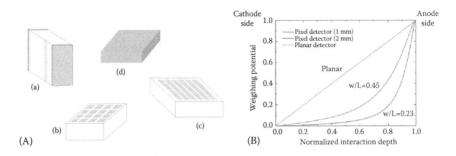

FIGURE 4.5 **(See color insert)** (A) Single-charge collection electrode configurations widely used in CdTe and CdZnTe detectors: (a) parallel-strip Frisch grid, (b) pixel, (c) strip, and (d) multiple electrodes. (B) Weighting potential of a pixel detector, compared with a planar detector. It is possible to improve the unipolar properties of pixel detectors by reducing the *w/L* ratio (i.e., pixel size to detector thickness), according to the theory of small pixel effect.

than holes. Single-charge carrier-sensing techniques are widely employed for CdTe and CdZnTe detectors (unipolar detectors), both by using electronic methods (pulse rise time discrimination [22], biparametric analysis [23,24]) and by developing careful electrode design (Frisch grid [25,26], pixels [27,28], coplanar grids [29], strips [30,31], multiple electrodes [32–34]). Figure 4.5 shows some unipolar electrode configurations widely used in CdTe and CdZnTe detectors.

The first single-charge carrier-sensing technique was implemented in gas detectors by Frisch [35] to overcome the problem of slow drift and loss of ions. A simple semiconductor Frisch grid detector and a derivate structure known as the coplanar-grid configuration can be built by using parallel metal strips on the opposite faces of the detector (yellow strips of Figure 4.5a) connected in an alternating manner to give two sets of interdigital grid electrodes. Pixels and strips on the anode electrode of detectors (Figure 4.5b, c), useful for their imaging properties, are also characterized by unipolar properties. The small-size anode electrode and the multiple ring electrodes (drift electrodes) on the anode surface of the detectors, as shown in Figure 4.5d, optimize charge collection, minimizing the effect of the hole trapping on the measured spectra.

In general, the almost unipolar characteristics of these detector configurations are due to the particular shape of the weighting potential: it is low near the cathode and rises rapidly close to the anode. According to this characteristic, the charge induced on the collecting electrode, proportional to the weighting potential, as stated by the Shockley–Ramo theorem, is mostly contributed by the drift of charge carriers close to the anode, that is, the electrons. On the contrary, the linear shape of the weighting potential of a planar detector makes the induced charge sensitive to both electrons and holes, as discussed above.

In particular, the introduction of coplanar-grid noncollecting electrodes in the anode-side design of the sensor provides an important additional feature that is fundamental to realizing 3-D sensing spectrometers, that is, the position information of the radiation interaction point inside the sensitive volume [29]. In fact, for these electrode configurations, the induced charge on the planar cathode Q_p increases roughly as a linear function of the distance D of γ-ray interaction location from the coplanar anodes ($Q_p \propto D \times E_\gamma$), because it is proportional to the drift time of electrons. On the other hand, the coplanar anode signal Q_s is only approximately proportional to the γ-ray deposited energy ($Q_s \propto E_\gamma$). Therefore, the interaction depth can be estimated by reading both Q_p and Q_s signal amplitude for each interaction through their ratio (also called depth parameter): $d = Q_p/Q_s, \propto D$ [36].

4.3   3-D CZT/CdTe SENSOR CONFIGURATIONS

A 3-D spectrometer is, in principle, a detector divided into volume elements (voxels), each operating as an independent spectroscopic sensor. The signal produced in each voxel by the interaction of an incoming x/γ photon must be able to be read and converted into a voltage signal proportional to the energy released. If the readout electronics of the detection system implements a coincidence logic, it will be possible to determine to some extent the history of the incident photon inside the sensitive volume by associating the energy deposits in more voxels with the same incident photon.

The need for this type of sensor comes from application requirements. For example, in the field of hard x- and soft γ-ray astrophysics (10–1000 keV), there are promising developments of new focusing optics operating for up to several hundreds of kiloelectronvolts through the use of broadband Laue lenses [37] and a new generation of multilayer mirrors [38]. These systems make it possible to push the sensitivity of a new generation of innovative high-energy space telescopes to levels far higher (100–1000 times) than current instrumentation. To obtain the maximum return from this type of optics up to the megaelectronvolt level requires the use of focal plane detectors with high efficiency (>80%) even at higher energies, and with that the ability to measure the energy spectrum with good spectroscopic resolution and also to localize accurately (0.1–1 mm) the point of interaction of the photons used for the correct attribution of their direction of origin in the sky.

In fact, the realization of 3-D spectrometers by a mosaic of single CdTe/CZT crystals is not as easy as in the case of bidimensional imagers, mainly due to the small dimension of each sensitive unit, necessary to guarantee the required spatial resolution, and also the intrinsic difficulty of packaging in 3-D sensor units, in which each one requires an independent spectroscopic readout electronics chain. A solution is offered by the realization of a stack of 2-D spectroscopic imagers [39,40]. This configuration, while very appealing for large-area detectors, has several drawbacks for application, requiring fine spatial resolution in three dimensions and compactness. Indeed, the distance between 2-D layers of the stack limits the accuracy and the sampling of the third spatial coordinate, and passive materials are normally required for mechanical support.

To solve this kind of problem, in the last 10–15 years several groups have focused their activity on the development of sensors based on high-volume (1–10 cm^3) crystals of CZT/CdTe capable of intrinsically operating as 3-D spectrometers and therefore able to meet the requirements for certain applications, or to make more efficient and easy the realization of 3-D detectors based on matrices of these basic units. The main benefits are: a limited number of required readout channels to achieve the same spatial resolution; packing optimization; and reduction of passive material between sensitive volumes. In these developments, a key role is played by the adopted electrode configuration. As already seen in Section 4.2.2, various electrode configurations have been proposed and studied to improve both the spectroscopic performance and the uniformity of response of CZT/CdTe detectors compensating for and correcting problems related to trapping and low mobility of the charge carriers in these materials. In fact, these electrode configurations, with the implementation of appropriate logical reading of the signals, are intrinsically able to determine the position of interaction of the photon in the direction of the collected charge (depth sensing) and therefore are particularly suited to the realization of 3-D monolithic spectrometers without requiring a drastic increase of the electronics readout chains.

In the following sections, we will review a couple of configurations currently proposed and in development for the realization of 3-D spectrometers based on single large-volume crystals of CdTe/CZT. Even though there are other developments in this direction [41–44], the present choice is mainly dictated by application to hard x-rays and soft γ-rays, which requires good spatial resolution (at millimeter level or better) in all three dimensions coupled with fine and uniform spectroscopic response.

4.3.1 PIXEL SPECTROMETERS WITH COPLANAR GUARD GRID

By combining a pixelated anode array, already performing with good energy resolution because of the small pixel effect introduced in Section 2.2, and an interaction depth-sensing technique for electron-trapping corrections, it is possible to build CdZnTe γ-ray spectrometers with intrinsic 3-D position-sensing capability over quite large-volume (1–3 cm^3) bulk crystals. This configuration was proposed in 1998 by He et al. [45].

The first prototype was based on a $10 \times 10 \times 10$ cm^3 CZT crystal with an 11×11-pixel anode array and a single cathode electrode on the opposite surface [46]. The 2-D sensing of γ-ray interactions is simply provided by the pixel (x, y) anode, where electrons are collected on the anode surface. Instead of using an array of simple square pixel anodes, each square collecting anode is surrounded by a common noncollecting grid (Figure 4.6a and b). In the first configuration, the pixel pitch has dimensions of 0.7×0.7 mm^2, with a collecting anode of 0.2×0.2 mm^2 at the center surrounded by a common noncollecting grid with a width of 0.1 mm. Since the noncollecting grid is biased at lower potential relative to that of the collecting anodes, electrons are forced toward the collecting pixel anodes. Even more important, the dimension of the pixel collecting anode is small with respect to the anode–cathode distance and smaller than the geometrical pixel dimension, enhancing the small pixel effect and minimizing any induced signal from the holes' movement. To guarantee good electron collection, for this configuration the bias between anodes and the planar cathode is in the 1500–2000 V range, while the voltage difference between anodes and the noncollecting common grid is typically a few tenths of a volt (30–50 keV).

The γ-ray interaction depth between the cathode and the anode is obtained from the ratio between the signal readout by the cathode and the anode, respectively. With a simple coincidence logic between cathode and anode signals, this technique

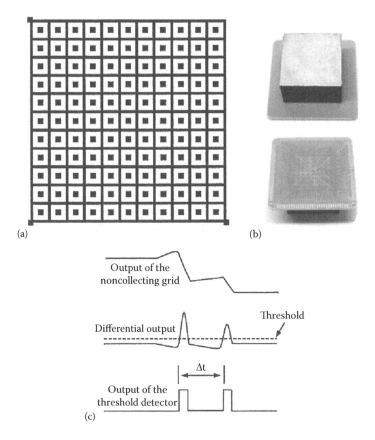

(a)   (b)

(c)

FIGURE 4.6   (a) Scheme of the anode side of the $10 \times 10 \times 10$ mm^3 CZT prototype. (b) Photos of the detector ($15 \times 15 \times 10$ mm^3) with the ceramic substrate facing up (top) and with the cathode facing down (bottom), where, through the thin ceramic substrate, the anode bonding pad array is visible. (c) Scheme of the depth-sensing logic used for multiple-site event handling.

can provide the depth (z) of the photon interaction for single-site events, and only the centroid depth for multiple-site interactions (e.g., Compton-scattered events). In order to identify individual hit depths for multiple-site events, the signal from the noncollecting grid is also read out using a charge-sensitive preamplifier. When electrons generated by an energy deposition are detected toward the collecting pixel anode near the anode surface, a negative pulse is induced on the noncollecting grid, as shown in Figure 4.6c. This signal is differentiated, generating positive pulses corresponding to the slope inversion points of the noncollecting grid signal. Finally, a threshold circuit uses the differential output to provide a logic pulse when it is above a defined threshold [47]. These logic pulses provide start and stop signals to a time-to-amplitude converter (TAC) that measures the interval of drifting times of electrons.

By combining the centroid depth, pulse amplitudes from each pixel anode, and the depth interval between energy depositions derived from the measure of electron drifting time, the depth of each individual energy deposition can be obtained. Although the differential circuit could identify multiple hits of the same incoming photon, the TAC limits the number of interactions to two. Therefore, the original system was able to provide interaction depths for only single and two-site (double) events. Events having more than two energy depositions can be identified by the number of anode-triggered pixels, but only the centroid interaction depth can be obtained. While the single-event low-energy threshold was low (~10 keV), the threshold for double events was relatively high, because their detection depended on the noncollecting grid signal threshold, which was, in the first measurements, in the order of 100 keV. The reconstructed interaction depth using this technique becomes worse with decreasing energy [48] and is ~0.25 mm for single events and ~0.4 mm for double ones at 662 keV.

In fact, since the first prototypes, several improvements have been realized by the same group on both the CZT sensor configuration and the readout and processing electronics, allowing the sensitive volume of each CZT device to be increased up to 6 cm³ (i.e., $2 \times 2 \times 1.5$ cm³) [49,50]. With these new sensors, very impressive spectroscopic performance can be achieved (Figure 4.7a) for all event types. These

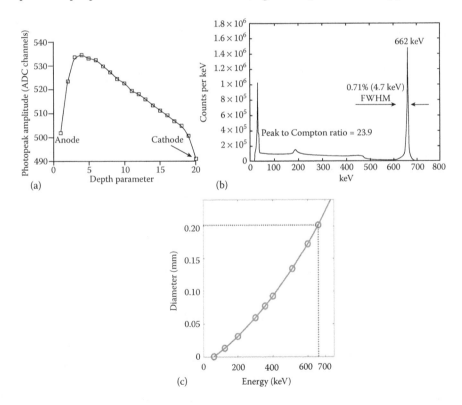

(a)

(b)

(c)

FIGURE 4.7 (a) The typical dependence of the centroids of ^{137}Cs photopeak from the interaction depth parameter (1 depth step = 0.5 mm) for one pixel. (b) ^{137}Cs spectrum measured by one pixel after compensation for interaction depth for all event multiplicity. (c) Diameter of the electron cloud generated by photon interaction vs. energy.

sensor units have been proposed as 3-D spectrometers for energy up to several megaelectronvolts. One of the main problems with operating in this energy region (>500 keV) is that the electron cloud, generated at each photon interaction point inside the sensitive volume, becomes larger than the pixels' lateral size (>1 mm) as the energy deposit increases. This effect, of course, tends to degrade the spatial resolution, because transient signals are collected by several anode pixels around the central one (charge sharing), which increases the uncertainty of depth reconstruction in the direction of charge collection.

Even if the geometrical spatial resolution in the anode plane of the larger CZT pixel sensor was at millimeter scale (1.8 mm pitch), with a custom-designed digital readout scheme, able to handle the signal coming from the eight pixels neighboring the triggered pixel, it has been demonstrated that a Δx of 0.23–0.33 mm can be achieved for a 662 keV single interaction [51].

Similar sensor configurations have been proposed, where the anode segmentation in a pixel array is replaced by a grid of orthogonal coplanar electrodes [52], as shown in Figure 4.8a, or by a parallel coplanar strip set [53] (Figure 4.8b), while maintaining in both cases a planar cathode on the opposite surface. The evident advantage of this detector scheme is the reduction of the required readout channels ($2N$) to achieve the same geometrical spatial resolution in the anode sensor plane (x, y) compared with an equivalent pixel array segmentation (N^2). In this case, the signal readout from the pixel lines gives one coordinate on the plane, while the orthogonal one is derived from analysis of the bipolar signals from noncollecting strips. As before, the third dimension (i.e., the interaction depth) can be inferred by the ratio between anode and cathode signals, but, due to the coplanar electrode configuration, the cathode signal can be replaced by the sum of all anode pixel line and strip signals. For a CZT detector of $10 \times 10 \times 10$ mm^3 with an orthogonal coplanar strip anode, the measurements have demonstrated that, despite a geometrical spatial resolution of

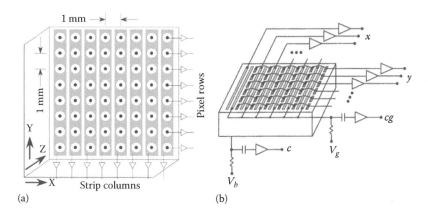

(a)   (b)

FIGURE 4.8 (a) Orthogonal coplanar-strip 3-D detector scheme: single-sided strip detector with collecting (row) and noncollecting (column) contacts on the anode surface. The cathode on the hidden side is planar. (b) Parallel coplanar-grid 3-D detector. The energy readout is accomplished by measuring the induced charge on a single set of interconnected anode strips, which are biased in order to collect the generated electrons.

1 mm in the (x, y) plane, the achievable 1σ values are between 0.2 and 0.4 mm for 122 and 60 keV, respectively, and ~0.4 mm in the interaction depth at 122 keV [54].

4.3.2 PTF MICROSTRIP WITH DRIFT CONFIGURATION

Another way to build 3-D spectroscopic sensors relies on the use of CZT crystals in the PTF configuration. The drawback of the PTF irradiating geometry is that all the positions between the collecting electrodes are uniformly hit by impinging photons, leading to a stronger effect of the difference in charge-collection efficiency and thus in spectroscopic performance with respect to the standard irradiation configuration through the cathode (PPF). Therefore, worse spectroscopic performance can be expected in PTF with respect to the standard PPF irradiation configuration [55]. In order to recover from this spectroscopic degradation and to improve the CZT sensitive unit performance, an array of microstrips in a drift configuration can be used instead of a simple planar anode (Figure 4.9): the anode surface is made of a thin collecting anode strip surrounded by guard strips that are biased at decreasing voltages. This anode configuration [56] allows the detector to become almost a single-charge carrier device, avoiding the degradation of the spectroscopic response by charge loss due to hole trapping and providing a more uniform spectroscopic response (i.e., independent of the distance of the interaction from the collecting electrodes), as shown in Figure 4.10a [57]. The spectroscopic resolution of this type of sensor ranges from 6% at 60 keV down to 1.2% at 661 keV, without any correction for the interaction depth. In fact, similarly to the other configuration presented above, it will be possible to perform compensation of the collected charge signals using the photon interaction position within the metalized surface, which can be inferred by the ratio between the cathode and the anode strip signals [58].

The achievable spatial resolution in this direction is, of course, a function of energy, because the dimensions of the charge-generated cloud, up to 500 keV, have been measured [59] to be around 0.2 mm (Figure 4.10b). In order to obtain 3-D sensitivity for the photon interaction position, the cathode can also be segmented into strips in the direction orthogonal to the anode ones. Of course, with the described configuration, the spatial resolution along the anode surface is defined geometrically by the collecting anode and cathode strip pitch.

Both anode and cathode strips are read by standard spectroscopic electronics chains, and therefore the segmentation of both cathode and anode surface will set the number of readout channels. In fact, ongoing developments on this sensor configuration are demonstrating that, with a readout logic able to weight the signal between strips, the achievable spatial resolution also across the anode and the cathode strip sets can be better than the geometrical strip pitch. For a sensor unit similar to the one shown in Figure 4.11a, in which the cathode is segmented into 2 mm pitch strips, the final spatial resolution can be as low as 0.6 mm (up to 600 keV), weighting the cathode strip signals. In fact, also along the anodic strip set the effective resolution can be improved to a small fraction (1/5–1/10) of geometrical pitch between collecting strips by implementing an appropriate readout of the drift strip signal, similar to the one suggested by Luke et al. [60] for 3-D coplanar grid detectors [61].

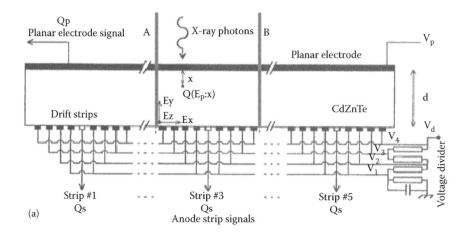

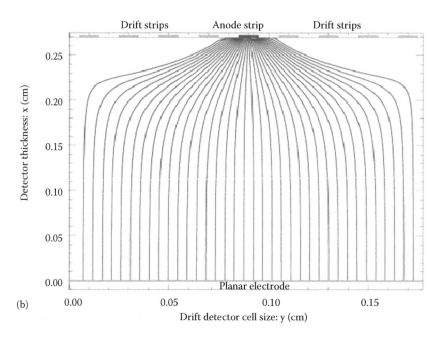

FIGURE 4.9 (a) Schematic configuration of a CZT drift-strip sensor. A drift-strip detector cell is shown between the two thick vertical lines marked "A" and "B." The drift-strip electrodes and the planar electrode are biased in such a way that the electrons move to the anode collecting strips (central white strips). (b) The shape of the charge-collecting field lines calculated for an anode cell on a CZT sensor with drift-strip electrode configuration (see Figure 4.8): the anode–cathode bias is set at 150 keV, and the difference between adjacent strips in pairs (ΔV) = –25 V.

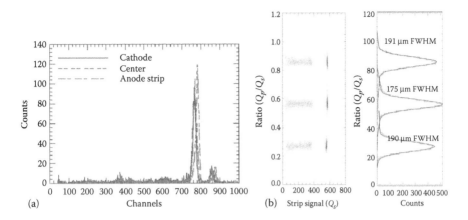

FIGURE 4.10 **(See color insert)** (a) Spectra of a collimated (0.6 mm spot) ^{57}Co source at three different positions between the collecting electrodes; the variation between the full energies peak and the corresponding energy resolution is within a few percent. (b) On the left, the biparametric distributions of the ratio between the planar electrode signal (Q_p) and the anode collecting-strip signal (Q_s) vs. Q_s for three positions of a 500 keV monochromatic x-ray beam; on the right, the corresponding measured depth resolution for these three different beam positions. The y-axis extension (ratio) is representative of the sensor interelectrode distance. The beam was collimated at 50 μm.

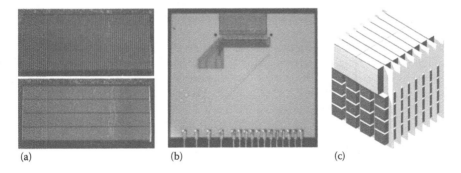

FIGURE 4.11 (a) Drift-strip CZT sensor (18 × 8 × 2.5 mm^3): (top) anode side with a set of 64 (0.15 mm wide) strips, (bottom) cathode side with four (2 mm wide) strips. (b) Linear module prototype seen from the anode side: this constitutes the basic element for building a large-volume 3-D sensor. (c) A suitable packaging scheme of eight linear modules, each supporting two CZT drift 3-D sensors of 20 × 20 × 5 mm^3 to obtain a spectroscopic imager of 32 cm^3 sensitive volume.

Because of the use of the PTF configuration, the dimensions of the 3-D sensor unit can reach up to 20–30 mm in lateral size and up to 5 mm in charge-collecting distance, making it possible to limit the high bias voltage required to completely deplete the sensitive volume to values below 500 V. A particular implementation of these types of sensor is shown in Figure 4.11a, in which a CZT sensor of 18 × 8 × 2.4 mm^3, with a fine-pitch μ-strip pattern on the CZT anode and a cathode segmented into four 2 mm wide orthogonal strips, is mounted on a 1 mm thick alumina support. In this sensor prototype, there are four collecting anode strips, with an overall pitch of 2.4 mm, each one surrounded by four strips on each side, used as drift electrodes. While the bias between cathode and anode is typically 100 V mm^{-1}, the drift strips, to be effective in shaping the charge-collection field to minimize dead volume, are biased at decreasing relative voltage with respect to the cathode strips of $\Delta V = 20$–30 V. These values, of course, depend in particular on the thickness (distance between cathode and anode surfaces) of the sensor tile, and the best bias scheme must be carefully studied using charge transport models. Using such sensor units [62], a large-volume 3-D spectrometer can be built by packaging several units (as shown in Figure 4.11b and c), in which CZT 3-D sensors are bonded on a thin high-resistivity support layer (e.g., Al_2O_3), forming linear modules that provide the electrical interface for both readout electronics and bias circuits.

4.4   CONSIDERATIONS ON 3-D CZT/CdTe SPECTROMETER APPLICATIONS

The development of CZT/CdTe spectrometers with high 3-D spatial resolution and fine spectroscopy represents a real challenge to the realization of a new class of high-performance instruments able to fulfil current and future requirements in several application fields.

The possibility of also achieving a very good detection efficiency even at high energy (up to a few megaelectronvolts) [63], because of the sensitive volumes that can also be obtained by mosaics or stacks of 3-D sensor units, without significant loss of spectroscopic performance and response uniformity, together with their capability to operate at room temperature, makes them really appealing for application in radiation monitoring and identification [64] and homeland security, as well as in industrial noninvasive controls and, in the research field, for new hard x- and soft γ-ray astronomy instrumentation.

Furthermore, the fine spectroscopic resolution (a few percent at 60 keV and <1% above 600 keV) and the high 3-D spatial resolution (0.2–0.5 mm) that these devices can guarantee, coupled with high-performance readout electronics, allow operation not only in full energy mode but also as Compton-scattering or pair detectors if equipped with appropriate electronics providing a suitable coincidence logic to handle multihit events. These possibilities imply that these sensors are suitable to realize wide-field detectors for localization and detection of γ-ray sources (>100 keV) in both ground and space applications [65]. Evaluation using a single 3-D CZT sensor (Section 3.1) as a 4π Compton imager has demonstrated the possibility of obtaining an angular resolution ~15° at 662 keV. This is an excellent result in the small distance scale used to reconstruct Compton event kinematics, and can be achieved only because of the good 3-D and spectroscopic performance of the CZT proposed sensor units.

The possibility of operating 3-D spectrometers as Compton-scattering detectors gives the appealing opportunity to utilize these devices for measurements of hard x-ray and soft γ-ray polarimetry. Today, this type of measurement is recognized as being of fundamental importance in high-energy astrophysics, and is one of the most demanding requirements for instrumentation for the next space mission in this energy range (10–1000 keV). The presence of linearly polarized photons in the incoming flux from a cosmic ray source determines a modulation in the azimuthal direction of Compton-scattered events [66]. A 3-D spectrometer able to properly handle scattered events is intrinsically able to measure these modulations, that is, to operate as a scattering polarimeter [67]. The quality (modulation factor) of a scattering polarimeter is strictly dependent on its spatial resolution and spectroscopic performance. Several experimental measurements [68,69] and simulation models have demonstrated that a detector allowing a good selection of events using both the spectroscopic and position information of each hit can achieve a very high modulation factor. In particular, the ability to select events within a thin layer of the sensitive volume, thanks to the intrinsic 3-D segmentation of the detector (i.e., close to 90° scattering direction), drastically improves the modulation factor and therefore the reliability of the polarimetric measurements.

REFERENCES

1. Pantazis, T., et al. (2010), The historical development of the thermoelectrically cooled x-ray detector and its impact on the portable and hand-held XRF industries, *X-Ray Spectrometry*, 39, 90.
2. Lechner, P., et al. (2004), Novel high-resolution silicon drift detectors, *X-Ray Spectrometry*, 33, 256.
3. Eberth, J. and Simpson, J. (2006), From Ge(Li) detectors to gamma-ray tracking arrays: 50 years of gamma spectroscopy with germanium detectors, *Progress in Particle and Nuclear Physics*, 60, 283.
4. Owens, A. and Peacock, A (2004), Compound semiconductor radiation detectors, *Nuclear Instruments and Methods in Physics Research A*, 531, 18.
5. Sellin, P.J. (2003), Recent advances in compound semiconductor radiation detectors, *Nuclear Instruments and Methods in Physics Research A*, 513, 332.
6. Lebrun, F., et al. (2003), ISGRI: The INTEGRAL soft gamma-ray imager, *Astronomy & Astrophysics*, 411, L141.
7. Lee, K., et al. (2010), Development of X-ray and gamma-ray CZT detectors for homeland security applications, *Proceedings of SPIE*, 7664, 766423–1.
8. Ogawa, K. and Muraishi, M. (2010), Feasibility study on an ultra-high-resolution SPECT with CdTe detectors, *IEEE Transactions on Nuclear Science*, 57, 17.
9. Limousin, O., et al. (2011), Caliste-256: A CdTe imaging spectrometer for space science with a 580 μm pixel pitch, *Nuclear Instruments and Methods in Physics Research A*, 647, 46.
10. Watanabe, S., et al. (2009), High energy resolution hard x-ray and gamma-ray imagers using CdTe diode devices, *IEEE Transactions on Nuclear Science*, 56, 777.
11. Berger, M.J., et al. (2010), XCOM: Photon cross sections database, National Institute of Standards and Technology, http://www.nist.gov/pml/data/xcom/index.cfm.
12. Cavalleri, G., et al. (1971), Extension of Ramo theorem as applied to induced charge in semiconductor detectors, *Nuclear Instruments and Methods in Physics Research*, 92, 137.
13. He, Z. (2001), Review of the Shockley–Ramo theorem and its application in semiconductor gamma ray detectors, *Nuclear Instruments and Methods in Physics Research A*, 463, 250.
14. Eskin, J.D., et al. (1999), Signals induced in semiconductor gamma-ray imaging detectors, *Journal of Applied Physics*, 85, 647.
15. Hecht, K. (1932), Zum Mechanismus des lichtelektrischen Primärstromes in isolierenden Kristallen, *Zeitschrift für Physik*, 77, 235.

16. Zanichelli, M., et al. (2013), Charge collection in semi-insulator radiation detectors in the presence of a linear decreasing electric field, *Journal of Physics D: Applied Physics*, 46, 365103.
17. Sato, G., et al. (2002), Characterization of CdTe/CdZnTe detectors, *IEEE Transactions on Nuclear Science*, 49, 1258.
18. Devanathan, R., et al. (2006), Signal variance in gamma-ray detectors: A review, *Nuclear Instruments and Methods in Physics Research A*, 565, 637.
19. Kozorezov, A.G., et al. (2005), Resolution degradation of semiconductor detectors due to carrier trapping, *Nuclear Instruments and Methods in Physics Research A*, 546, 207.
20. Casali, F., et al. (1992), Characterization of small CdTe detectors to be used for linear and matrix arrays, *IEEE Transactions on Nuclear Science*, 39, 598.
21. Caroli, E., et al. (2008), A three-dimensional CZT detector as a focal plane prototype for a Laue Lens telescope, *Proceedings of SPIE*, 7011, 70113G.
22. Jordanov, V.T., et al. (1996), Compact circuit for pulse rise-time discrimination, *Nuclear Instruments and Methods in Physics Research A*, 380, 353.
23. Richter, M. and Siffert, P. (1992), High resolution gamma ray spectroscopy with CdTe detector systems, *Nuclear Instruments and Methods in Physics Research A*, 322, 529.
24. Auricchio, N., et al. (2005), Twin shaping filter techniques to compensate the signals from CZT/CdTe detectors, *IEEE Transactions on Nuclear Science*, 52, 1982.
25. McGregor, D.S., et al. (1998), Single charge carrier type sensing with a parallel strip pseudo-Frisch-grid CdZnTe semiconductor radiation detector, *Applied Physics Letters*, 12, 192.
26. Bolotnikov, A.E., et al. (2006), Performance characteristics of Frisch-ring CdZnTe detectors, *IEEE Transactions on Nuclear Science*, 53, 607.
27. Barrett, H.H., et al. (1995), Charge transport in arrays of semiconductor gamma-rays detectors, *Physical Review Letters*, 75, 156.
28. Kuvvetli, I., and Budtz-Jørgensen, C. (2005), Pixelated CdZnTe drift detectors, *IEEE Transactions on Nuclear Science*, 52, 1975.
29. Luke, P.N. (1995), Unipolar charge sensing with coplanar electrodes: Application to semiconductor detectors, *IEEE Transactions on Nuclear Science*, 42, 207.
30. Shor, A., et al. (1999), Optimum spectroscopic performance from CZT γ- and x-ray detectors with pad and strip segmentation, *Nuclear Instruments and Methods in Physics Research A*, 428, 182.
31. Perillo, E., et al. (2004), Spectroscopic response of a CdTe microstrip detector when irradiated at various impinging angles, *Nuclear Instruments and Methods in Physics Research A*, 531, 125.
32. Lingren, C.L., et al. (1998), Cadmium-zinc telluride, multiple-electrode detectors achieve good energy resolution with high sensitivity at room-temperature, *IEEE Transactions on Nuclear Science*, 45, 433.
33. Kim, H., et al. (2004), Investigation of the energy resolution and charge collection efficiency of Cd(Zn)Te detectors with three electrodes, *IEEE Transactions on Nuclear Science*, 51, 1229.
34. Abbene, L., et al. (2007), Spectroscopic response of a CdZnTe multiple electrode detector, *Nuclear Instruments and Methods in Physics Research Section A*, 583, 324.
35. Frisch, O. (1944), British Atomic Energy Report, BR-49.
36. He, Z., et al. (1997), Position-sensitive single carrier CdZnTe detectors, *Nuclear Instruments and Methods in Physics Research A*, 388, 180.
37. Frontera, F., et al. (2013), Scientific prospects in soft gamma-ray astronomy enabled by the LAUE project, *Proceedings of SPIE*, 8861, 886106–1.
38. Della Monica Ferreira, D., et al. (2013), Hard x-ray/soft gamma-ray telescope designs for future astrophysics missions, *Proceedings of SPIE*, 886, 886116–1.

39. Watanabe, S., et al. (2002), CdTe stacked detectors for gamma-ray detection, *IEEE Transactions on Nuclear Science*, 49, 1292.
40. Judson, D.S., et al. (2011), Compton imaging with the PorGamRays spectrometer, *Nuclear Instruments and Methods in Physics Research A*, 652, 587.
41. Cui, Y., et al. (2008), Hand-held gamma-ray spectrometer based on high-efficiency Frisch-ring CdZnTe detectors, *IEEE Transactions on Nuclear Science*, 55, 2765.
42. Bale, D.S. and Szeles, C. (2006), Design of high performance CdZnTe quasi-hemispherical gamma-ray CAPture™ plus detectors, *Proceedings of SPIE*, 6319, 63190B.
43. Owens, A., et al. (2006), Hard x- and γ-ray measurements with a large volume coplanar grid CdZnTe detector, *Nuclear Instruments and Methods in Physics Research A*, 563, 242.
44. Dish, C., et al. (2010), Coincidence measurements with stacked (Cd,Zn)Te coplanar grid detectors, *2010 IEEE Nuclear Science Symposium Conference Record*, 3698.
45. He, Z., et al. (1999), 3-D position sensitive CdZnTe gamma-ray spectrometers, *Nuclear Instruments and Methods in Physics Research A*, 422, 173.
46. Stahle, C.M., et al. (1997), Fabrication of CdZnTe strip detectors for large area arrays, *Proceedings of SPIE*, 3115, 90.
47. Li, W., et al. (1999), A data acquisition and processing system for 3-D position sensitive CZT gamma-ray spectrometers, *IEEE Transactions on Nuclear Science*, 46, 1989.
48. Li, W., et al. (2000), A modeling method to calibrate the interaction depth in 3-D position sensitive CdZnTe gamma-ray spectrometers, *IEEE Transactions on Nuclear Science*, 47, 890.
49. Zhang, F., et al. (2004), Improved resolution for 3-D position sensitive CdZnTe spectrometers, *IEEE Transactions on Nuclear Science*, 51, 2427.
50. Zhang, F., et al. (2012), Characterization of the H3D ASIC readout system and 6.0 cm 3-D position sensitive CdZnTe detectors, *IEEE Transactions on Nuclear Science*, 59, 236.
51. Zhu, Y., et al. (2011), Sub-pixel position sensing for pixelated, 3-D position sensitive, wide band-gap, semiconductor, gamma-ray detectors, *IEEE Transactions on Nuclear Science*, 58, 1400.
52. Macri, J.R., et al. (2002), Study of 5 and 10 mm thick CZT strip detectors, *2002 IEEE Nuclear Science Symposium Conference Record*, 2316.
53. Luke, P.N. (2000), Coplanar-grid CdZnTe detector with three-dimensional position sensitivity, *Nuclear Instruments and Methods in Physics Research A*, 439, 611.
54. Macri, J.R., et al. (2003), Readout and performance of thick CZT strip detectors with orthogonal coplanar anodes, *2003 IEEE Nuclear Science Symposium Conference Record*, 468.
55. Auricchio, N., et al. (1999), Investigation of response behavior in CdTe detectors versus inter-electrode charge formation position, *IEEE Transactions on Nuclear Science*, 46, 853.
56. Gostilo, V., et al. (2002), The development of drift-strip detectors based on CdZnTe, *IEEE Transactions on Nuclear Science*, 49, 2530.
57. Caroli, E., et al. (2010), Development of a 3D CZT detector prototype for Laue Lens telescope, *Proceedings of SPIE*, 7742, 77420V.
58. Kuvvetli, I., et al. (2010), CZT drift strip detectors for high energy astrophysics, *Nuclear Instruments and Methods in Physics Research A*, 624, 486.
59. Kuvvetli, I., et al. (2010), Charge collection and depth sensing investigation on CZT drift strip detectors, *2010 IEEE Nuclear Science Symposium Conference Record*, 3880.
60. Luke, P.N., et al. (2000), Coplanar-grid CdZnTe detector with three-dimensional position sensitivity, *Nuclear Instruments and Methods in Physics Research A*, 439, 611.

61. Kuvvetly, I., et al. (2014), A 3D CZT high resolution detector for x-and gamma-ray astronomy, presented at High Energy, Optical, and Infrared Detectors for Astronomy VI, SPIE Astronomical Telescopes and Instrumentation Conference, 22–27 June 2014, Montréal, Quebec, Canada.

62. Auricchio, N., et al. (2012), Development status of a CZT spectrometer prototype with 3D spatial resolution for hard X ray astronomy, *Proceedings of SPIE*, 8453, 84530S.

63. Boucher, Y.A., et al. (2011), Measurements of gamma rays above 3 MeV using 3D position-sensitive $20 \times 20 \times 15$ mm^3 CdZnTe detectors, *2011 IEEE Nuclear Science Symposium Conference Record*, 4540.

64. Wahl, C.G. and He, Z. (2011), Gamma-ray point-source detection in unknown background using 3D-position-sensitive semiconductor detectors, *IEEE Transactions on Nuclear Science*, 58, 605.

65. Xu, D., et al. (2004), 4π Compton imaging with single 3D position sensitive CdZnTe detector, *Proceedings of SPIE*, 5540, 144.

66. Lei, F., et al. (1997), Compton polarimetry in gamma-ray astronomy, *Space Science Reviews*, 82, 309.

67. Xu, D., et al. (2005), Detection of gamma ray polarization using a 3-D position-sensitive CdZnTe detector, *IEEE Transactions on Nuclear Science*, 52, 1160.

68. Curado da Silva, R.M., et al. (2011), Polarimetry study with a CdZnTe focal plane detector, *IEEE Transactions on Nuclear Science*, 58, 2118.

69. Antier, S., et al. (2014), Hard X-ray polarimetry with Caliste, a high performance CdTe based imaging spectrometer, *Experimental Astronomy*, (in press).

5 CdTe Pixel Detectors for Hard X-Ray Astronomy

Olivier Limousin, Aline Meuris,
Olivier Gevin, and Sébastien Dubos

CONTENTS

5.1 NEEDS FOR NEW HARD X-RAY DETECTION SYSTEMS FOR SPACE SCIENCE

5.1.1 Status of X-Ray Astronomy and Solar Physics

Because our atmosphere is opaque to UV, x-rays, and soft gamma rays (50% attenuation at 30 km altitude for 1 MeV photons), high-energy astrophysics and solar physics are rather young scientific fields that emerged with the development of space telescopes. At the end of the twentieth century, novel imaging techniques for x-rays and gamma-rays were invented and implemented in the place of classical optics transparent to these radiations. In 1952, Hans Wolter proposed the technique of grazing incidence mirrors to focus x-rays after a double reflection on hyperboloid and paraboloid mirrors [1]; the type I configuration he described was first implemented onboard the Einstein U.S. satellite in 1979. In 1978, Fenimore and Cannon proposed the multiplexing technique of coded-aperture masks with uniformly redundant arrays to detect and localize gamma rays [2]; this technique was implemented in the Sigma telescope of the Granat Russian satellite in 1989 to map the gamma-ray sky for the first time. Other multiplex collimation schemes based on modulation collimators were considered to image solar flares in hard x-rays and gamma rays [3], and were implemented on the Japanese mission Hinotori in 1981.

The beginning of the twenty-first century saw the deployment of powerful x-ray and gamma-ray observatories—still active today—that provide quantitative information on accretion and particle acceleration mechanisms taking place in various stellar objects at the end of their evolution, in active galaxy nuclei (AGN), and at the sun's surface. With an angular resolution of 2 arcsec at 1 keV, the National Aeronautics and Space Administration (NASA) mission Chandra obtained impressive images of young supernova remnants and jets in pulsars. With a total effective area of 1000 cm² at 1 keV, the European Space Agency (ESA) mission XMM-Newton could observe many galaxy clusters and AGNs (whose high energy emission comes from the gravitational energy produced during matter accretion around a supermassive black hole). With a sensitivity better than 1 mCrab at 20 keV in 100 ks exposure time (10^{-5} photons s^{-1} cm^{-2} keV^{-1}), the Imager on-Board INTEGRAL Satellite (IBIS) telescope onboard the International Gamma Ray Astrophysics Laboratory (INTEGRAL) mission was able to detect x-ray binaries formed by a compact object accreting its massive companion star and obtain the first gamma-ray image of the galactic center, revealing the unexpectedly weak activity of its supermassive black hole Sagittarius A*. With its 5200 cm² cadmium zinc telluride (CZT) detection area, the NASA Swift mission has detected and observed more than 200 gamma-ray bursts since 2002. With an angular resolution down to 2.3 arcsec from 3 to 17 MeV, the Fourier transform telescope of the NASA Reuven Ramaty High Energy Solar Spectroscopic Imager (RHESSI) mission has obtained since 2002 numerous hard x-ray and gamma-ray images of solar flares resolved in time and energy.

5.1.2 New Space Missions and New Technological Challenges

What is the trend for new high-energy observatories? There is definitely a huge gap in terms of sensitivity (two or three orders of magnitude) between focusing techniques below

10 keV and indirect imaging techniques beyond this limit. Focusing hard x-rays appears as a scientific breakthrough to study transitions of states in black holes, to observe the Compton hump in AGN, to localize gamma-ray bursts, and to observe nucleosynthesis nuclear lines in young supernova remnants and several harmonics of cyclotron lines in neutron stars. That implies, on the one hand, the development of superpolished mirrors and multilayer coating to extend the reflectivity at high energy and, on the other hand, the development of focal planes with small pitch and high-Z material. This path was opened by the NASA Nuclear Spectroscopic Telescope Array (NuSTAR) mission [4], launched in 2012, and will be followed by the Japan Aerospace Exploration Agency (JAXA) Astro-H mission [5], to be launched in 2015. New instrumental concepts should also include polarization capabilities to improve our understanding of the physics of compact objects. Polarimetry measurements would help to discriminate between emission coming from the jets and the corona in microquasars and validate or invalidate some models of prompt hard x-ray emission in gamma-ray bursts.

Another modern scientific topic is the concept of "space weather." Studying the sun, in particular its interaction with the Earth, is one of the tools for addressing the problem of climate change and global warming and their impacts on our society. That partly explains the recent decisions for new solar missions to be launched between 2017 and 2022: Solar Orbiter by ESA [6], Solar Probe Plus by NASA, and Interhelioprobe by Roskosmos [7]. For these missions, the challenge is to bring close to the sun small instruments (less than 10 W of power and 10 kg in mass) that withstand high temperatures and high radiation levels. Other satellites will complete the sentinel: SOLAR-C by JAXA [8] and the Solar Polar ORbit Telescope (SPORT) of the Chinese Academy of Sciences [9]. Imaging solar flares in hard x-rays would improve our understanding of energy-conversion and release processes at the sun's surface.

This chapter presents one instrumental concept based on pixel CdTe detectors for new space instruments, from x-rays to soft gamma rays.

5.2 CALISTE TECHNOLOGY FOR COMPACT CAMERAS

5.2.1 INTRODUCTION TO CALISTE R&D PROGRAM FOR NEW CdTe-BASED IMAGING SPECTROMETER HYBRIDS

In 2004, an R&D program based on CdTe semiconductor detectors was set up in France by the Institute of Research into the Fundamental Laws of the Universe of the French Alternative Energies and Atomic Energy Commission (CEA/Irfu, Saclay, France) with the support of CNES (the French space agency) to build large focal planes for hard x-ray astronomy telescopes using focusing optics (from 5 to 100 keV) with excellent spatial and energy resolution. The performance of such arrays relies on the quality of the CZT crystals and associated electrode technologies, the noise performance of the front-end electronics, and the bonding technologies to interconnect them. Several approaches have been considered worldwide to realize CdTe-based hybrid detectors with imaging spectroscopy capability: flip-chip bonding of a CZT pixel detector on top of a two-dimensional front-end application-specified integrated circuit (ASIC) was realized for the NuSTAR focal plane, whereas CdTe double-sided strip detectors were wire-bonded to ASICs on both sides for the Astro-H hard

TABLE 5.1

Features of Caliste Devices Designed and Produced from 2005 to 2011 for Hard X-Ray Imaging Spectroscopy

Parameters	Caliste-64	Caliste-256	Caliste-HD
Years of development	2005–2007	2008–2009	2010–2011
Pixel array	8×8	16×16	16×16
Pixel pitch μm)	900	580	625
Guard ring width (μm)	900	100	20
Front-end electronics	IDeF-X V1.1 (16 channels)	IDeF-X V2 (32 channels)	IDeF-X HD (32 channels)
Number of ASICs	4	8	8
Pin grid array	7×7	7×7	4×4
Power consumption (mW)	200	800	200
Energy range	2–250 keV	1.5–250 keV	1.5 keV–1 MeV
Dimensions[a] (mm^3)	$10 \times 10 \times 18.6$	$10 \times 10 \times 20.7$	$10 \times 10 \times 16.5$

Note: From Caliste-64 to Caliste-256, the number of spectroscopic channels was multiplied by four in the same volume. From Caliste-256 to Caliste-HD, the guard ring was reduced to 20 μm, and the power consumption and electrical interface were drastically reduced thanks to the new ASIC generation.

[a] The height of the module is given from the pixelized anode surface to the 3-D module bottom side, without the pin grid length or the CdTe detector thickness, which is chosen according to the user's needs, typically from 0.5 to 2 mm thick or more.

x-ray imager. The concept presented here, Caliste, is based on three-dimensional (3-D) integration of the CdTe crystal and its associated front-end electronics into a system-in-package hybrid modular component. Caliste hybrid detectors have been developed as elementary units of a focal plane, which can be placed side by side to form a mosaic of the required size. The main drivers for the design of these devices are as follows: (a) the CdTe pixel detector is 1 cm² to maximize the production yield of uniform high-quality crystals; (b) the front-end electronics are made of several one-dimensional (1-D) multichannel ASICs placed perpendicular to the detection plane to minimize the input stray capacitance; (c) the electrical interface is fully implemented into the right cylinder volume underneath the CdTe crystal surface to result in four-side-buttable hybrid detectors. Several generations of Caliste devices based on these criteria have been designed, produced, and tested to reach optimal performance step by step [9–11]. The main properties are summarized in Table 5.1.

5.2.2   IDeF-X HD ASIC: Imaging Detector Front End for X-Rays with High Dynamic Range

5.2.2.1   Architecture, Operation, and General Features

IDeF-X HD is the latest generation of low-noise and radiation-hard multichannel front-end ASICs designed by CEA/Irfu for high-resolution CdTe or CZT detectors, in particular to be integrated into Caliste-HD [12]. X-ray interaction in the semiconductor detector produces a number of pairs of electrons and holes proportional to the deposited energy. The circuit includes 32 analog channels to convert the photogenerated charge into an amplified pulse-shaped signal, and a common part for slow control and readout communication with a specific controller. The first stage of the analog channel consists of a charge-sensitive preamplifier (CSA) based on a folded cascode amplifier with an inverter input amplifier (see Figure 5.1). It integrates the incoming charge on a feedback capacitor and converts it into a voltage; the feedback capacitor is discharged by a continuous reset system realized with a p-channel metal-oxide-semiconductor (PMOS) transistor. The increase of drain current in this transistor during the reset phase is responsible for a nonstationary noise; to minimize the impact of this noise on the equivalent noise charge, a so-called nonstationary noise suppressor was implemented for the first time in this chip version using a low-pass filter between the CSA output and the source of the reset transistor to delay this noise. The second stage is a variable-gain stage to select the input dynamic range from 10 fC (250 keV with CdTe) up to 40 fC (1 MeV in CdTe). The third stage is a pole zero cancellation stage (PZ), implemented to avoid long-duration undershoots at the output and also used to perform a first integration. The next stage of the analog channel is a second-order low-pass filter (RC^2) with adjustable shaping time. To minimize the influence of the leakage current on the signal baseline, a so-called baseline holder circuitry (BLH) was implemented by inserting a low-pass filter in the feedback loop between the output of the RC^2 filter and the input of the PZ stage. The DC level at the output is consequently stabilized for leakage current up to 7 nA per channel. The output of each single analog channel is driven both to a discriminator and to a stretcher. The discriminator

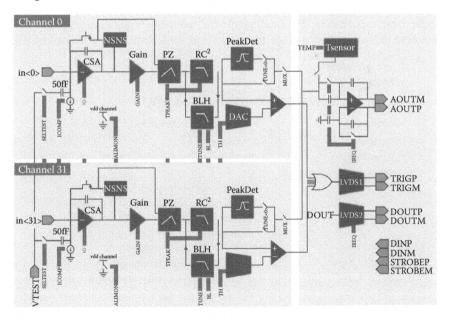

FIGURE 5.1 Architecture of IDeF-X HD front-end ASIC, including 32 analog channels, one digital part for slow control and readout communication, and one temperature sensor.

compares the signal amplitude with an adjustable in-pixel reference low-level threshold used for event triggering, while the stretcher consists of a peak detector and a storage capacitor that samples and holds the maximum amplitude of the signal, the latter being proportional to the integrated charge and hence to the impinging photon energy. In addition, each channel can be powered off by a slow control command to minimize the total power consumption of the chip when the application does not require the use of all the channels.

The slow control interface signals, as well as the analog output drivers and power supply lines, were designed to be connected together for up to eight ASICs. Thus, each ASIC of a group is addressed individually using a reduced number of wires. This optimization allowed the electrical interface to be reduced from 49 pins in Caliste 256 to 16 pins in Caliste-HD, keeping the same number of channels and using a fully differential electrical interface. When an event is detected in at least one channel, a global trigger signal (TRIG) is sent out of the chip. The controller starts a readout communication sequence using three digital signals (DIN, STROBE, and DOUT) to obtain the address of the hit ASIC and then the hit channels. Then the amplitudes stored in the peak detectors of the hit channels are multiplexed to a differential output buffer (AOUT). The whole readout sequence for one ASIC lasts between 5 and 20 μs, according to the set delays and clock frequencies and the number of channels to be read out. The main features of the chip are summarized in Table 5.2.

5.2.2.2 Noise Performance

Figure 5.2 shows noise characteristics of the ASIC expressed in terms of equivalent noise charge (ENC). Noise at short peaking times is dominated by series noise, which depends on the main charge preamplifier features (power and geometry) and on the input capacitance. The small pixel area and the hybridization technique limit the input capacitance in a range from 1 to 2 pF in Caliste-HD. Without any dark

---

TABLE 5.2
Main Features of IDeF-X HD Front-End ASIC

Parameters	IDeF-X HD
Die area	3.5×5.9 mm^2
Technology	Austriamicrosystems 0.35 μm CMOS 3.3 V
Number of analog channels	32
Input stage	DC coupling, negative polarity (for connection to detector anode), optimized for 2–5 pF input capacitance, 1 pA to 7 nA dark current (CdTe small pixels)
Peaking time	0.7–10.7 μs (16 possible settings)
Dynamic range	10–40 fC (4 possible gain settings)
Output stage	Single-ended or differential output buffer (0.4–2 mA)
Typical power consumption	27 mW in total, that is, 0.8 mW per channel
Radiation hardness	Latchup free, tolerant to total ionizing dose > 200 krad

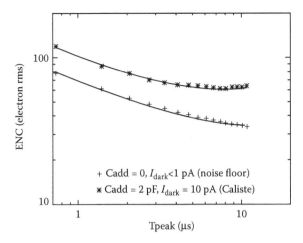

FIGURE 5.2 Equivalent noise charge characteristics of IDeF-X HD vs. peaking time for two input capacitances and leakage current values corresponding to the ASIC noise floor and the ASIC integrated in Caliste-HD.

current, noise at long peaking times is dominated by 1/f noise, which depends on the geometry and technology of the charge preamplifier and on the input capacitance. With dark current, it is the parallel noise that dominates. Dark current in small pixels can be limited to a few tens of picoamps in CdTe Schottky diodes when the crystal is cooled below 0°C. The expected best performance in the Caliste-HD configuration is an ENC between 50 and 60 electrons rms. That corresponds to an energy resolution between 550 and 650 eV FWHM at 14 keV with a CdTe detector.

5.2.2.3 Radiation Hardness

The IDeF-X family has been designed to be radiation hard with respect to the space environment. The radiation hardness of IDeF-X HD was evaluated during irradiation campaigns. Heavy ion tests were performed at the heavy ion cyclotron facility (HIF) of Louvain-la-Neuve (Belgium), and no latchup was noticed up to the maximal available linear energy transfer of 110 MeV mg^{-1} cm^{-2}. This was possible due to a radiation-hardened design of the digital circuitry of the chips. Total ionizing dose tests were performed with a ^{60}Co source at the CoCase facility (Saclay, France); circuit parameters, including noise performance, were monitored during the campaign and are reported in Figure 5.3. No noise degradation was noticed up to 40 krad(Si) total ionizing dose. Tens of kilorads is the typical level that would be expected for payload equipment in a spacecraft during a mission if the instrument is properly shielded. The chip can actually work properly until a total ionizing dose of 200 krad or more, with very satisfactory performance and no modification of the power consumption.

5.2.3 Caliste HD: Last Generation of 256-Pixel 1 cm² CdTe Microcamera

The Caliste-HD detection system consists of eight 32-channel IDeF-X HD ASICs hybridized with a 256-pixel CdTe sensor of 1 cm² surface area and typically 1 or 2 mm in thickness. The ASICs are mounted on microprinted circuit boards, tested,

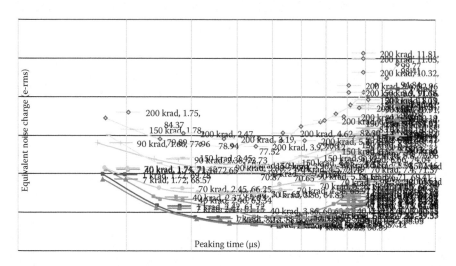

FIGURE 5.3 Equivalent noise charge characteristics vs. the peaking time setting during total ionizing dose test campaign with a ^{60}Co source, for a leakage current set to 20 pA.

stacked in eights with high alignment precision, and molded into an epoxy resin, according to the technology of the 3D-Plus company. The channels of an ASIC are connected to two adjacent rows of 16 pixels. The resulting block is cut and metalized with a gold coating. A laser ablation process is used to create tracks on the lateral sides of the block to make electrical connections between the eight ASICs, which share all supply voltages and input and output signals (but have individual addresses). The bottom interface includes a filtering stage for power supplies and a grid array of 16 pins. The top interface of the so-called Caliste body is then prepared by a laser-assisted polymer bump bonding technique to realize a matrix of 16×16 conductive epoxy bumps with a pitch of 625 µm and two extra bumps at the periphery for the connection of the guard ring [13]. The CdTe pixel detector is connected to the Caliste body by a flip-chip technique. This full hybridization process is performed at fairly low temperature to keep the CdTe below 80°C, while the bumps are high enough to withstand the thermal cycling required for space qualification. The resulting device, with a 1 mm thick detector, has a volume of $10 \times 10 \times 22.5$ mm^3, including the pin grids, and a weight of 6 g (see Figure 5.4). Performance results are presented in Section 5.3.

5.2.4 MACSI: Modular Assembly of Caliste Spectroscopic Imagers

By design, Caliste-HD is a four-side-buttable imaging spectrometer device with a 1 cm^2 CdTe sensor and can be used for the realization of large cameras. A prototype with an 8 cm^2 detection plane, called MACSI, was designed and realized with an array of 2×4 Caliste-HD units. This geometry was chosen with a view to the realization of a 64 cm^2 CdTe focal plane integrating eight MACSI detection sectors next to each other for space astronomy. In this context, the only way to cool down the CdTe sensors is by thermal conduction from the base plate and through

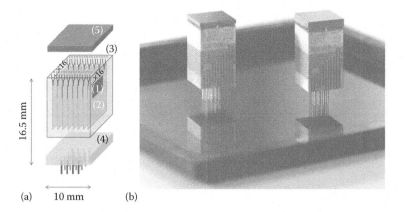

FIGURE 5.4 (a) Principle of Caliste-HD hybridization with IDeF-X ASIC (1) on flex substrate (2); the channels are connected to the 16×16 conductive epoxy bumps at the top surface, and all the interface signals are driven to the bottom interface in a 4×4 pin-grid array (4); the CdTe pixel detector is connected to the Caliste body by flip-chip bonding (5). (b) Picture of two Caliste HD units, with 1 or 2 mm thick CdTe detector, fitting in a volume of $10 \times 10 \times 25$ mm³. (From A. Meuris, O. Limousin, O. Gevin, et al., *Nucl. Sci. Symp. Conf. Rec.* 2011, 4485–4488.)

the Caliste bodies. To optimize this, the printed circuit board hosting the Caliste units is a flex circuit realized with thick ground planes and glued to a mechanical drain made of aluminum. The 20 cm flex lead allows routing in a focal plane enclosure with an anticoincidence shielding; it is terminated with a small printed circuit board, integrating power-supply regulators and filters, and could possibly include amplifier buffers, low-voltage differential signaling (LVDS) drivers, and receivers for other applications. All components and processes were chosen to meet space quality standards. After manufacture of this assembly, presented in Figure 5.5a, the Caliste-HD units are placed and manually soldered, with a gap of ~400 µm between them. In this way, using an electrode pattern of the CdTe detector anode with a thin 20 µm wide guard ring, we guarantee a dead zone no larger than one pixel pitch between two pixels of adjacent Caliste-HD units. The high voltage is connected by wire bonding between the high-voltage filtering stage and the platinum planar cathodes of the two closest Caliste-HD units; the other units are biased by wire bonds coming from their adjacent Caliste-HD neighbors, as shown in Figure 5.5b.

The MACSI ends with two 37-pin micro-D connectors to be connected to the warm electronics, including the controller and analog-to-digital converters (ADCs). The first prototype was tested in a vacuum chamber using an interface board with LVDS drivers and receivers and test points. The acquisition board, placed outside the vessel, includes a field programmable gate array (FPGA) that controls the configuration and the readout sequence of the eight Caliste-HD units (i.e., 64 ASICs), controls the conversion of the amplitudes with one 12-bit ADC, generates event frames, and sends the data stream to a computer through a Spacewire link. When a photon interacts in one CdTe crystal, all the ASICs of the corresponding Caliste-HD

(a)   (b)

FIGURE 5.5 Picture of the first MACSI realization. (a) MACSI is the assembly of CdTe hybrid detectors on an electrical and thermal interface terminated with a flex lead to be connected to warm electronics outside a focal plane enclosure. (b) Close-up view of the 8 cm^2 CdTe detection plane with 2048 individual spectroscopic channels made of eight Caliste-HD units placed side by side.

are temporarily disabled during the readout sequence of the corresponding event, but the other Caliste-HD remains in observation mode.

5.3 PERFORMANCE OF FINE PITCH PIXEL CdTe DETECTORS

5.3.1 SPECTROSCOPIC PERFORMANCE

Caliste-HD units were tested individually before being integrated in a MACSI system. Twenty-five defect-free units were fabricated, with excellent spectral performance. Figure 5.6 shows the ^{241}Am spectrum obtained after calibrating the

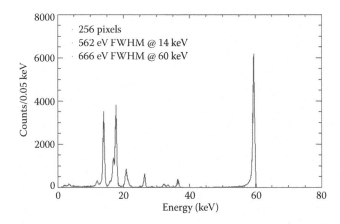

FIGURE 5.6 ^{241}Am spectrum obtained by summing the 256 calibrated spectra of the Caliste-HD matrix when the device is cooled to −5°C. The 1 mm-CdTe detector is biased at −300 V. The energy resolution is 0.56 keV FWHM at 14 keV, and the low-level threshold is 1.5 keV.

256 pixels of a sample and summing the 256 histograms by energy bins of 50 eV. The resulting energy resolution is 562 eV FWHM at 14 keV and 666 eV FWHM at 60 keV. This corresponds to a mean equivalent noise charge of 50 electrons rms. This performance can be obtained due to the low-noise IDeF-X HD ASIC, the hybridization principle that minimizes the input capacitance, and the Al-Schottky CdTe diodes from Acrorad Co. Ltd (Japan), which show very low leakage current (typically 1 nA cm^{-2} at $-5°C$, 300 V). The pixel configuration compared with double-sided strip detectors is favorable for imaging spectroscopy, since the capacitance and the leakage current that drive ASIC performance are proportional to electrode area.

The architecture of the low-level discriminator in IDeF-X HD, with individual settings, makes it possible to reach low-energy thresholds down to 1 keV on individual channels and 1.5 keV on a Caliste-HD unit. This capability was exploited to determine the real detection efficiency of the CdTe detectors at low energy with a platinum electroless planar cathode [14]. Caliste-HD was placed into the SOLEX facility (Saclay, France), which generates a flux-calibrated monoenergetic x-ray collimated beam in a range from 2 up to 12 keV. Images and spectra were recorded down to 2.13 keV with Caliste-HD; the energy resolution at this energy is 441 eV FWHM for a single pixel and 560 eV after reconstruction of the sum spectrum. We derived the detector efficiency by computing the ratio between detected photons and incoming photons. The impinging photon rate on Caliste-HD was determined by measuring time-interval statistics between two consecutive events, fitting a Poisson-law distribution (typically 1000 counts s^{-1}).

From our measurements, we derived the total photopeak efficiency of Caliste equipped with a Schottky type Pt//CdTe//Al detector as a function of the impinging photon energy. We found the efficiency to be 97% at 7 keV and above, and we successfully measured up to 40% at 2.13 keV. This result is surprisingly high and demonstrates that, despite the nature of the Pt heavy-metal contact at the entrance of our crystals, it is worth pushing down the discrimination threshold as low as possible. From this observation, it is also possible to derive the detection-efficiency limiting factors at low energy, that is, at the entrance. For this purpose, as shown in Figure 5.7, we fitted our model to single events only for simplicity—getting rid of subtle effects due to charge sharing, photon escapes, and discrimination-threshold cutoff—and found the data compatible with a 34 nm thick Pt electrode layer superimposed on a 440 nm thick dead layer, showing the same attenuation properties as CdTe. These layers only affect the response at very low energies.

5.3.2 Imaging Performance

To illustrate the imaging capability in x-rays, we illuminated MACSI with a ^{241}Am source using a tungsten mask with a spiral-shaped aperture between the source and the detectors (see Figure 5.8a). Cooling the system to 0°C, we successfully operated 2045 pixels out of 2048. The pixels show a normal statistics count rate thanks to excellent uniformity of the sensing material. The dead zones of the camera, appearing in black in the image, are due to the required space between two neighboring Caliste

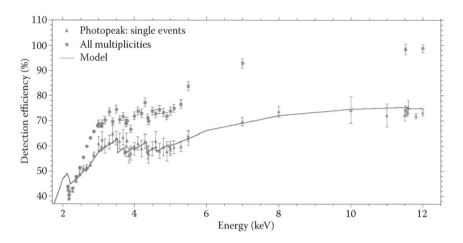

FIGURE 5.7 X-ray detection photopeak and total efficiencies at low energy measured with Caliste-HD equipped with 1 mm CdTe detector from Acrorad Co. Ltd. (Japan). The response modeling that matches experimental data corresponds to an entrance window made of 34 nm platinum with 440 nm interface dead zone.

units in the MACSI design for mounting tools. This dead space is limited to the width of one pixel and appears every 16 pixels. For a 64 cm² focal plane made of eight MACSI sectors, the total dead zone would correspond to 9% of the total surface. Thanks to impressive progress in Schottky CdTe production in terms of uniformity and yield since 2005, it is now possible to design new space-imaging spectrometers with 2–4 cm² CdTe detectors using the same hybridization concept to reduce the dead-area fraction.

Despite the presence of the dead region between Caliste modules, note that the entire volume of the pixelated CdTe crystal is sensitive to x-rays: when a photon interacts in a pixel gap (100 μm pixel gap for 525 μm pixel size in Caliste-HD geometry), the charge cloud induces charges in all the neighboring channels with very little charge loss (less than 1% at 60 keV) and creates split events. All MACSI channels are independent and include their own discriminator. Consequently, multiple hits are read out and associated in the data with a unique time-tag but distinct energy and position coordinates. Therefore, it is possible to analyze the data and efficiently correct the split events between neighbor pixels. The split events are mostly due to charge sharing (two, three, or four pixels) between pixels or fluorescence escapes (double hits). Assuming that a large majority of split events are caused by photons impinging into the interpixel gaps, their localization is assigned to a 100×525 μm rectangle between two triggered pixels for double hits or assigned to a 100×100 μm square at the center of four neighboring pixels in the case of triple or quadruple hits. The flexibility of MACSI architecture allows selection of only single events to obtain a high-resolution image associated with excellent spectral performance, as well as the possibility of refining interaction areas by visualizing split events. In this case, the improvement of spatial resolution

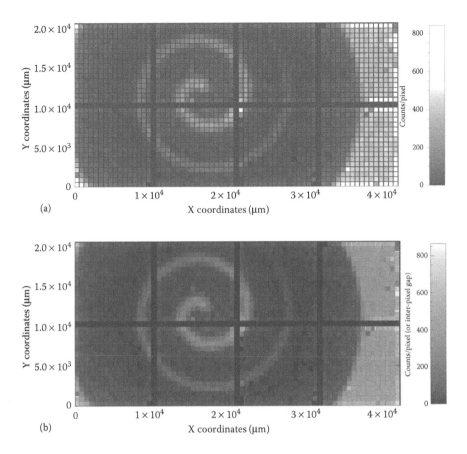

FIGURE 5.8 X-ray image of the first MACSI detection plane of 8 cm² CdTe (2048 pixels) by illuminating the matrix with a ²⁴¹Am source through a 1 mm thick tungsten mask. The round mask has a spiral-shaped opening. (a) Image of counts obtained for each single pixel (raw image). (b) Image of counts obtained for each single pixel and in interpixel gaps (processed image). Respective counts measured in interpixel gaps were weighted to their surface.

comes at the expense of a slight degradation of energy resolution. As a matter of fact, the energy resolution results from the quadratic sum of the electronic noise of each channel. The energy resolution is degraded by $\sqrt{2}$ for double hits.

Regarding the imaging refinement, it is not sufficient to compute the geometrical ratio of the interpixel gap surface to the total sensitive surface in order to estimate the fraction of multiple to single events. For instance, in the image displayed in Figure 5.7, the fraction of split events is found to be 23%, while the interpixel regions represent 28% of the whole detection surface. Apart from the fluorescence, which plays a role when the impinging photon energy is high enough with respect to Cd and Te K edges (a few percent of the double events), the operating conditions must be taken into account as well: high voltage and operating temperature affect the probability of charge sharing occurring. Moreover, the low-level discriminator threshold settings affect the capability to detect multiple events. In fact, a high discrimination level will cause the single event rate to be artificially higher than expected. Consequently, a rigorous image reconstruction, in which single and multiple events are properly assigned to pixels and interpixel areas, respectively, requires some careful corrections. An "effective pixel surface," likely to be larger than the metalized surface, must be defined to flatten the images. The effective pixel surface is a correction coefficient matrix that depends on the photon energy (e.g., charge-cloud size), the operating conditions, and the detector geometry. Nevertheless, for simplicity and as a first approximation, Figure 5.8b is obtained by weighting the pixel and interpixel counts to their respective geometrical surfaces, and looks flat and uniform.

In addition to the imaging and spectrometric capabilities, each event of MACSI is time-tagged with an accuracy better than 1 μs. Since photon events are processed in 20 μs or less, such a detection system can produce x-ray movies by integrating flux every second, or less often if the source is bright.

5.3.3 Caliste Used as Hard X-Ray Polarimeter

Caliste devices are not only imaging spectrometers resolved in time but can be used in hard x-rays as Compton polarimeters. The polarimetric performance of our detector is primarily determined by the fundamental physics underlying the Compton-scattering process for linearly polarized photons. In this case, the Compton-scattering differential cross section, per unit of elementary solid angle $d\Omega$, is defined as follows, after Klein-Nishina [15]:

$$\frac{d\sigma}{d\Omega} = \frac{r_0^2}{2} \left(\frac{E'}{E} \right) \left[\frac{E'}{E} + \frac{E}{E'} - 2\sin^2\theta \cos^2\varphi \right] \tag{5.1}$$

where:

$r_0 =$ classical electron radius
E and $E' =$ energies of the impinging and scattered photons

The scattered photon is deviated from its original direction by θ. The azimuthal deviation angle φ corresponds to the angle formed by the scattering plane (defined by the initial direction and the scattered direction) and the incoming photon polarization plane (defined by the photon direction and its polarization vector, that is, the photon electric field component), as shown in Figure 5.9a.

Considering linearly polarized photons, this equation provides the azimuthal (φ) dependency for the Compton-scattered photons. The Compton scattering of polarized photons generates a nonuniformity in its azimuthal angular distribution (Figure 5.9b). From Equation 5.1, fixing all parameters except the azimuthal angle φ, the probability of interaction reaches its minimum and maximum for orthogonal directions, $\varphi = 0°$ and $\varphi = 90°$, respectively. Moreover, the relative difference of the two extrema presents its maximum for a given scattering angle θ_M, which is always around 90° in the soft gamma and hard x-ray energy range (100 keV–1 MeV).

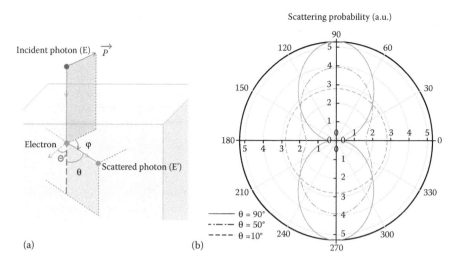

FIGURE 5.9 (a) Scheme of Compton scattering of a polarized photon. (b) Azimuthal angle φ probability distribution for a given Compton scattering angle Θ of linearly polarized photons at E = 200 keV. The direction of the polarization is parallel to the horizontal axis of the polar plot.

The polar plot in Figure 5.9b shows the probability distribution against the azimuthal angle φ for linearly polarized photons at 200 keV, after they Compton scatter at a given angle θ (10°, 50°, or 90°). It appears clearly that the asymmetry of this distribution increases with the scattering angle. Note that the asymmetry is almost invisible at θ = 10°, while for θ = θ_M = 90° the probability of a photon scattering perpendicularly to the polarization plane is roughly five times higher than the probability of scattering along the polarization plane. Finally, the asymmetrical shape of the azimuthal probability is enhanced with the scattering angle, making the distribution look like a bow tie (cf. Figures 5.9b and 5.10a).

If a primary photon, whose direction is perpendicular to the surface, Compton scatters into the detector at roughly 90° and stops further away inside the same detector, but in a distinct pixel, it is therefore possible to record the statistical distribution of the relative position of double events. Building an image of the double-event counts will draw the bow tie in question. Analyzing the data allows the degree of polarization and the polarization direction of the incident radiation to be derived by evaluating the symmetry in the count distribution.

At this point, it is possible to define a figure of merit for a polarimeter: the polarimetric modulation factor Q. For a pixel detector, Q is written as

$$Q = \frac{d\sigma(\varphi = 90) - d\sigma(\varphi = 0)}{d\sigma(\varphi = 90) + d\sigma(\varphi = 0)} = \frac{N_y - N_x}{N_y + N_x}$$

(5.2)

where N_x and N_y are the double events integrated over two orthogonal directions (x and y) defined over the detector plane.

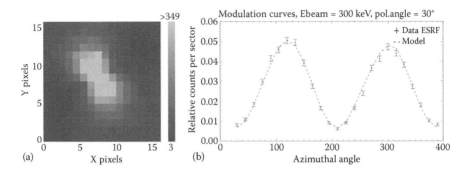

FIGURE 5.10 Polarimetry measurements realized at ESRF Grenoble by selecting Compton events in one Caliste-256 detector (scattering around 90°). A 300 keV photon beam with 98% polarization fraction and a polarization angle of 30°. (a) Compton photon distribution over the CdTe 256-pixel matrix. (b) Count-rate modulation: comparison of experimental data and modeling. A Q factor of 0.76 ± 0.02 can be obtained in these conditions.

We used a Caliste-256 device in the form of a Compton polarimeter at the European Synchrotron Radiation Facility (ESRF, Grenoble, France), where we exposed the detector to a highly polarized and monoenergetic beam at 300 keV, for different angles and fractions of polarization [16]. At these energies, it is very likely that a Compton-scattering process occurs in a CdTe detector. Deviating by 90°, the scattered photon energy is 189 keV and has a mean free path of 5.42 mm, well suited to the 1 cm² of Caliste. Figure 5.10a shows an example of a map of photoelectric absorption in the energy range from 181 keV up to 197 keV, the beam—300 keV, 98% linearly polarized with an angle of 30°—being focused in the middle of the bow-tie figure. The data are corrected for fluorescence line escape, charge sharing, and detector alignment. and the modulation curve is derived as shown in Figure 5.10b. We find the modulation factor to be $Q = 0.76 \pm 0.02$. Our model is superimposed on the data, and the match is almost perfect. Note that this is not a model fitting with free parameters but a superimposition of the theoretical response computing the cross sections at 300 keV (in Equation 5.1) and projections of the photoelectric probabilities of the scattered photons on the pixel pattern. Model and real data sets are analyzed in the same way, in the same energy range, to build up the modulation curves.

5.3.4 Timing Performance

With a focusing technique for hard x-ray imaging, the count rate expected for bright astronomical sources is in the order of 100–1000 counts s⁻¹ spread over a few millimeters on the detection place (surface defined by the angular resolution and the focal length of the telescope). Despite this rather low count-rate requirement, CdTe detection planes based on Caliste hybridization with IDeF-X front-end ASIC are well suited for high count-rate applications, due to the self-triggered architecture. Each Caliste of a detection array can be connected to an ADC, and hence can be operated fully independently of its neighbor. When the controller detects a trigger signal from a Caliste unit, it has to wait for a latency time at least equal to the time for the pulse to reach its maximum value, which is stored in the peak detector and is proportional to the incident energy. During this period, a second photon arriving in the same pixel would provoke a pileup event, or a "misfit" event if it interacted in another pixel (good count estimate but bad energy estimate). Then the controller starts the readout sequence and the whole device is locked and blind. That corresponds to a dead time for the instrument: events occurring in this period in the same Caliste unit are missed. A mathematical model has been built to estimate the timing response of such a system. The input events are generated by Monte Carlo simulation of a monoenergetic bunch of photons interacting in a 1 mm thick CdTe detector and 100 nm Pt entrance window, and time tagged using a Poisson-law distribution. A basic model of induction in the semiconductor detector is assumed to take charge sharing into account according to the pixel pattern of the anode: for Caliste-HD, only interactions occurring in the 100 µm pixel gap provoke split events. The events are then processed sequentially using simulation data of the ASIC response; events are tagged as detected, misfit, in pileup, or missed. The results obtained after several simulation runs with different source count rates are shown in Figure 5.11 for a typical configuration of Caliste-HD. For a peaking time of 2 µs and a latency time of 2.5 µs (corresponding to the time to summit) and a low-level threshold of 3 keV, the fraction of misfit events, including pileup, can be inferior to 2% until 10,000 photons s^{-1} cm^{-2}. At that rate, the dead time is 9% of the observation time, but the actual number of missed events due to the readout of the device is only 3%. It is actually possible to operate the device up to 10^5 photons s^{-1} cm^{-2} and to correct the recorded counts knowing the dead time and using such simulation tools to properly estimate the incident flux. Another efficient way to estimate this flux is to study the statistics of time intervals between two events and to model this by the Poisson law.

For applications for which time-tagging accuracy is important, for instance an instrument surrounded by an active shielding with a veto system, the time of arrival of the photon can be precisely estimated if the trigger time of the channel, the signal amplitude, and the peaking time of the filter stage are known. The time uncertainty comes from the jitter of the trigger due to electronic noise and is higher at low energy. Time resolution of 15 ns rms was measured at 50 keV for 5 µs peaking time.

5.4 DERIVATIVE PRODUCTS AND APPLICATIONS

5.4.1 SOLAR FLARE OBSERVATION

The spectrometer telescope imaging x-rays on board the Solar Orbiter ESA mission will embed 32 Caliste-SO spectrometer units to detect hard x-rays from 4 to 150 keV produced by electrons accelerated in solar flares. These devices are inherited from the Caliste R&D program (see Figure 5.12a). They take advantage of the compact, low-power, low-noise, and radiation-hard design of Caliste-HD. The pixel pattern is adapted for the indirect imaging technique based on a Fourier transform of the image. The anode is divided into four bands to measure the amplitude and the phase of the Moire pattern realized by the collimator placed in front of each Caliste-SO

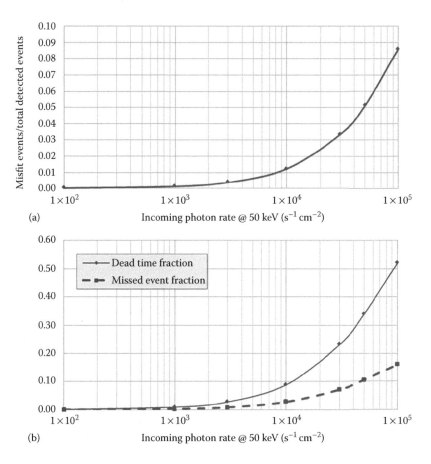

(a)

(b)

FIGURE 5.11 Simulation of timing properties of Caliste-HD for a 50 keV photon beam, a Poisson distribution of events, a shaping time of 2 μs, a readout time of 9 μs for single events and 10 μs for double events, and a low-level threshold of 3 keV. (a) Fraction of misfit events (photons incoming during the latency) as a function of the source count rate. (b) Fraction of dead time and missed events as a function of the source count rate.

(responsible for the measurement of one unit of visibility of the image). Each band is divided into two large pixels for redundancy and one small pixel to precisely measure source intensity and energy in the case of very bright events (lower pileup-event probability). A guard ring representing 12% of the detector area surrounds the device. The resulting adopted pixel pattern is shown in Figure 5.12b. Caliste-SO uses the same ASIC as for Caliste-HD, but only one per device. The room left inside the Caliste body is used to integrate space-qualified passive parts for power-supply local filtering and to route the high-voltage signal from the bottom electrical interface (small outline package) to the detector cathode at the top by wire bonding. It is possible in this application to build a Caliste body slightly larger than the CdTe crystal and to use a surface-mounting interface, since it is not required to put several Caliste units next to each other. STIX will benefit from the count-rate capability of IDeF-X HD, illustrated in Section 5.3.4. The Caliste-SO units will be

(a)   (b)

FIGURE 5.12 (a) Picture of Caliste-SO spectrometer unit for Solar Orbiter/STIX. (b) Picture of the patterned Schottky anode side of the CdTe detector on top of Caliste-SO.

grouped in twos (fully sharing their electrical interface and ADC); consequently, reading out a single event in one of the Caliste-SO locks two units in dead time. For simplicity, the onboard system will allow the readout of single events only, and the duration of the readout sequence is set to a fixed value to facilitate accurate dead-time calculation aboard the spacecraft. By using the simulation tool presented in Section 3.4 with the STIX configuration (3 keV threshold, 2.5 µs latency time, 10 µs readout time, trigger common to two units, single-hit processing), we estimate the pileup fraction to be 0.2%, the missed event fraction to be 5%, and the fraction of processed events to be 62% for a mean source rate of 20,000 counts s^{-1} cm^{-2}. More precise results can be obtained by using a real solar-flare spectrum as input source, by modeling the imaging system with a collimator made of pairs of tungsten grids above the sensors, and by implementing a more realistic model of charge induction in CdTe.

5.4.2 NUCLEAR IMAGING UP TO MEGAELECTRONVOLT RANGE

Thanks to its high performance and its compact and modular design, Caliste is particularly well suited for use in handheld, low-size and weight-imaging spectrometers for nuclear inspection and radiation monitoring. After the major accident at the Fukushima nuclear power plant following the devastating tsunami in 2011, the international community reinforced its efforts toward nuclear plant safety, including the development of high-precision portable imaging spectrometers [17,18]. For these reasons, our group has initiated the development of a new camera using Caliste to enable high-precision identification, localization, and activity measurement of radioactive isotopes dispersed after an accident. A first prototype of a miniature system, WIX, equipped with Caliste-HD, has been recently designed, assembled, and tested (see Figure 5.13a). Caliste-HD is placed into a hermetic enclosure, with a sealed beryllium entrance window, ideal for transparency to x-rays. The detector housing is depressurized. Small thermoelectric coolers are placed against two lateral faces of Caliste-HD and enable cooling of the CdTe

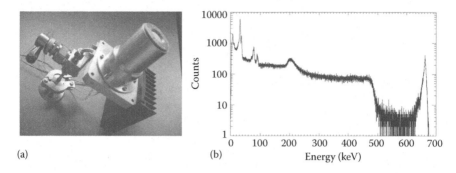

(a)   (b)   Energy (keV)

FIGURE 5.13 (a) WIX, a first prototype of portable nuclear imaging spectrometer: the hermetic housing embeds a Caliste-HD detector. (b). ^{137}Cs spectrum obtained with Caliste-HD. The spectrum is the sum of the single events spectra for all calibrated channels. In this configuration, Caliste-HD is equipped with a 2 mm thick CdTe (Schottky). The energy resolution is 4.1 keV at 662 keV, and the low energy threshold is set at 2 keV.

crystal to ~−5°C to guarantee stable operation and good spectral performance. The heat flux dissipated by the ASICs is evacuated by means of a passive radiator placed on the camera side. The full system fits into a volume of $8 \times 8 \times 12$ cm^3. The detector is directly connected to the Spacewire acquisition board developed for MACSI. In the near future, the interface will be simplified and reduced in size, making use of a standard USB driver interface. WIX will eventually be fully autonomous and will have wireless data-transfer capability. The acquisition system will be redesigned to fit into the small volume right behind the detector head. The system will include preprocessing for calibration, spectroscopy, Compton event reconstruction, and source localization by means of an adequate coded-mask aperture, placed at the front of the camera. Caliste-HD was originally optimized for the observation of faint celestial point sources in the hard x-ray range up to ~250 keV. However, its electronics readout system is able to perform fine spectroscopy in a much broader energy range, up to 1 MeV. Combining double events allows the range to be extended accordingly, which is mandatory in the nuclear-imaging application domain. For purposes of illustration, we show in Figure 5.13b preliminary results obtained with the system illuminated by a ^{137}Cs radioactive source. The spectrum is the sum of all calibrated pixels for single events only. The energy resolution is 4.1 keV FWHM (0.6%) at 662 keV, and the peak-to-valley ratio between counts at 662 keV and 650 keV is 11:1. In this case, Caliste-HD is equipped with a 2 mm thick CdTe pixel detector, which limits the photopeak efficiency to 4% at 662 keV. However, Caliste-HD will eventually be equipped with a thicker detector, possibly in CZT. Note that, even when working with the highest dynamic range setting of the camera (1 MeV), the low energy threshold remains at 2 keV.

5.5 CONCLUSION

Recent developments in advanced pixel detectors, for hard x-rays particularly, emphasize the use of CdTe semiconductors worldwide, associated with high performance and full-custom-readout ASIC chips. Our group at CEA has contributed to playing a major role in this field in the last decade, creating and promoting the Caliste product line, an original configuration of CdTe-based pixel detectors using 3-D technology assembly (system in package) for the integration of IDeF-X readout chips. 3-D technology enables low noise; uniform response; and modular, reliable, and compact design, particularly suitable for application in space science, astronomy, or solar physics. We demonstrated our ability to integrate the same technology in various configurations with contradictory requirements: from x-rays to gamma rays, from low to high rates, with any shape and size of pixels, but always with high spectral response. The maturity of the Caliste product line is good enough to foresee applications in other fields of application, such as portable devices for radiation monitoring.

ACKNOWLEDGMENTS

The authors warmly thank CEA-Irfu and CNES for supporting and funding the R&D programs that enabled the creation of Caliste detectors. Our team is grateful for cooperative and innovative work performed at 3D-Plus (France) for Caliste manufacturing, Acrorad Co. Ltd. (Japan) for high-quality CdTe detector manufacturing, Systronic (France) for advanced PCB manufacturing, and Resa (France) for the assembly of MACSI technological demonstrators. The authors acknowledge the professionalism and the key contributions of their colleagues to the design, fabrication, and performance tests: S. Antier, C. Blondel, M. Donati, L. Dumaye, F. Ferrando, A. Goetschy, I. Le Mer, F. Lugiez, J. Martignac, F. Nico, and F. Pinsard.

The WIX portable device is initiated in the frame of the ORIGAMIX RSNR research program "Investissement d'Avenir," referenced ANR-11-RSNR-0016, supported by the French Government and managed by the Agence Nationale de la Recherche.

REFERENCES

1. H. Wolter, Spiegelsysteme streifenden Einfalls als abbildende Optiken für Röntgenstrahlen, *Annalen der Physik* 10 (1952): 94–114.
2. E. E. Fenimore and T. M. Cannon, Coded aperture imaging with uniformly redundant arrays, *Applied Optics* 17(3) (1978): 337–347.
3. T. A. Prince, G. J. Horford, H. S. Hudson, et al., Gamma-ray and hard x-ray imaging of solar flares, *Solar Physics* 118 (1998): 269–290.
4. F. A. Harrison, W. W. Craig, F. E. Christensen, et al., The nuclear spectroscopic telescope array, *Astrophysical Journal* 770(2) (2013): 103.
5. T. Takahashi, K. Mitsuda, R. Kelley, et al., The ASTRO-H mission, *Proceedings of SPIE* 7732 (2010): 77320Z.
6. D. Müller, R. G. Marsden, O. C. StCyr, et al., Solar Orbiter: Exploring the sun–heliosphere connection, *Solar Physics* 285 (2013): 25–70.
7. V. Kuznetsov, The interhelioprobe mission for the study of the sun and inner heliosphere, *39th COSPAR Scientific Assembly* E2.2–3-12 (2012) 1010.
8. T. Shimizu, S. Tsuneta, H. Hara, et al., The SOLAR-C mission: Current status, *Proceedings of SPIE* 8148 (2011): 81840C.

9. A. Meuris, O. Limousin, F. Lugiez, et al., Caliste 64, an innovative CdTe hard x-ray micro-camera, *IEEE Transactions on Nuclear Science* 55(2) (2008): 778–784.

10. O. Limousin, F. Lugiez, O. Gevin, et al., Caliste 256: A CdTe imaging spectrometer for space science with a 580 μm pixel pitch, *Nuclear Instruments and Methods in Physics Research Section A* 647 (2011): 46–54.

11. A. Meuris, O. Limousin, O. Gevin, et al., Caliste HD: A new fine pitch Cd(Zn)Te imaging spectrometer from 2 keV up to 1 MeV, *Nuclear Science Symposium Conference Record* (2011): 4485–4488.

12. O. Gevin, et al., Imaging X-ray detector front-end with dynamic range: IDeF-X HD, *Nuclear Instruments and Methods in Physics Research Section A* 695 (2012): 415–419.

13. O. Limousin and F. Soufflet, Procédé d'interconnexion par retournement d'un composant électronique, Institut National de la Propriété Intellectuelle, patent INPI-11/00719.

14. S. Dubos, O. Limousin, C. Blondel, et al., Low energy characterization of Caliste HD, a fine pitch CdTe-based imaging spectrometer, *IEEE Transactions on Nuclear Science* 60(5) (2013): 3824–3832.

15. O. Klein and Y. Nishina, Über die Streuung von Strahlung durch freie Elektronen nach der neuen relativistischen Quantendynamik von Dirac. *Zeitschrift für Physik* 52 (1952): 853–869.

16. S. Antier, O. Limousin, P. Ferrando, et al., Hard x-ray polarimetry with Caliste, a high performance CdTe based imaging spectrometer, to be published in *Experimental Astronomy*, 2014.

17. D. Matsuura, K. Genba, Y. Kuroda, et al., ASTROCAM 7000HS, radioactive substance visualization camera, *Mitsubishi Heavy Industries Technical Review* 51(1) (2014): 68–75.

18. F. Carrel, K. Khalil, S. Colas, et al., GAMPIX: A new gamma imaging system for radiological safety and homeland security purposes, *IEEE Nuclear Science Symposium Conference Record* (2011): 4739–4744.

6 Technology Needs for Modular Pixel Detectors

Paul Seller

CONTENTS

6.1 INTRODUCTION

Area-array imaging detectors are ubiquitous in our everyday lives and in science instrumentation. For imaging IR, visible, and UV photons, the usual approach taken is to use focusing optics to concentrate the light onto a small imaging area and to use the smallest pixels possible to reduce the cost of the imaging detector. This is usually true in science experiments as well as commercial cameras. The modality for detecting x-rays is very different. In most cases it is impossible to focus the x-rays, and large-area detectors are required. This is true for x-ray photon and particle detectors in high-energy physics, space science, and synchrotron applications. Many applications can use large flat panels [1] with deposited scintillators or scintillating screens. Amorphous silicon screens have been very successful in medical and security applications as well as science but do have limitations in speed and several other performance criteria [2]. Complementary metal–oxide–semiconductor (CMOS) sensors have recently surpassed charge-coupled devices (CCDs) for many applications and can now be constructed on a wafer scale [3]. Even with these wafer-scale sensors, there is a requirement to tile these to create larger arrays. Up to a certain size, this is relatively easy, as the detectors can be three-side-butted and connected to scintillator screens. The readout can be performed on one edge using

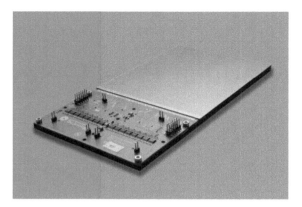

FIGURE 6.1   Three-side-buttable LASSENA CMOS sensor of 140×140 mm^2 with 50 μm pixels mounted on readout card.

conventional wire bonding and the sensor area connected to a stable substrate which also allows cooling if required (Figure 6.1).

These CMOS sensors are also used in this configuration for the direct-conversion mode of detecting x-rays using diodes integrated in the CMOS technology to convert the x-rays to electrons. This is usually restricted to low-energy x-rays (<10 keV) and can suffer from radiation damage effects. Overall, flat-panel detectors and CMOS sensors have been very successfully integrated into arrays using large-area modules. The situation with direct-conversion semiconductor detectors using thicker conversion layers is very different. This requires the use of separate conversion layers of thick silicon or high-atomic number material bonded to the readout electronics in a so-called hybrid configuration. The process technologies and interconnect for these hybrid sensors are much more complicated.

6.2   HYBRID SEMICONDUCTOR X-RAY IMAGING DETECTOR MATERIALS AND CONTACTS

X-rays interact inelastically with matter in three ways: photoelectric, Compton, and pair production [4]. These energy-loss mechanisms all eventually produce thermalized electron–hole pairs in the conduction and valence bands of the semiconductors. The basic conversion process consists of the electric field of the x-ray photon transferring energy to a bound electron in the material. This can, for instance, be a k-shell electron, which causes it to be ejected out of the atom. This resulting photoelectron (and also subsequent Auger electron) interacts with the electric field in the crystal lattice, elevating electrons to the conduction band and leaving holes in the valence band of the semiconductor. This is a statistical effect, producing electron–hole pairs by multiple photon–phonon interactions. On average, one electron–hole pair is produced by 3.6 eV of the photoelectron energy in silicon. These mobile carriers drift in an applied electric field across the detector. In semiconductor detectors, a large voltage can be applied across the crystal to cause these charge carriers to drift to the anode and cathode sides of the detector. As the charge carriers move, they induce a charge on electrodes on the anode and cathode sides of the detector. It is this induced signal that is measured by the preamplifiers on application-specific integrated circuits (ASICs), as explained in Section 6.3. Detectors can have thousands or millions of contacts and similar numbers of amplifier channels. In most cases, the amplifiers are fabricated on silicon ASICs, with each amplifier connected directly to the detector contact, as explained in Section 6.4.

Silicon has been used as a good conversion material to convert x-rays into charge carriers. It has a large bandgap, so very few electron–hole pairs are thermally produced at room temperature. Blocking contacts can be produced, which further reduce unwanted leakage currents. The bandgap is sufficiently low that the photoelectron produces many carriers. Good crystals with low trapping/ recombination centers are produced, so that the charge carriers produced can drift large distances in the material. The advanced technology allows nearly any geometry of pixels and contact structures. The photoelectric effect dominates in most materials up to a few tens of kiloelectronvolts. High-resistivity silicon detectors are typically limited to about 500 mm thick material (at reasonable full-depletion voltages), and this thickness can absorb most photons up to about 15 keV by the photoelectric effect. Unfortunately, above this energy silicon becomes essentially transparent. Photoelectric absorption increases as the fourth power of the atomic number, so higher-atomic number elements can absorb x-rays more efficiently. What we would like is a high-Z material (for good absorption) with a high bandgap (for low leakage) and good crystal quality with good charge transport (high mobility–lifetime product). As one would expect, the perfect high-Z material does not exist. Germanium is nearly a perfect material, with a reasonable atomic number and superb charge transport, but it has a low bandgap of 0.66 eV, which requires cooling to cryogenic temperatures to stop unacceptable leakage current. GaAs and HgI have been proposed for a long time [5], but it has not been possible to produce reliable material with unpopulated bandgaps. Another continuing problem with epitaxial GaAs has been the need to grow detector-grade material on thick substrates, which then have to be removed. This process has been a technical barrier to development of material with good conversion efficiency. Currently, highly doped bulk material has acceptable performance in some applications [6]. CdTe and CdZnTe are currently the most favored crystals for this application. The bandgap is about 1.4 eV, increasing with increasing zinc concentration. The advantage of a wider bandgap is higher resistivity and lower leakage currents. As a rule of thumb, the energy to produce an electron–hole pair in semiconductor material is about three times the bandgap, and this holds for Cd(Zn)Te. The charge transport over a few millimeters is acceptable, and growth techniques have now improved to give reliable material [7–9]. The x-ray absorption [10] is shown in Figure 6.2: several millimeters of material can efficiently convert 100 keV photons and above.

Area-array detectors built with CdTe or cadmium zinc telluride (CZT) [11,12] use typically 1–10 mm thick material, depending on the energy of the incident photons and the efficiency required. Cd(Zn)Te detector material is now grown with good quality. Bulk material in boules 4 in. in diameter can be obtained by several Bridgeman techniques, the traveling heater method, or multitube physical vapor

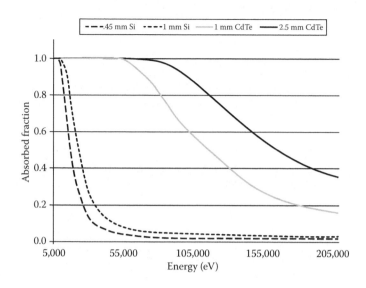

FIGURE 6.2   X-ray absorption against photon energy for different thicknesses of CdTe and Si.

transport. Resistivities of 10^{10} to 10^{11} ohm-cm and mobility–lifetime products of 1×10^{-2} to $3 \times 10^{-2}\,\mathrm{V\,cm^{-1}\,s^{-1}}$ are achieved (Chapter 3). There is now good availability of high-quality detector material, but there are still many academically interesting and technically challenging issues to be solved in creating reliable electrical contacts to the material.

Indium and aluminum contacts (cathode and anode, respectively) are used for good-quality Shottky blocking contacts on CdTe [13]. The advantage of these is reduced leakage current, but blocking the thermally generated and signal generated carriers causes issues due to so-called polarization or field collapse below one of the contacts in the bulk of the material [14]. This can be controlled by pulsing the bias supplies if the problem is not too pronounced. Gold and platinum are used as more ohmic contacts to both CZT and CdTe material. There are still problems of uniformity, injection of current, and stability of contacts. At the present time, single-element or large pad (1–2 mm) detectors are available with good uniformity from material suppliers and other fabricators. Small pixel detectors with lithographic features below 250 μm are very much more difficult to obtain [15]. Detectors with small pixels (50–250 μm pitch) are required due to their imaging resolution, but also because the small-pixel effect allows single carrier-type readout due to the high weighting field around the pixels. This gives very good spectroscopic capability and reduces the effects of the poor hole mobility in Cd(Zn)Te. In CZT and CdTe, these small pixels are hard to obtain commercially. GaAs detectors are produced with small pixels due to the more advanced processing technology. This modality is one area where GaAs detectors have an advantage over other high-Z detector materials. Germanium has historically not been available with pixelated contacts, but in the last few years strip and pixel-contacted detectors have been developed [16]; however, there is still the huge issue of running the detector at liquid nitrogen temperature and readout electronics at a higher temperature.

6.3   ASIC TECHNOLOGY FOR HYBRID PIXEL DETECTOR READOUT

Clearly, for large-area imaging detectors with many imaging pixels a high level of segmented multichannel readout is needed. Several decades ago, the only way to achieve this was via massive fan-out schemes to route signals to discrete low-density electronics. At the present time, CMOS technology allows us to build very dense low-power electronics with many channels that can be bonded directly to the conversion/sensing medium [17]. There are different requirements for the CMOS technology used for the analog front-end signal processing, as opposed to that for the digital signal processing and off-chip communication. For the analog part of the electronics, there is a requirement for a robust technology that has low electronic noise in terms of thermal noise, junction leakage and capacitance, and flicker noise. Also, there is a need for large power-supply voltages (3.3–5 V) to accommodate a wide dynamic range. For digital signal processing and communication, there is a requirement for very high-speed, high-density, low-power technologies that are compatible with the more modern low-voltage supply, deep submicron processes.

There seems to be a technology optimum at present at around 0.35–0.25 μm minimum feature size for the analog requirements. These CMOS processes have robust insulated-gate field-effect transistor (FET) oxides so that input protection circuits can be optimized for noise performance, and also so that they have wide dynamic range and low noise coefficients. The additional advantage of these technologies is that they are relatively cheap, costing less than EUR 100,000 for an engineering run of a new design and EUR 2000 per wafer. This technology is also relatively easy to design with and has acceptable die sizes of about 20×20 mm^2. Obviously, the large feature size limits the complexity of circuitry that can be integrated in a pixel, but the device performance compensates for this. Surprisingly, even with the relatively thick gate oxides, the total dose radiation hardness of these technologies can be acceptable [18], and, with suitable consideration of substrate biasing, latchup-free designs can be achieved.

In comparison, the digital signal-processing side of the measurement chain can benefit from the rapid development of deep submicron processes. Science applications have used 0.25, 0.18, and 0.135 μm, and are now using 65 nm [19] and proposing 22 nm technologies for digital components. These technologies are well suited to high-speed analog-to-digital converter (ADC) architectures and to very fast data manipulation for data sparsification and packaging. The technologies have their own limitations in terms of gate-oxide thickness, noise, and cost. The nanometer gate oxides and small interdevice distances reduce the power supply range and can introduce noise problems. A major practical problem is the cost of a mask set and engineering wafer batch, and also the design complexity, which increases design engineer costs. A full ASIC development can easily cost over EUR 1,000,000 and extend over 2 years, which can be a significant problem for the budget of some scientific experiments.

These high-density processes will be important for new devices, and SiGe and multigate FET processes will expand the range of processing architectures we can integrate close to the front end of detector systems. Also, new auto-gain switching front-end circuits [20] allow wide dynamic range without needing large power-supply voltages. In Section 6.9, we will discuss how we can use mixed technologies to optimize the use of the strong points of these different technologies.

6.4 AREA INTERCONNECT

Pixel detectors require a direct connection from the pads on the sensor material to the bond pads on the ASICs. Figure 6.3a shows a simple pad geometry for connection to the detector where the pixel pitch on the readout ASIC is the same as the detector pixel pitch. It is also possible to fan out the connections on the detector with multilevel metal routing on the detector, as shown in Figure 6.3b. This has to be done very carefully, as there is huge scope for signal cross talk. With integrating readout and synchronous input signals [21], where the signal is totally removed from the detector, this might not be a problem, but transients can still upset thresholds in these systems.

The pitch of x-ray imaging systems currently ranges from about 50 μm to 1 mm. Bump bonding is used to connect the detector pixels to the ASICs. There are many different technologies to do this, depending on the requirements of the detectors and environmental constraints.

The industry-standard area bump bonding method is to deposit solder onto underbump metalization on the pads of the detector and ASIC and then to align the two and heat them to reflow the solder. Various solders are used, including lead–tin, bismuth–tin, indium alloys, and silver alloys, depending on the temperature

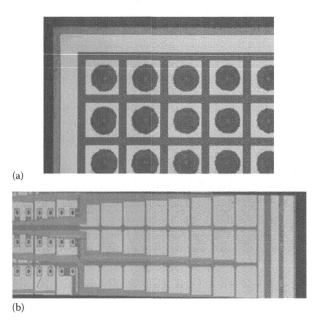

(a)

(b)

FIGURE 6.3 **(See color insert)** (a) Hexitec CdTe 250 μm pitch pad geometry with silver glue dots. (b) XFEL LPD redistributed bonding pads $250 \times 400 \ \mu m^2$ pitch (left) to the 500 μm pitch pixels on a two-layer interconnect silicon detector.

to reflow and the operating temperature required [22]. Typically, these materials require 240°C–140°C to reflow. Indium is used in either a lower-temperature reflow process or straight compression bonding [23], and gives good results, but cannot be used if high operating temperatures will ever be experienced. With all these processes, either fluxes or special gas environments, including nitrogen, hydrogen, or formic acid, have to be used to ensure good contact between parts. These processes can enable the use of 10 μm bond diameters, which allows 25–50 μm pixels to be connected [24]. Another method, used particularly for larger-pixel (150–250 μm) CZT detectors, is gold stud and silver-loaded epoxy dots. The gold studs are either bonded or deposited on the ASIC part, and the silver-loaded epoxy is screen printed on the detector. A process for this is used by our group at Science & Technology Facilities Council (STFC) for connection of CZT. After alignment, the unit is heated to set the epoxy. This can be done at low temperatures of 45°C to 120°C so that the detectors are not affected. CZT is a very sensitive crystal, and it has been seen that temperatures of around 150°C can cause movement of impurities and damage sites, which affect detector performance. This method uses 50 μm gold studs, which gives large distances between the components and is good for capacitance and stress relief. For all these processes, there is the option to use an underfill after bonding to strengthen the connection and seal against corrosion.

Other area-interconnect methods have been used, such as direct gold-to-gold and indium pillar contacts. Also, solid liquid interface diffusion (SLID) bonding can be used. This Cu–Sn process leaves a bond that is robust to subsequent processing up to 400°C. This is being used in three-dimensional integrated circuits (3DIC) [25] and silicon-interposer technology, but could be a prospect for high-strength detector-to-ASIC connection.

A serious issue with hybrid detector systems is stress produced in the detector material and to the interconnection surfaces [26]. One inherent issue is the different coefficient of thermal expansion of the different components. The bump bonding interconnection processes above require the module to be heated to 100°C–200°C and above. The device might then be required to operate at 0 to −10°C temperature for best performance (with CdTe). With large detectors of 20 mm length and above, this causes a considerable strain to be set into the interconnection face. With a CZT detector connected to silicon, this is several microns of linear change over a few centimeters of detector with 100°C temperature change. Either the bumps have to accommodate this or the material is stressed and the structure could bend like a bimetallic strip or break. None of these is good, but considering using the minimum processing temperature and allowing the bumps to take the deformation is the least bad option. With CZT, there is certainly an additional problem with inbuilt stress, as this can cause residual internal electric fields around the pixels as well as affecting the long-term reliability of the fragile electrical contact interfaces.

Another problem is the connection of the ASIC to the cooling/alignment block used to support the ASIC to the instrument. Aluminum would be a very convenient mounting substrate, but has a very different CTE from silicon. Molybdenum is often used, as it has a good CTE match to silicon (even better to CZT) and is also a good thermal conductor. It is, however, quite expensive. A good option we have found is to accept the mismatch with a metallic mounting plate but use a compliant heat transfer tape. We had a spectacular failure of a 54×54 mm^2 area CMOS detector when hard glued to a copper block but complete success mounted with compliant tape. With smaller ASICs, the epoxy glue is quite adequate, but it can still stress the silicon and affect reliability.

6.5   NOISE

There are several intrinsic noise processes that limit the performance of pixelated detectors. These can be split into sources related to the detector and processes related to the way the electronics reads out the detector [27]. These processes define the ultimate limit of performance, but there are many ways that the architecture can be designed inappropriately so that these are not the dominant effect.

Assuming all environmental noise is eliminated, the fundamental front-end-related noise can be summarized as equivalent noise charge (ENC) given by Equation 6.1:

$$ENC^2 = ENC^2_{\text{thermal}} + ENC^2_{\text{shot}} + ENC^2_{\text{flicker}}$$

$$ENC^2 = \left(b_t c^2_{tot} kT \frac{2}{3gm} + b_s iq + b_f c^2_{tot} \frac{Q_f}{gm^2} \right) (\text{coulombs}^2) \tag{6.1}$$

where:

(b_t, b_s, b_f) = coefficients calculated for a particular shaping filter
c_{tot} = capacitance of the detector and tracking
i = leakage current of the detector
gm = transconductance of the input FET of the amplifier
T = temperature
k = Boltzmann's constant

Definition of terms are fully explained in [27].

The ENC is the signal charge you would have to inject onto the detector in order to produce an output signal equivalent to the rms noise you measure at the output of the system. The terms for the individual ENC components are added in quadrature to give the total noise. This method of addition is required as statistically independent noise sources add as powers so we add the squares of the voltages. These noise contributions from the electronics affect the accuracy with which the signal can be measured. The signal produced in the detector has an intrinsic statistical noise. In semiconductor detectors, ionizing radiation produces a number of charge carriers, as described in Section 6.2. With a stochastic process, one would expect the noise on the number of carriers to be the square root of the number of carriers. In fact, because of the carrier-formation process, the signal noise is better than this square root by a factor known as the Fano factor [28]. This term is the ultimate or Fano limit of resolution of the detector, but has to be added in quadrature to the electronic noise contributions. Lowering the bandgap increases the number of carriers, which improves the accuracy of measurement; this is one reason why Ge detectors have good resolution. The down side of a low bandgap is that the thermally generated leakage current can offset this advantage. High-Z conversion materials all have other noise mechanisms and inhomogeneity, which produce signal uncertainty problems. In CZT, the major problem areas can be small inclusions [29] and other bulk and contact nonuniformities, which cause very localized changes in charge-carrier transport efficiency. This creates fixed-pattern noise, which unfortunately cannot always be removed in subsequent processing. Other problems occur due to charge sharing across pixels [30]. This can cause additional signal uncertainty problems related to the exact detailed system parameters and subsequent signal processing.

If one inspects the basic ENC formula, one can obtain several design objectives. As the thermal noise (bt) decreases with shaping time and leakage current (shot) noise (bs) increases with shaping time [31], there is an optimum shaping time for a detector:

$$ENC^2 = \left(\frac{e^2}{2\tau} c_{tot}^2 kT \frac{2}{3gm} + \frac{e^2\tau}{4} iq + \frac{e^2}{2} c_{tot}^2 \frac{Q_f}{gm^2} \right) \left(\text{coulombs}^2 \right) \left(\text{for } CR - RC \right) \quad (6.2)$$

Because of the formula for the thermal noise of an insulated-gate FET (IGFET):

$$ENC = (c_d + c_{in}) \sqrt{b_t kT \frac{2L}{\sqrt[3]{2\mu I_D C_{in}}}} \quad \text{(coulomb)} \quad (6.3)$$

one can optimize against c_{in} and obtain an optimum input FET gate capacitance of 1/3 to 1 times the total input detector capacitance [32]. This so-called matching condition sets a good criterion for the input circuit design.

Further inspection of the ENC formula shows an even more important design driver. If the leakage current is reduced, the leakage noise reduces. If the detector capacitance reduces, the thermal noise reduces. The way to achieve both of these together is to segment the detector into smaller pixels. A limit is reached where capacitance actually increases, but to first order having many pixels improves measurement accuracy. Also, as there are more amplifier channels, the possible bandwidth to measure signal photons is increased. Another practical limitation is reached when the pixel size is so small that charge-carrier clouds spread across multiple pixels, and this charge-sharing effect then has to be resolved by more complicated signal processing. Overall, fine pixel systems have the potential to be lower noise and faster than single-channel systems.

Several direct-conversion x-ray imaging systems have been built using a photon-counting technique, including Medipix, Pilatus, and XPAD [33]. Typically, this involves the front-end electronics amplifying the charge signal and then using a discriminator to count the number of hits in the pixel with a value above a set level. Some ASICs have the capability to have several discriminators and several counters to get a rudimentary spectroscopy [34]. To achieve a true spectrum, the threshold has to be scanned during repeated experiments. Another method used by STFC is to amplify the signal, output the analog signal for every photon incident in the pixel, and perform an ADC conversion on this value [35]. The advantage of this is that the position and energy of every charge cloud are measured, allowing an effective improvement in efficiency and good correction of charge sharing. The advantage of this hyperspectral data with very good spectral resolution is that there is a new richness in the data to allow advanced imaging techniques [36].

6.6   COOLING

An important technology aspect of array detectors is the close integration of the electronics to the semiconductor detector material. As we have seen, this is important for capacitance and other noise-injection possibilities. With wire-bonded detectors, it is possible to have thermal breaks between the electronics and the detectors [37].

The intimate connection of the electronics to the detectors in array detectors means that the electronics and detector are essentially thermally coupled. The low-noise front ends and complex electronic processing generate a lot of heat: 0.5 W cm^{-1} is certainly possible. CZT semiconductor detectors will work at room temperature, but the leakage increases rapidly with elevated temperature. In order to operate the detectors efficiently, some form of heat removal is required. In most systems, this is only achieved by attaching the ASICs to a cooling substrate. The mechanics and materials used for this are different in different systems, including circulating binary ice, high-pressure evaporative liquids [38], Peltier devices, forced air cooling, heat pipes, and off-detector cooling. The technologies vary, but the consequence is the same: the back of the detector becomes very congested. The cooling takes place in exactly the space where one would like to route interconnections to control and take data from the ASICs. An example of the cooling is shown in Figure 6.4, but many other configurations are used [39].

6.7   DEAD AREAS IN MODULAR SYSTEMS

As already stated, large-area hybrid detector systems have to be built from multiple modules. There is a general requirement to reduce the dead areas between modules for cosmetic reasons in the image as well as necessary science issues for full image coverage. Several hybrid systems are built using conversion material directly bump bonded to ASICs. In order to read out these ASICs, at least one edge of the ASIC has to be exposed (not covered by conversion material) so that wire bonds can be made to pads on the surface. An image of this is shown in Figure 6.5.

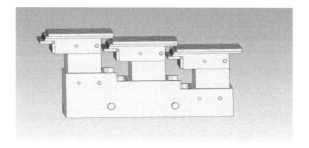

FIGURE 6.4   Cooling block of the multimodule Hexitec system.

FIGURE 6.5   Wire bonds connecting the ASIC to the readout board. The detector (top) cannot be close to these wire bonds, so there is an area of the ASIC not covered by the detector. This causes a large dead area between this detector and the next.

A gap has to be left between detectors to allow room for these wire bonds and the connection board; typically this is about a minimum of 1–2 mm. This has two consequences: there is a dead area for the detector, and also the ASIC is directly exposed to radiation, which can have radiation-damage consequences. One possibility is to have a roof-tile geometry for the detectors, shown in Figure 6.4. This allows the wire-bonded area of one ASIC to be covered by an active detector area of the next detector. This is very effective but produces a detector surface which is not a flat surface normal to the incident x-rays. Section 6.8 describes a new method to allow full coverage of the detector plane and also to have a flat detector. All direct-conversion detectors need a high-voltage bias to be connected to the back of the detector. This can also be done using the same method as the wire bond to the ASIC I/O pads, as shown in Figure 6.6.

FIGURE 6.6   Wire bonds to the exposed I/O pads on the fourth side of the ASIC, with single HV connection to the back of the detector.

The other three edges of the ASIC, not required for bonding, can be covered by the detector material, so that the detectors can be spaced as close as mechanically allowed. The edges of the detector also cause dead areas. Semiconductor detectors have to be cut from a larger wafer, which nearly always causes the edge material to have significantly different properties from the bulk material. In silicon, a diced edge has crystal damage, which shorts the top and bottom of the edge together. So-called edgeless or active-edge silicon detectors have been built by very specifically fabricating the edges [40]. With GaAs and CdTe detectors, usually there is a guard band on the edges of the detector to stop the edge leakage current from entering the edge pixels. STFC have been working on active-edge CdTe detectors, which have shown that good counting efficiency and spectroscopy can be achieved without guard bands, allowing very close packing of detectors [41]. Similar results are shown on Medipix detectors with very small intermodule dead regions [42].

6.8   THROUGH-SILICON VIA (TSV) FOR I/O CONNECTION NEEDED FOR LOW DEAD AREAS

STFC have built an x-ray imaging spectroscopy readout system based on CdTe detector material. A $20 \times 20 \times 1$ mm^3 crystal is bonded to a Hexitec ASIC with gold stud bonds to 80×80 pixels of the detector. Each pixel is 250×250 µm and has an amplifier and a peak-track-and-hold circuit. This allows the energy of every photon to be recorded and measured with subkiloelectronvolt full width at half maximum (FWHM) accuracy from 4 to 200 keV and above [43]. The first versions of the ASIC allow the detector modules to be butted on three edges with only a 170 µm gap between detectors. The fourth edge of the ASIC has wire bonds, which precludes butting the detectors in a flat geometry. A similar restriction has compromised the design of the popular Dectris Pilatus detector. The only way to avoid these wire bonds restricting the coverage of the ASIC is to read out the ASIC from the back of the detector. Several methods are proposed, but STFC has chosen an approach whereby the wire bond pads are shielded with a metal plane on the top of the ASIC to stop electromagnetic injection into the detector. The I/O bond pads are then read out with through-silicon vias (TSVs) [44]. The process involves thinning the wafer to 100 µm so that it is still mechanically robust and then etching a TSV from the back to connect to the back of the metal bond pad. The TSV is plated with metal, and a pad of the same metal is fabricated on the back of the ASIC. Then an almost identical, but mirrored, wire-bonding method is used to connect to the back of the "4S" ASIC, as shown in Figure 6.7.

With this geometry, it is possible to extend the active CdTe over the whole of the front surface of the ASIC. This allows large-area coverage with four-side butting and only very small dead regions on all edges of the detector modules.

6.9   3DIC TECHNOLOGY ADVANTAGES

CMOS ASIC technology has progressed over the decades with the relentless Moore's law. However, as we approach the 22 nm feature-size regime it is believed that simple dimension scaling will not be sufficient. This is where so-called 3DIC technology will

FIGURE 6.7 Wire bonding to redistributed pads on the back of a "4S" ASIC. Each pad has a TSV through the thinned 100 μm ASIC.

be an important next-generation technology [45]. The objective of 3DIC technology is to distribute the signal processing on several layers of silicon and then connect the layers by TSV technology as shown in Figure 6.8. The advantage comes partly from simply having multiple layers of silicon but mostly from the immense connection density. This mimics, as a biological analogy, the animal nervous system, with the immense processing capability produced by the billions of three-dimensional neural connections. The drive for this technology is to scale processing power, but this will be also extremely useful for area-array detectors. As we have seen, different CMOS processes have advantages for either analog or digital functions. Connecting them together allows the best technology to be used for the analog process and the best for the digital process. Typically, connections would be made between the layers within each pixel. The signal from the analog layer is amplified and transferred using a small via from this top layer through the thinned (20–50 μm) silicon substrate to the digital layer below.

Several demonstration prototypes of this technology are being trialed in the science community [46,47]. The 3DIC technology is only starting to be used in

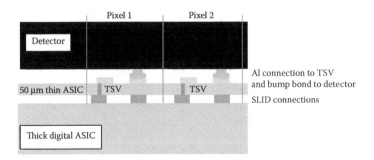

FIGURE 6.8 3DIC stacking showing bump bond from detector to active surface of the thin ASIC, and the TSV through to the SLID connection to the active layer of the thick ASIC. In this case, there is a SLID support under the bump bond.

the commercial sector [48], and the first large investments in plant are only now being made, to achieve capability in the next few years. STFC has a program to investigate the feasibility of 3DIC technology for imaging detectors [49]. We have used the analog front-end part of the Hexitec CZT readout ASIC as the pixel for the analog layer. The amplified output from each 250 µm pixel is then sent via a TSV to a peak-track-and-hold and Wilkinson ADC circuit on each pixel of a lower digital ASIC. The aim of this is to digitize all the signals from the detector on one 3DIC ASIC. The analog and digital ASICs are both fabricated on an Austrian Micro Systems (AMS) 0.35 µm CMOS process to reduce the cost of the trial. The tungsten via technology and the SLID interconnection between the wafers are performed by EMFT in Munich [50]. This technology has rather large TSVs of 3×10 µm and requires a large connection pitch due to the large size of the SLID pad technology, shown in Figure 6.9. The advantage of the process is that it is a via-last (or back-end-of-line) process, so that any CMOS process can be used, and possibly different technologies for the two layers. TSV processing can induce large strain in the silicon, which can affect the active circuitry, particularly with modern processes that rely on strain in the FET for high-performance operation. The relatively large space around the via in this technology is an advantage in this respect.

The aim of 3DIC technology for imaging detectors would be to eventually reduce the vias to a few microns and have pixel sizes in the range of 50–100 µm. This would allow a fine-pitch detector with smart signal processing in each multilayer pixel. In order to do this, we will probably have to wait for extremely thin silicon layers and the introduction of front-end-of-line processes, which have the vias inserted in the silicon, before active devices are fabricated. These processes will be very expensive and will probably be driven by the digital processing and communications markets. This will mean that the technologies might not be ideal for all science applications. Until then, via-last and silicon-interposer [51] technologies will be used, which will allow mixed technologies and cheaper cost, if not ideal in other respects. The alternative silicon-interposer technology has been used in the large pixel detector

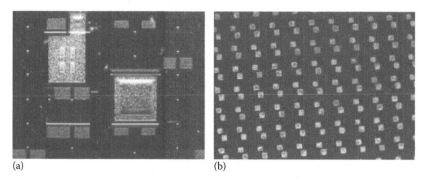

(a)   (b)

FIGURE 6.9   (a) The four 3×10 µm² vias are shown in the 3DIC via area. The aluminum contact to this can be seen entering from the top. The other large square is the bump bond pad for connection to the detector. (b) The SLID bond pairs connecting to the digital ASIC. One of the pair is the TSV connection and the other is a pad under the bump bond pad for support.

(LPD) for the x-ray free electron laser (XFEL) European x-ray source [52]. This is a useful technology but is widely recognized as a stepping stone to full 3DIC technology for these imaging applications.

6.10 FLATNESS AND STRESS

Silicon wafers have to be very flat in order that photolithographic and other processes can be reliably performed on the active surface. This dictates the use of very thick wafers of 700–1000 µm for dimensional stability. These thick wafers are perfect for subsequent bump bonding processes to detectors that require a relatively relaxed flatness of a few microns across the ASIC. In conventional solder bump bonding of silicon to detectors, the bonding can be done at the wafer level or at the ASIC level. When the wafer is thinned, the highly stressed processed side of the ASIC causes the whole wafer to distort. In 3DIC processing, the only way to obtain very thin and flat wafers is to attach them to handle wafers and perform all the grinding and interconnect steps while the thin wafer remains on the handle wafer.

If we want to use these devices for imaging detectors, the final stack has to have at least one thick substrate to remain flat enough for the subsequent bonding to detectors. The stress in the stack still remains, but at least the ASICs will be flat enough to be bonded.

An even more difficult challenge is when TSVs are used on a single-layer device to redistribute the I/O connection from the ASIC bond pads to the back of the wafer, as described in Section 6.8. This is particularly useful for four-side-bonded area-array detectors with very small dead areas. At STFC, we have used single layers thinned to 100 µm and seen that the stress in the ASIC causes them to distort over an area of 20 mm. The solution to this might be to firmly bond them to the detector. The detector and the bonding will then form the support for the structure.

REFERENCES

1. S. Kasap, Amorphous and polycrystalline photoconductors for direct conversion flat panel X-ray imaging sensors, *Sensors* (2011), 11, 5112–5157.
2. J. A. Seibert, Flat panel detectors: How much better are they? *Pediatric Radiology* (2006), 36(Suppl. 2), 173–181.
3. I. Sedgwick, LASSENA: A 6.7 MegaPixel, 3 sides buttable wafer-scale CMOS sensor using a novel grid-addressing architecture, in *Proceedings of the 2013 International Image Sensor Workshop*, June 12–16, Snowbird, UT, (2013).
4. G.F. Knoll, *Radiation Detection and Measurement*. Wiley, New York.
5. B.E. Patt, Development of mercuric iodide detector array for medical imaging applications, *Nuclear Instruments and Methods in Physics Research Section A* (1995), 366, 173–182.
6. O.P. Tolbanov, GaAs structures for X-ray imaging detectors, *Nuclear Instruments and Methods in Physics Research Section A* (2001), 466, 25–32.
7. H. Chen et al., Characterization of large cadmium zinc telluride crystals grown by traveling heater method, *Journal of Applied Physics* (2008), 103, 014903.
8. J.T. Mullins et al., Vapor-phase growth of bulk crystals of cadmium telluride and cadmium zinc telluride on gallium arsenide seed. *Journal of Electronic Materials* (2008), 37, 1460.

9. H. Shiraki et al., Improvement of the productivity in the THM growth of CdTe single crystal as nuclear radiation detector, *IEEE Transactions on Nuclear Science* (2009), 56, 1717–1723.

10. E. Gullikson, X-ray interactions with matter, The Center for X-Ray Optics. http://henke.lbl.gov/optical_constants/.

11. P. Seller et al., Pixellated Cd(Zn)Te high energy X-ray instrument, *Journal of Instrumentation* (2011), 6, C12009.

12. W. Kaye et al., Calibration and operation of the Polaris 18-detector CdZnTe array, in *2010 IEEE Nuclear Science Symposium Conference Record (NSS/MIC)*, October 30–November 6, Knoxville, TN. IEEE, (2010), pp. 3821–3824.

13. T. Takahashi et al., High-resolution Schottky CdTe diode detector, *IEEE Transactions on Nuclear Science* (2002), 49, 1297–1303.

14. A. Cola, The polarization mechanism in CdTe Schottky detectors, *Applied Physics Letters* (2009), 94, 102113.

15. M.C. Veale, S.J. Bell, P. Seller, M.D. Wilson and V. Kachkanov, X-ray micro-beam characterization of a small pixel spectroscopic CdTe detector, *Journal of Instrumentation* (2012), 7, P07017.

16. Semikon Detector GmbH, Germany.

17. R. Ballabriga, Medipix3: A 64 k pixel detector readout chip working in single photon counting mode with improved spectrometric performance, *Nuclear Instruments and Methods in Physics Research Section A* (2011), 633 (Suppl. 1).

18. F. Faccio, Total dose and single event effects (SEE) in a 0.25 μm CMOS technology. in *4th Workshop on Electronics for the LHC* (CERN-LHC-98-036), September 21–25, Rome. CERN, (1998), p. 105.

19. S. Bonacini, P. Valerio, R. Avramidou, R. Ballabriga, F. Faccio, K. Kloukinas and A. Marchioro, Characterization of a commercial 65 nm CMOS technology for SLHC applications, *Journal of Instrumentation* (2012), 7, P01015.

20. J. Becker, Challenges for the adaptive gain integrating pixel detector (AGIPD) design due to the high intensity photon radiation environment at the European XFEL, *Proceedings of Science Vertex* (2013), 2012, 012.

21. European XFEL LPD detector, Hamburg. http://www.xfel.eu/.

22. W.K. Choi et al., Development of low temperature bonding using In-based solder, in *58th Electronic Components and Technology Conference*, May 27–30, Lake Buena Vista, FL. IEEE, (2008), pp. 1294–1299.

23. H. Heikkinen et al., Indium-tin bump deposition for the hybridization of CdTe sensors and readout chips, in *2010 IEEE Nuclear Science Symposium Conference Record (NSS/MIC)*, October 30–November 6, Knoxville, TN. IEEE, (2010), pp. 3891–3895.

24. P. Norton, HgCdTe infrared detectors, *Opto-Electronics Review* (2002), 10(3), 159–174.

25. P. Ramm, 3D-IC fabrication challenges for more than Moore applications, in *Workshop Manufacturing and Reliability Challenges for 3D ICs using TSVs*, September 26, San Diego, CA. Sematech, (2008).

26. W. Sang, Primary study on the contact degradation mechanism of CdZnTe detectors, *Nuclear Instruments and Methods in Physics Research Section A* (2004), 527, 487–492.

27. P. Seller, Noise analysis in linear electronic circuits, *Nuclear Instruments and Methods in Physics Research Section A* (1996), 376, 229–241.

28. A. Owens, On the experimental determination of the Fano factor in Si at soft x-ray wavelengths, *Nuclear Instruments and Methods in Physics Research Section A* (2002), 491, 437–443.

29. A.E. Bolotinkov, Internal electric-field-lines distribution in CdZnTe detectors measured using X-ray mapping, *IEEE Transactions on Nuclear Science* (2009), 56(3), 791–794.

30. Study of charge sharing in Medipix3 using a micro-focused synchrotron beam, 2011 JINST 6 C01031, *12th International Workshop on Radiation Imaging Detectors*.

31. P. Seller, Summary of thermal, shot and flicker noise in detectors and readout circuits, *Nuclear Instruments and Methods in Physics Research Section A* (1999), 426(2–3), 538–543.

32. P. Seller, The matching condition for optimum thermal noise performance of FET charge amplifiers, RAL-87-063, internal publication.

33. P. Pangaud et al., XPAD3-S: A fast hybrid pixel readout chip for x-ray synchrotron facilities, *Nuclear Instruments and Methods in Physics Research Section A* (2008), 591(1), 159–169.

34. C. Ponchut, MAXIPIX, a fast readout photon-counting x-ray area detector for synchrotron applications, *Journal of Instrumentation* (2011), 6, C01069.

35. L. Jones, P. Seller, M. Wilson, A. Hardie, HEXITEC ASIC—A pixellated readout chip for CZT detectors, *Nuclear Instruments and Methods in Physics Research Section A* (2009), 604, 34.

36. P. Seller et al., Pixellated Cd(Zn)Te high energy x-ray instrument, *Journal of Instrumentation* (2011), 6, C12009.

37. J. Headspith, First experimental data from XH, a fine pitch germanium microstrip detector for energy dispersive EXAFS (EDE), *Nuclear Science Symposium Conference Record* (2007), 4.

38. A. Nomerotski, Evaporative CO_2 cooling using microchannels etched in silicon for the future LHCb vertex detector. arXiv.org > physics > arXiv:1211.1176. February (2013).

39. C. Broennimann et al., The PILATUS 1M detector. *Journal of Synchrotron Radiation* (2006), 13, 120–130.

40. G.F. Dalla Betta, Recent developments and future perspectives in 3D silicon radiation sensors, *Journal of Instrumentation* (2012), 7, C10006.

41. D.D. Duarte, et al. Edge effects in a small pixel CdTe for x-ray imaging, *Journal of Instrumentation* (2013), 8, P10018.

42. T. Koenig et al., Imaging properties of small pixel spectroscopic x-ray detectors based on cadmium telluride sensors, *Physics in Medicine and Biology* (2012), 57, 6743–6759.

43. J.W. Scuffham, M.D. Wilson, P. Seller, M.C. Veale, P.J. Sellin, S.D.M. Jacques, R.J. Cernik, A CdTe detector for hyperspectral SPECT imaging, *Journal of Instrumentation* (2012), 7, P08027.

44. P. Seller, S. Bell, M.D. Wilson and M.C. Veale, Through silicon via redistribution of I/O pads for 4-side butt-able imaging detectors, in *IEEE Proceedings of Nuclear Science Symposium and Medical Imaging Conference (NSS/MIC)*, October 27-November 3, Anaheim, CA. IEEE, (2012), pp. 4142–4146.

45. *3DIC Proceedings*. IEEE, October 2–4, (2013). San Francisco.

46. E. Ramberg, 3-Dimensional ASIC development at Fermilab, Vienna Instrumentation Workshop. February 14, (2013).

47. A. Macchiolo, SLID-ICV vertical integration technology for the ATLAS pixel upgrades, arXiv.org > physics > arXiv:1202.6497 (2012).

48. *Proceedings of European 3D TSV Summit*, January (2013), Grenoble, France.

49. AIDA, Advanced European Infrastructures for Detectors at Accelerators. EU Framework 7. http://arxiv.org/abs/1202.6497. Cornell University Library.

50. L. Sung Kyu, *Design for High Performance, Low Power, and Reliable 3D Integrated Circuits*. Springer, Berlin.

51. M. Rimskog, Through wafer via technology for MEMS and 3D integration, in *2007 Electronic Manufacturing Technology Symposium, 32nd IEEE/CPMT International*, 3–5 October, San Jose, CA. IEEE, (2007), pp. 286–289.

52. M. Hart et al. Development of the LPD, a high dynamic range pixel detector for the European XFEL, in *IEEE Nuclear Science Symposium and Medical Imaging Conference (NSS/MIC)*, October 27–November 3, Anaheim, CA. IEEE, (2012), pp. 534–537.

7 Single Photon Processing Hybrid Pixel Detectors

Erik Fröjdh

CONTENTS

7.1 INTRODUCTION

This chapter will introduce the concept of single photon processing hybrid pixel detectors before moving on to characterize hybrid pixel detectors using synchrotron radiation and in particular monoenergetic pencil beams. The focus will be on the sensor side and especially cadmium telluride (CdTe) sensors with small pixels, but application-specific integrated circuit (ASIC) aspects will also be covered. A monochromatic pencil beam is a very useful tool for detector characterization since both the position and the amount of energy deposited are then known. Together with a hybrid pixel detector that can resolve the response from a single photon, effects in the sensor layer can be studied in detail. The fine segmentation, down to 55 μm, also gives the ability to map how parameters change over the area and, in combination with a titled beam, it is possible to generate a three-dimensional (3-D) map over the sensor volume.

Most of the measurements in this chapter have been performed with the Timepix [1] or the Medipix3RX [2] chip and therefore an introduction to these detectors is included. Methods and results can, however, be generalized to other detector systems.

7.1.1 HYBRID PIXEL DETECTORS

A hybrid pixel detector consists of two layers, a sensor layer and an electronics layer as shown in Figure 7.1. The advantages of this detector are that the different layers can be optimized for a specific application ex. using a high-Z sensor material to increase x-ray photon absorption and that there is no dead area where the electronics are situated as in monolithic sensors. However, each pixel in the sensor layer must be connected to the corresponding pixel in the read-out layer thus adding a complex processing step and increasing the cost. For particle physics applications, the extra electronics layer and the bump bonds also add extra material causing the particles to scatter. Due to their high performance, hybrid pixel detectors are still preferred over monolithic sensors.

The pixel pitch varies between applications and detectors ranging from approximately 0.25–1 mm for medical computed tomography (CT) detectors [3] to about 50–100 μm for state-of-the-art multipurpose photon counting (PC) chips [1,2,4–6] down to 25 μm for the newest prototypes [7].

7.1.1.1 Timepix

The Timepix [1] chip is a hybrid pixel detector developed in the framework of the Medipix2 collaboration [8]. The pixel matrix consists of 256×256 pixels with a pitch of 55 μm, which gives a sensitive area of about 14×14 mm^2. Timepix is designed

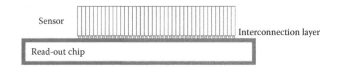

FIGURE 7.1 Schematic of a hybrid pixel detector.

FIGURE 7.2 Timepix chip with a 1 mm thick CdTe sensor.

in a 0.25 μm complementary metal–oxide–semiconductor (CMOS) process and has about 500 transistors per pixel (Figure 7.2).

The chip has one analog threshold and it can be operated in PC, time over threshold (ToT), or time of arrival (ToA) mode. The principles of the different operating modes are shown in Figure 7.3. In the PC mode, the counter is incremented once for each pulse that is over the threshold, while in the ToT mode the counter is incremented as long as the pulse is over the threshold, and in the ToA mode the pixel starts to count when the signal crosses the threshold and keeps counting until the shutter is closed. In the measurements described, only the PC and the ToT modes have been used but the ToA mode is very useful and has applications in mass spectrometry [9,10] and particle tracking [11] among other areas.

7.1.1.2 Medipix3RX

While Timepix is a general-purpose chip, the Medipix3RX [2] is much more aimed at x-ray imaging. It can be configured with up to eight analog thresholds per pixel and features analog charge summing over dynamically allocated 2×2 pixel clusters. The intrinsic pixel pitch of the ASIC is 55 μm and if bump bonded in this mode, which is called the fine pitch mode, the chip can then be run with either four thresholds per pixel in single pixel mode or with two thresholds per pixel in charge summing mode. Optionally, the chip can be bump bonded with a 110 μm pitch then combining counters and thresholds from four pixels. Operation is then possible in single pixel mode with eight thresholds per pixel or in charge summing mode having four thresholds and a summing charge of 220×220 μm^2 in area.

As the Medipix3RX is a very versatile and configurable chip there is also a possibility to utilize the two counters per pixel and run in continuous read/write mode where one counter counts while the other one is being read out. This eliminates the read-out dead time but comes at the cost of losing one threshold since both counters need to be used for the same threshold.

The charge summing mode is a very important feature to combat contrast degradation by charge sharing in hybrid pixel detectors with small pixels. The effect of the charge summing mode is showed in Figure 7.4 where it is compared with the single

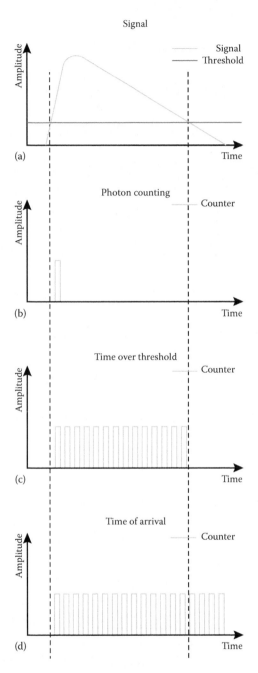

FIGURE 7.3 Schematic of the three different operating modes in the Timepix chip showing (a) the signal after the preamplifier, (b) the response in photon counting mode where each pulse is counted once, (c) response in time over threshold mode where the length of the pulse (in clock cycles) is counted, and (d) time of arrival mode where the counter starts counting when the pulse goes over the threshold and then counts until the shutter is closed.

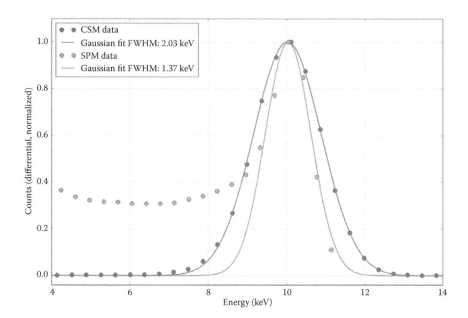

FIGURE 7.4 Comparison between the single pixel mode and the charge summing mode of Medipix3RX.

pixel mode for a Medipix3RX chip illuminated by 10 keV photons at the ANKA synchrotron. In this case, the sensor was a 300 μm thick p-on-n silicon sensor with a 55 μm pixel pitch.

7.2 CALIBRATION OF HYBRID PIXEL DETECTORS

7.2.1 THRESHOLD EQUALIZATION

Before calibrating a hybrid pixel detector, the chip has to be equalized in order to minimize the threshold dispersion between pixels. This is because the threshold that the pixel sees is applied globally but the zero level of the pixel can be slightly different due to process variations affecting the baseline of the preamplifier.

The equalization is done with a threshold adjustment digital-to-analog converter (DAC) in each pixel. The resolution of the adjustment DAC is usually in the range of 4 bits as in the Timepix to 6 bits as in the Eiger chip [6]. The standard way to calculate the adjustment setting for each pixel is by scanning the threshold and finding the edge of the noise, then aligning the noise edges. This adjusts correctly for the zero level of the pixel but gain variations can still deteriorate the energy resolution at a given energy. To correct for the gain mismatch, either test pulses or monochromatic x-ray radiation has to be used for the equalization, thereby equalizing at the energy of interest instead of the zero level. In Figure 7.5 the threshold dispersion before equalization and after equalization with both noise and test pulses is shown for the Medipix3RX chip. A method for the equalization and calibration of the Eiger chip using monochromatic x-ray radiation at several energies can be found in a paper by Kraft et al. [12].

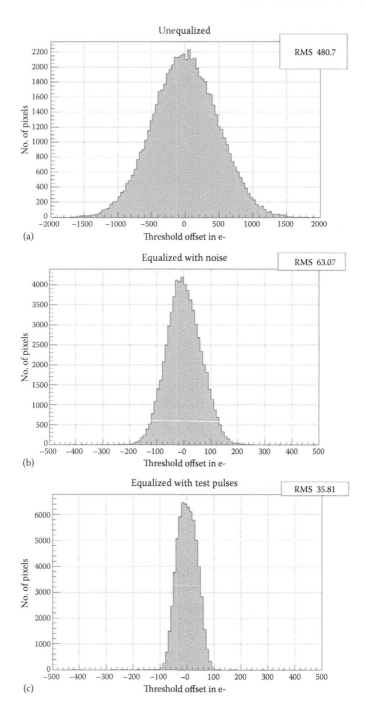

FIGURE 7.5 Threshold dispersion at (a) noise level before equalization and at 8 keV after (b) noise and (c) test pulse equalization for the Medipix3RX chip in single pixel super high gain mode. Note the different scales on the x-axis.

7.2.2 ENERGY CALIBRATION

Depending on the chip, two types of energy calibration are necessary: calibration of the analog threshold and calibration of the ToT response. For a strictly PC chip such as Medipix3RX, Pixirad, or Eiger, the only calibration is the calibration of the analog threshold, whereas in a ToT measuring chip, such as Timepix and Clicpix, the ToT response has to be calibrated as well.

7.2.2.1 Energy Calibration of Analog Threshold

To calibrate the threshold, monochromatic photons or at least radiation with a pronounced peak is required. These can be obtained from radioactive sources, by x-ray fluorescence, or from synchrotron radiation. To find the corresponding energy for a certain threshold, the threshold is scanned over the range of the peak and an integrated spectrum or S-curve is obtained. The data are then either directly fitted with an error or sigmoid function or are first differentiated and then fitted with a Gaussian function. From this fit, the peak position and the energy resolution can be extracted. Repeating the procedure for multiple peaks, the result can then be fitted with a linear function and the relation between the threshold setting in DAC steps or millivolts (mV) and the deposited energy in the sensor can be found. A calibration of a Medipix3RX chip in charge summing mode is shown in Figure 7.6 with the corresponding information on the fluorescence peak that was used to generate the radiation.

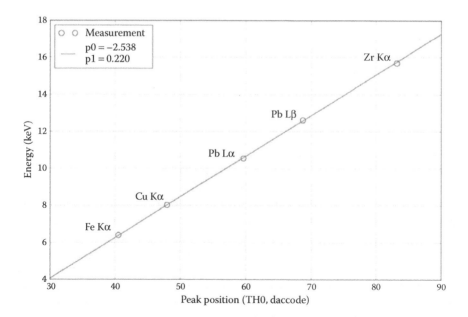

FIGURE 7.6 Energy calibration of a Medipix3RX chip in charge summing mode. The labels show the corresponding fluorescence peak used and p0 and p1 are the coefficients for the linear fit.

7.2.2.2 Energy Calibration of Time over Threshold Mode

For detectors with the ToT capability, for example the Timepix, the ToT response also has to be calibrated. This is slightly more difficult since the ToT to energy relation is not linear but rather a surrogate function. This is because while the amplitude of the pulse is proportional to the energy of the incoming photon, the length of the pulse deviates from linear behavior at low pulse heights due to the shape of the pulse. Equation 7.1 shows the relation between ToT, $f(x)$, and the energy of the incoming photon, x. The equation is taken from [13], where Jan Jakubek presents a calibration procedure for the Timepix chip.

$$f(x) = ax + b - \frac{c}{x-t} \tag{7.1}$$

In Figure 7.7, the fitted calibration function for several pixels is plotted and in Figure 7.8 the energy spectrum from a ^{241}Am source is shown before and after calibration for a few pixels in the detector. This highlights the importance of per-pixel calibration to achieve an optimal energy response.

7.3 PERFORMANCE AND LIMITATIONS

Single photon processing hybrid pixel detectors offer excellent performance; however, it is important to understand the applications and the limitations of the current detectors. By applying a threshold and counting each photon, the noise from

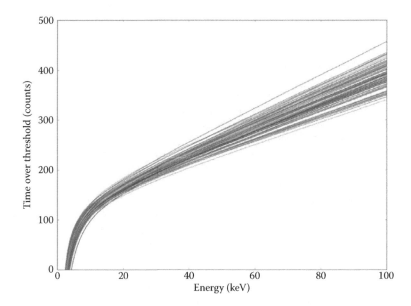

FIGURE 7.7 (**See color insert**) Calibration functions for several pixels in a Dosepix chip.

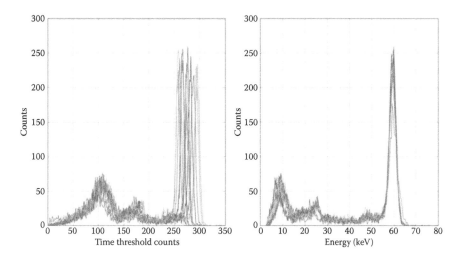

FIGURE 7.8 **(See color insert)** Measured photon energy spectrum for a ^{241}Am source for a few pixels in a Dosepix detector before and after calibration.

integrating the leakage current seen in charge-integrating devices is removed. This is especially important in low-flux measurements where the signal could be completely drowned in noise.

Also looking at the contrast-to-noise ratio, in a charge-integrating device the photons are weighted according to their energy, $W \propto E$, which means that for a low-contrast imaging task the high-energy photons will mask the more important low-energy photons. A study by Giersch et al. [14] has shown that the signal-to-noise ratio in a low-contrast imaging task such as mammography CT can be increased by 30% by using PC, thus weighting photons $W = 1$ by 90% by introducing optimal weighting where each photon is weighted by $W \propto E^{-3}$.

7.3.1 Sensor Material

To minimize the dose to the patient, a high quantum efficiency is very important in medical imaging. To this end, there is a need to move away from silicon because although it is a very well understood material and is available in excellent quality, its atomic number is too low to reach a high efficiency. Figure 7.9 shows the x-ray absorption for silicon (Si), gallium arsenide (GaAs), and CdTe of 1 mm thickness up to 140 keV, which are the three main candidates for x-ray imaging applications. GaAs for medium energies and CdTe/cadmium zinc telluride (CdZnTe) for high energies. Both materials still have their problems, such as point defects and trapping, but their quality is improving. Later in the chapter, the properties of CdTe will be more closely examined and its defects will be characterized.

An intrinsic problem with all high-Z sensor materials is x-ray fluorescence. When a primary photon is absorbed, an electron in the atom is excited and the atom then emits a fluorescence photon as it is de-excited. The range of this photon depends on

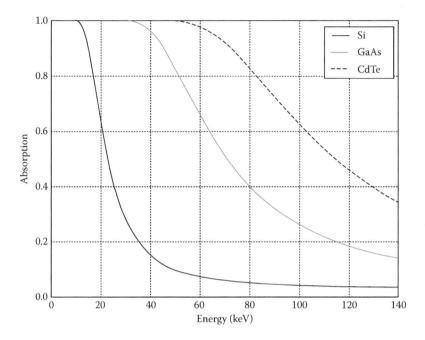

FIGURE 7.9   X-ray absorption for a 1 mm thick sensor of Si, GaAs, and CdTe. (Generated with XOP2.3 [15].)

the material and the electron layer containing the excited electron. Both the range and the yield of the fluorescence photons increase with the atomic number of the material. In GaAs, the fluorescence photons have a mean free path of 16 and 40 μm, respectively, and in CdTe the ranges are 120 and 60 μm for the Kα photons.

If the distance that the fluorescence photon can travel is comparable or larger than the pixel pitch, this will lead to a degradation of the energy response and a blurring of the picture. This can, to some extent, be corrected for by charge summing, either in the ASCI as for the Medipix3RX or off-line provided that the energy and time information are available from, for example, ToT measurements with Timepix. However, one portion of the fluorescent photons will always escape from the sensor volume and therefore will not be corrected for.

7.3.2   CHARGE SHARING

When the pixel size starts to approach the size of the charge cloud, the signal from more and more photons is subjected to charge sharing. Charge sharing creates a characteristic low-energy tail and leads to a reduced contrast and distorted spectral information. To counteract this problem there are two possibilities, either to use larger pixels and then reduce the spatial resolution or to implement charge summing on a photon-by-photon basis.

For lower rates and with detectors that store the energy information in each pixel, using either ToT such as Timepix or a peak-and-hold circuit and an ADC such as

Hexitec [16] the charge summing can be done off-line. This requires, however, that a second hit does not occur in the same pixel before the read out. Using this approach, charge that is below the threshold is lost.

Another approach is to sum the charge in the detector as implemented in Medipix3RX, where the analog charge is summed in a 2×2 cluster before being compared with the threshold. The advantage of this approach is that it can handle much higher interaction rates and also that even charge below the threshold is summed as long as one pixel is triggered. However, since this has to be implemented in the ASIC, it complicates the design and is less flexible. There is also a practical limit to how many pixels over which the summation can be done. The importance of charge summing when using thick sensors with small pixels is highlighted by Figure 7.10 where the energy response of a 2 mm thick CdTe sensor with 110 μm pixels bump bonded to Medipix3RX is shown both in single pixel mode and in charge summing mode.

7.3.3 PILEUP

Given that the processing of each photon takes a finite time, during high interaction rates there will be problems with pileup especially for medical CT where the photon flux can reach 10^9 photons mm^{-2} s^{-2} in the direct beam. This is caused when a second photon arrives in the same pixel before the first photon is processed. Depending on the detector, the second photon could either be lost or added to the signal of the first photon. The result will be a deviation from linear behavior for the count rate as shown in Figure 7.11. This can, however, be corrected for up to a certain limit, but more problematic is the spectral distortion as an effect of pileup, which is an effect that cannot be easily corrected for.

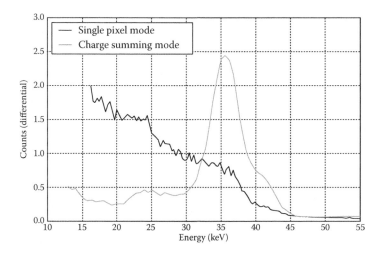

FIGURE 7.10 Energy response of a 2 mm thick CdTe sensor with 110 μm pixels bump bonded to Medipix3RX in both single pixel mode and charge summing mode (CSM). The source is x-ray fluorescence from praseodymium. Kα and Kβ lines are visible but not clearly resolved.

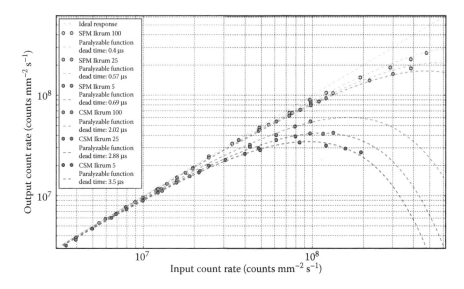

FIGURE 7.11 (See color insert) Count-rate behavior for Medipix3RX in single pixel mode (SPM) and charge summing mode (CSM) for different settings.

Different detectors will have different responses and it is important that the detector used is characterized and suitable for the flux in a specific application. Since the flux is measured per area, smaller pixels are an advantage leading to less photons per second per pixel.

7.4 CHARACTERIZATION OF HYBRID PIXEL DETECTORS

Many synchrotron facilities offer microfocused pencil beams, with a beam size in the range of a few microns. This section describes practical considerations for the setup and information gained from the tests available.

7.4.1 PRACTICAL CONSIDERATIONS

7.4.1.1 Aligning the Sensor

In an effort to combine high spatial resolution with high quantum efficiency many hybrid pixel detectors use a small pixel pitch combined with a thick sensor. For example, both 55 and 110 μm pixel pitches with a 1 mm thick sensor have been tested with Timepix [17] and recently a 110 μm pixel pitch with a 2 mm thick sensor was used with Medipix3RX [18]. Another detector, Pixirad, uses a 60 μm pitch with 650 μm thick sensors [5]. An aspect ratio, in the range of 1:10–1:20, puts very high demands on the alignment of the sensor in respect of the incoming beam because even a slight misalignment would illuminate several pixels and distort the results as shown in Figure 7.12.

For simplicity, the rotation of a sensor in respect of a perfectly aligned sensor with a horizontal x-axis, a vertical y-axis, and a z-axis in the beam direction will be

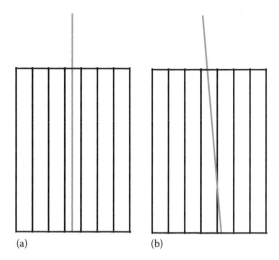

FIGURE 7.12   (a) Good alignment and (b) the effects of a 5° misalignment with a 1:10 aspect ratio.

defined, as shown in Figure 7.13. In order to align the sensor, a small visible laser is installed at the beamline. The laser is aimed in the direction of the beam and the reflection of the laser spot of the sensor back to the laser is used for alignment. A large distance of, for example, 4 m, and a small spot in size guarantees that when the reflected spot hits the laser there is a very small rotation around either the x- or y-axis. Assuming that, using only the eye, the spot can be aligned with the laser to a 1 mm precision, then the biggest rotation possible is 0.014°, which could at most have the beam moving 0.25 μm in x or y while traversing a 1 mm thick sensor.

The alignment of the rotation around the z-axis is then done by positioning the beam in the center of one pixel at the edge of the chip and then moving it to the other side of the chip. This will reveal any rotation around the z-axis. Because of the charge sharing in the sensors used, a movement significantly smaller than one pixel can be determined.

7.4.1.2   Beam Hardening
The beams produced by synchrotrons are generally very clean and have excellent energy resolution. However, for some of the tests with hybrid pixel detectors, the beam has to be strongly attenuated using aluminum filters to be able to resolve

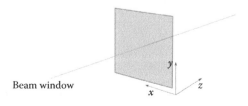

FIGURE 7.13   Sensor orientation.

individual photons. This causes beam hardening since the lower energies are attenuated more than the higher energies. As an example, in calculating the intensities of a 10 keV primary beam and a 30 keV second harmonic passing through a 2 mm aluminum filter, the primary beam is attenuated 150 times more than the harmonic.

So, after attenuation, otherwise nonvisible higher-order harmonics of a crystal monochromator might become visible. Figure 7.14 shows a range of harmonics measured with the Medipix3RX chip at B16 and Figure 7.15 shows the attenuation of a 25 keV tilted beam passing through a CdTe sensor bonded to a Timepix chip. With the tilted beam the signal is expected to decrease exponentially (a straight line in log scale) but the plot shows two different slopes, indicating that the primary 25 keV beam dominates in the upper layers and the second harmonic at 75 keV dominates in the lower layers.

The actual amount of harmonics in the beam will vary between beamlines and depends strongly on the filtration but it is something that has to be considered when planning and analyzing a measurement.

7.4.2 CHARACTERIZATION OF CdTe SENSORS BUMP BONDED TO HYBRID PIXEL DETECTORS

7.4.2.1 Sensors

To show the types of measurements possible and also report on the results of a new CdTe sensor with very small pixels, this section will go through the measurements performed on sensors with a 55 and 110 μm pixel pitch bump bonded to the Timepix ASIC. Both sensors were 1 mm thick. The CdTe used was grown by ACRORAD [19] using the traveling heater method and processing and hybridization was done at FMF Freiburg. To prevent the material properties from deteriorating, a low-temperature solder process was used. Contacts were intentionally ohmic, which should lead to less polarization and also offers the possibility to operate the same detector in either electron or hole collection mode, which is interesting when characterizing the material.

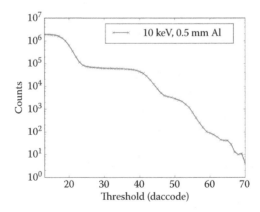

FIGURE 7.14 Integrated spectrum measured with Medipix3RX showing several harmonics. Monochromator set to 10 keV.

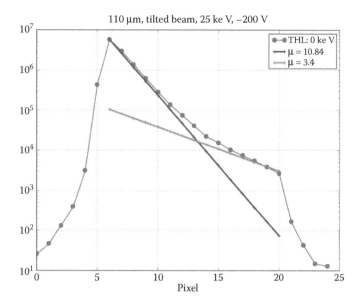

FIGURE 7.15 Attenuation during transmission of a 25 keV pencil beam at 30° through a CdTe sensor.

The $\mu\tau$ product was measured by Greiffenberg et al. [17] to be $\mu_e\tau_e = (1.9 \pm 0.6)10^{-3}$ cm^2 V^{-1} and $\mu_h\tau_h = (0.75 \pm 0.25)10^{-4}$ cm^2 V^{-1} for electrons and holes, respectively, using α particles. The possibility of collecting both electrons and holes is of course essential in this case and having a pixellated ASIC allows for a two-dimensional (2-D) map of the transport properties using a standard ^{241}Am laboratory source.

7.4.2.2 Verification of the Sensitive Volume and Depth-of-Interaction Effects

The depth-of-interaction dependence in the 1 mm thick CdTe sensor with a pixel pitch of 110 μm was investigated at the I15 Beamline at Diamond Light Source [20]. To do this, the sensor was illuminated with a 77 keV monochromatic pencil beam at an angle of 20° to the sensor surface. This meant that the beam passed through 25 pixels before exiting the sensor. First, to verify that the detector was working well and to verify the sensitivity of the whole volume and that there was no problem with the beam, an absorption profile was measured in PC mode. Then, to measure the energy response the Timepix chip was used in ToT mode and spectra were generated for each pixel thus representing a different depth in the sensor. For both of these measurements, the sensor was biased at −300 V to collect electrons.

Figure 7.16 presents the results at 20° and 25°. Much valuable information can be extracted from the plot. It shows that the detector is working correctly throughout the full thickness covering more than three orders of magnitude in count rate, which is expected but nevertheless an important confirmation. Since the y-axis is in logarithmic scale, the straight line of the profile confirms that the incoming beam

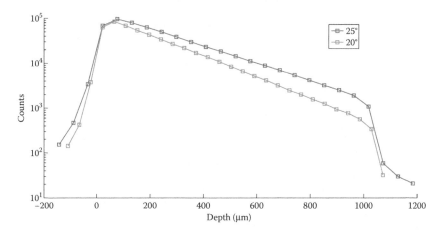

FIGURE 7.16 Absorption profile of a 77 keV pencil beam at 20° and 25° to the sensor surface.

is dominated by a single energy and the extracted linear attenuation coefficient is consistent with CdTe and 77 keV. Furthermore, having two different angles with a well-known difference between them (given by the rotation stage), it is possible to calculate the absolute angle and verify the alignment.

To investigate the material performance, a good area of the chip was chosen to eliminate the influence from localized defects. Analyzing interactions from individual photons it was also possible to look at the diffusion at different depths and the influence of fluorescence. To minimize the effects of trapping, the sensor was used in electron collection mode, which is the normal way to operate CdTe sensors.

In Figure 7.17, no depth-of-interaction effect is visible even for the lowest layer measured. This is consistent with a signal that is generated very close to the pixel and is completely dominated by electrons as the "small pixel effect" [21] for a detector W/L = 0.11 would predict. This also shows that the trapping of electrons is sufficiently low not to affect the signal on a level that can be measured with the Timepix chip in a 1 mm thick sensor.

7.4.2.3 Electrical Field Distortions around "Blob" Defects

Using a pencil beam it is possible to map distortions of the electrical field in a sensor. The simplest way is to illuminate the sensor with the beam perpendicular to the surface. The beam is first centered on a good pixel so that the initial position is known. Then, scanning over the defect, at each step, the position of the beam is known and compared with the position of the signal.

This method was used to scan a blob defect (Figure 7.18a) in a 1 mm thick sensor with a pixel pitch of 110 μm bump bonded to the Timepix chip. The experiment was carried out at the Extreme Conditions Beamline I15 at Diamond Light Source [20]. The photon energy was 79 keV and the beam size was measured to be 22.4 μm full width at half maximum (FWHM) by scanning the beam over one pixel and then deconvoluting the beam size from the pixel size. The defect was scanned in 110 μm steps putting the beam in the center of each pixel in the defect.

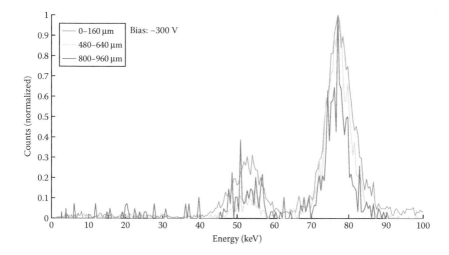

FIGURE 7.17 **(See color insert)** Time over threshold spectra at different depths in a 1 mm thick CdTe sensor bonded to Timepix. Clusters covering several pixels have been summed together and assigned to the pixel with the highest energy.

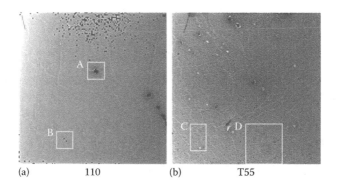

FIGURE 7.18 (a,b) Defects in CdTe sensors bonded to Timepix.

The Timepix detector was operated in PC mode using a low threshold of approximately 5 keV. For each step, the signal position was determined using a weighted centroid. Figure 7.19 presents the result of the measurement. An arrow is drawn from the known position of the beam to the position of the weighted centroid of the signal. This gives an indication of the electrical field in the defect as it shows how charge is flowing toward the center of the defect. If the electrical field were perpendicular to the sensor surface, the length of the arrow would be zero. The arrows indicate the movement of electrons so the orientations of the electrical field are in the opposite direction. The number of counts is independent of the beam position indicating that the main effect is a distortion of the field following displacement of the charge from the photon interactions. This distortion of the electrical field is probably due to an excess leakage current in the center of the defect. Further

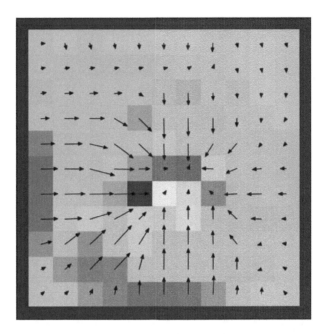

FIGURE 7.19 (**See color insert**) Arrowplot.

indications of the excess leakage current come from the Timepix chip itself. The chip features a preamplifier scheme proposed by Krummenacher [22] based on a cascoded differential CMOS amplifier operating with a bias current Ikrum. Global DACs control the front-end including the Ikrum current. The maximum leakage current that a single pixel can handle is Ikrum/2 [1]. Returning to the sensor, the area in the defect is unresponsive at normal Ikrum settings; however, by increasing the Ikrum current, and therefore the amount of leakage current that the pixel can handle, the response is recovered.

The limitation of this approach is of course that it does not give any information of the depth in the sensor at which the distortion in the field occurs. For that, a tilted beam has to be used, but such an approach requires more time and complex postprocessing. This method is faster and still gives an indication of the field and an accurate representation of how the image would be distorted in an x-ray imaging application.

7.4.2.4 Single Pixel Defects

Another common type of defect in CdTe hybrid pixel detectors is single pixel defects, where one pixel has low or no response while it is surrounded by pixels with a slightly higher than normal response. One such defect was scanned with the same kind of pencil beam as the blob defect but in this case the scan was done only in one row and in subpixel steps of 20 μm.

To investigate the charge transport, the cluster size at each position was studied. A cluster is defined as the pixels responding from a single photon interaction. To be able to resolve these the beam had to be strongly attenuated. The cluster size gives a good indication of the amount of diffusion during the charge transport. As shown in Figure 9.20a the cluster size increases at the edge of the pixels, which is expected since the charge cloud is then spread over two pixels, but it also increases when crossing the defect showing a maximum in the center of the defect pixel.

The total energy deposited in each cluster was also measured using the ToT mode of the Timepix chip. In Figure 7.20b, the total value for the cluster is presented as a function of its position. There seems to be no loss of charge when the beam passes the defect but rather the charge is spread over a larger area and collected in the neighboring pixels.

One possibility for this behavior is the high resistivity in the pixel contact or in the bump bond that connects the pixel. This would result in a lower electrical field and therefore more diffusion. The effect is also more visible with low-energy photons interacting high in the sensor, further strengthening this belief.

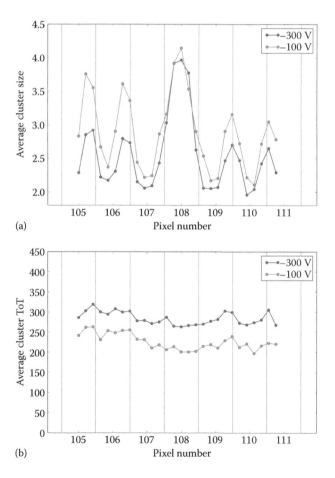

FIGURE 7.20 (See color insert) Line scan in 20 μm steps over a single pixel defect in the CdTe detector with 110 μm pixels. The vertical lines show pixel borders. (a) Cluster size; (b) cluster TOT.

7.4.2.5   Line Defects

The bright lines shown in Figure 7.18 are characteristic of pixellated CdTe detectors. A study by Hamann et al. [23] matched the lines to similar features in a back-reflection white-beam topography map. Further studies with the scanning rocking curve (SRC) showed that the borderlines correspond to mutually tilted regions in the crystal, revealing small-angle subgrain boundaries and/or dislocation networks.

In flood exposures (as Figure 7.18), the lines show a slightly higher count rate (10%) in electron collection mode and lower count rates in hole collection mode. The depth effect of the lines was investigated by scanning the lines with a tilted pencil beam and then reconstructing images from different depths in the sensor shown in Figure 7.21. The scan was also done with different threshold settings to extract information on the pulse height. For a threshold set at 6, 15, or 30 keV, the lines were between 1.1 and 1.3 times brighter than the rest of the matrix but with no significant depth dependence. Looking at the image of a 60 keV threshold there is a strong depth effect with the lines being as much as eight times brighter at the top of the sensor. Within the resolution of the Timepix chip, the lines remain in the same

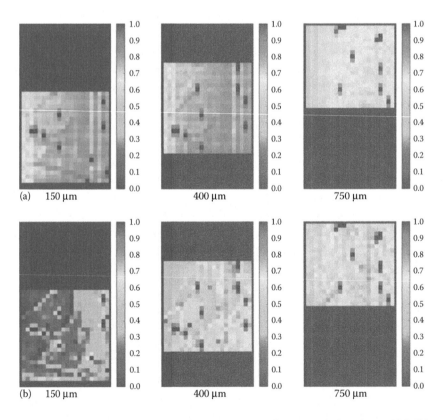

FIGURE 7.21   (**See color insert**) (a,b) Line defects at different depths in a 1 mm thick CdTe sensor for two different threshold settings: (a) 6 keV and (b) 60 keV.

place throughout the depth of the sensor as expected if they are low-angle subgrain boundaries.

The lines have been correlated with areas that have higher leakage current, but the use of a PC approach with leakage current compensation should not be affected by this. However, variations in leakage current will affect the electrical field in a high-resistivity material. Having the lines more visible even when the trapping should be increased along the subgrain boundaries further strengthens the theory that the effect of field distortion can be seen. This is also supported by the fact that the lines appear darker in hole collection mode since charge carriers of the opposite sign should be diffused instead of focused.

7.4.2.6 Circular Defects

Using the detector in hole collection mode immediately reveals the poor transport properties of holes in CdTe. Looking at the flood exposures in Figure 7.22, the sensor shows a large number of defects and inhomogeneities. Also, a strange circular defect appears in some detectors. It is almost perfectly circular and either unresponsive or with noisy pixels inside. Operating the same chip in electron collection mode the defect is not visible at all.

The defect has been scanned with a pencil beam at 30° in order to characterize it. In hole collection mode the charge is clearly pushed away from the center of the defect (Figures 7.23 and 7.24). The effect depends strongly on the depth of interaction as holes created in the top layer are affected by the electrical field over a longer

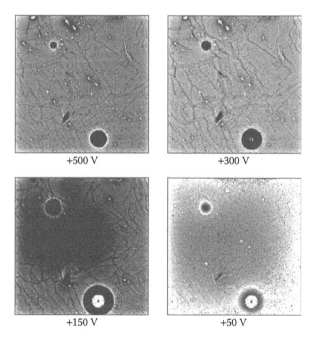

FIGURE 7.22 Flood exposure in hole collection mode at different bias for a 1 mm thick CdTe sensor with a 55 μm pixel pitch bump bonded to Timepix.

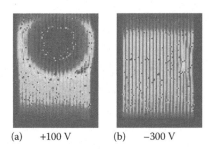

(a)   +100 V   (b)   −300 V

FIGURE 7.23 **(See color insert)** (a,b) Sum of all scan steps through the circular defect. Shown for both electron and hole collection mode.

time than holes created near the pixel contact. Repeating the same scan in electron collection mode does not reveal anything and the sensor is working correctly in the whole scanned region.

The fact that the effect is present throughout the whole depth and is not visible at all in electron collection mode points to it being located at the surface and injecting charge when positively biased. The large current affects the electrical field and repels the charge created from radiation interaction. Given that the defect is only visible in one polarity, it probably behaves as a diode, blocking the current in one direction. To fully understand this defect, further investigations are needed.

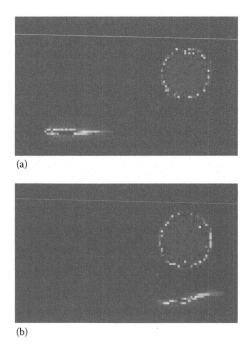

FIGURE 7.24 **(See color insert)** (a,b) A single scan step showing the displacement of the signal.

7.5 OUTLOOK AND FUTURE CHALLENGES

To be a viable alternative for medical imaging, single PC detectors must be available with a high-Z sensor material to provide sufficient absorption up to 140 keV. Currently, the best candidates are CdTe and CdZnTe but even though these materials have greatly improved in recent years, much research is still needed to produce large volumes of high-quality material at a reasonable price. Reducing point defect and polarization will be a key development. Silicon can still be useful in some medical imaging application using lower energies, for example, the edge on silicon strip detectors has shown good results in mammography [24].

On the ASIC side, the big challenge is the high flux in CT measurements since pileup affects both the count-rate linearity and the spectral response. One way to counteract that is to use smaller pixels but this is also a balance since smaller pixels will lead to more charge sharing. In this respect, the Medipix3RX chip offers an interesting combination of small pixels and good energy resolution. Also important for medical applications and imaging applications in general is the size of the detector. Currently, the largest produced module in the Medipix family detectors is 14×14 cm^2 [25] and even though larger modules are available with the Pilatus chip, these come at a steep price and with relatively large dead areas between sensors.

Efforts have been made to produce large-area detectors using edgeless sensors and through-silicon vias but the technology is still at an early stage. However, there is no doubt that the current movement in medical imaging is toward PC systems with spectral resolution.

REFERENCES

1. X. Llopart, et al., Timepix, a 65k programmable pixel readout chip for arrival time, energy and/or photon counting measurements, *Nucl Instrum Methods Phys Res A*, 581(1–2), (2007), 485–494.
2. R. Ballabriga, et al., The Medipix3RX: A high resolution, zero dead-time pixel detector readout chip allowing spectroscopic imaging, *J Instrum*, 8, (2013), C02016.
3. W. Barber, et al., Characterization of a novel photon counting detector for clinical CT: Count rate, energy resolution, and noise performance, *Proc SPIE*, 7258, (2009), 725824.
4. P. Pangaud, et al., XPAD3: A new photon counting chip for X-ray CT-scanner, *Nucl Instrum and Methods A*, 571(1–2), (2007), 321–324.
5. R. Bellazzini, et al., Chromatic X-ray imaging with a fine pitch CdTe sensor coupled to a large area photon counting pixel ASIC, *J Instrum*, 8, (2013), C02028.
6. R. Dinapoli, et al., EIGER: Next generation single photon counting detector for X-ray applications, *Nucl Instrum and Methods A*, 650(1), (2011), 79–83.
7. P. Valerio, et al., A prototype hybrid pixel detector ASIC for the CLIC experiment, *J Instrum*, 9, (2014), C01012.
8. Medipix, The Medipix2 and Medipix3 collaboration, Medipix, http://www.cern.ch/medipix.
9. V. Pugatch, et al., Metal and hybrid TimePix detectors imaging beams of particles, *Nucl Instrum and Methods A*, 650(1), (2011), 194–197.
10. J. Jungmann, et al., Biological tissue imaging with a position and time sensitive pixelated detector, *J Am Soc Mass Spectrom*, 23, (2012), 1679–1688.

11. A. Kazuyoshi, Charged particle tracking with the Timepix ASIC, *Nucl Instrum and Methods A*, 661(1), (2012), 31–49.
12. P. Kraft, et al., Characterization and calibration of PILATUS detectors, *IEEE Trans Nucl Sci*, 56(3), (2009), 758–764.
13. J. Jakubek, Precise energy calibration of pixel detector working in time-over-threshold mode, *Nucl Instrum and Methods A*, 633(1), (2011), S262–S266.
14. J. Giersch, et al., The influence of energy weighting on x-ray imaging quality, *Nucl Instrum and Methods A*, 531, (2004), 68–74.
15. M. Sánchez del Río and R.J. Dejus, XOP: A multiplatform graphical user interface for synchrotron radiation spectral and optics calculations, *Proc SPIE*, 3152, (1997), 148–157.
16. J. Lawrence et al., HEXITEC ASIC—A pixellated readout chip for CZT detectors, *Nucl Instrum and Methods A*, 604(1–2) (2009), 34–37.
17. D. Greiffenberg, et al., Energy resolution and transport properties of CdTe-Timepix-Assemblies, *J Instrum*, 6 (2011), C01058.
18. T. Koenig, et al., Charge summing in spectroscopic x-ray detectors with high-Z sensors, *IEEE Trans Nucl Sci*, 60(6), (2013), 4713–4718.
19. ACRORAD, CdTe, http://www.acrorad.co.jp/.
20. Diamond Light Source, I15: Extreme conditions beamline at Diamond Light Source, Diamond Light Source, http://www.diamond.ac.uk/Home/Beamlines/I16.html.
21. L.-A. Hamel, et al., Charge transport and signal generation in CdTe pixel detectors, *Nucl Instrum and Methods A*, 380(1–2), (1996), 238–240.
22. F. Krummenacher, Pixel detectors with local intelligence: An IC designer point of view, *Nucl Instrum and Methods A*, 305(3), (1991), 527–532.
23. E. Hamann, et al., Applications of Medipix2 single photon detectors at the ANKA synchrotron facility, in *2010 IEEE Nuclear Science Symposium Conference Record (NSS/MIC)*, Knoxville, TN, 30 October–6 November, pp. 3860–3863, IEEE.
24. M. Lundqvist, Evaluation of a photon counting X-ray imaging system, *IEEE Trans Nucl Sci*, 48(4), (2002), 1530–1536.
25. WidePix, http://www.widepix.cz.

8 Integrated Analog Signal-Processing Readout Front Ends for Particle Detectors

Thomas Noulis

CONTENTS

8.1 INTRODUCTION

The current trend in high-energy physics, biomedical applications, radioactivity control, space science, and other disciplines that require radiation detectors is toward smaller, higher-density systems to provide better position resolution. Miniaturization, low power dissipation, and low noise performance are stringent requirements in modern instrumentation, where portability and constant increase of channel numbers are the main streamlines. In most cases, complementary metal–oxide–semiconductor (CMOS) technologies have fully proven their adequacy for implementing data acquisition architectures based on functional blocks such as charge preamplifiers, continuous time or switch-capacitor filters, sample and hold amplifiers, analog-to-digital converters, etc., in analog signal processing for particle physics, nuclear physics, and x- or beta-ray detection [1–3].

Several motivations suggest that most of these applications can benefit from the use of application-specific integrated circuit (ASIC) readouts instead of discrete solutions. The most crucial motivation is that the implementation of readout electronics and semiconductor detectors onto the same die offers enhanced detection sensitivity thanks to improved noise performances [4–8]. Placing the very first stage of the front-end circuit close to the detector electrode reduces the amount of material and complexity in the active detection area and minimizes connection-related stray capacitances. This method allows the noise-optimization theory predictions [9,10] to be effectively satisfied, especially in the case of silicon sensors with very low anode capacitance, such as silicon drift detectors and charge-coupled devices (CCDs). On the other hand, the use of discrete transistors, with their relatively high (a few picofarads) gate capacitances, as front-end elements of hybrid circuits cannot comply with the stringent low-noise requirements. As a result, continuous efforts have been made to implement readout systems in monolithic form, and CMOS and SiGe bipolar CMOS (BiCMOS) technologies have been chosen due to their high integration density, relatively low power consumption, and capability to combine analog and digital circuits on the same chip.

8.2   READOUT FRONT-END ANALOG PROCESSING CHANNEL

8.2.1   PREAMPLIFIER–SHAPER STRUCTURE

The preamplifier–shaper structure is commonly adopted in the design of the above systems. A block diagram of such a detection system is shown in Figure 8.1. An inverse-biased diode (Si or Ge) detects radiation events, generating electron–hole pairs proportional to the absorbed energies. A low-noise charge-sensitive preamplifier (CSA) is used at the front end due to its low noise configuration and insensitivity of the gain to the detector capacitance. The generated charge Q is integrated onto a small feedback capacitance, which gives rise to a step voltage signal at the output of the CSA with an amplitude equal to Q/Cf. This is fed to a main amplifier, called a pulse shaper, where pulse shaping is performed to optimize the signal-to-noise ratio (SNR). The resulting output signal is a narrow pulse suitable for further processing.

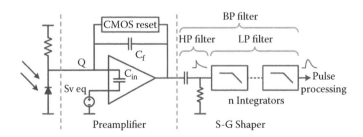

FIGURE 8.1   Preamplifier–shaping filter readout front-end system.

8.2.2   CSA: SHAPING FILTER SYSTEM NOISE ANALYSIS AND OPTIMIZATION

In order to analyze clearly the noise-matching mechanism in monolithic implementations, it is necessary to briefly review the noise characteristics of the metal-oxide-semiconductor field-effect transistor (MOSFET). In CMOS technology two major noise sources exist: thermal noise and flicker noise (or 1/f noise). Using a basic MOS transistor model and the Nyquist theory, the MOS drain-current thermal-noise spectral density, in the saturation regime, is given by Equation 8.1 [9,11,12]:

$$i_d^2(f) = \frac{8}{3} kT \mu C_{ox} \left(\frac{W}{L} \right) (V_{GS} - V_T) = \frac{8}{3} kT g_m \tag{8.1}$$

where:

k	= the Boltzmann constant
T	= the absolute temperature
μ	= carrier mobility
g_m	= the MOSFET transconductance
$W, L,$ and C_{ox}	= transistor's width, length, and gate capacitance per unit area, respectively

In contrast to channel thermal noise, which is well understood, the flicker noise mechanism is more complex, although its presence is surprisingly universal in all types of semiconductor devices [13–16]. A large number of theoretical and experimental studies show that the 1/f noise in MOSFET is caused by the random trapping and detrapping of the mobile carriers in the traps located at the Si–SiO$_2$ interface and within the gate oxide. On the basis of this model, the short circuit drain-current noise spectral density in the saturation region is given by Equation 8.2 [9,12]:

$$i_f^2(f) = \frac{K_F I_{DS}}{C_{ox} L^2 f} \tag{8.2}$$

where:

K_F = a technological dependent constant proportional to the effective trap density $N_t (F_n)$

I_{DS} = the drain current

Dividing Equation 8.2 by the square of the transconductance, the equivalent input 1/f noise can easily be calculated:

$$v_f^2(f) = \frac{K_F}{2\mu C_{ox}^2 WLf} = \frac{K_f}{C_{ox}^2 WLf} \tag{8.3}$$

The 1/f noise voltage depends only on the gate area, and its amplitude is dependent on the 1/f noise coefficient, whose value is variable in relation to the V_{gs} voltage of the MOS transistor [14,17].

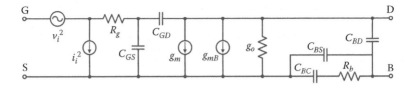

FIGURE 8.2 Equivalent input noise generator MOS small signal model. (Reprinted from Noulis, T., Siskos, S. and Sarrabayrouse, G., *IET Circuits Dev. Syst.*, 2, 324–334, 2008. With permission.)

In addition to the channel thermal and the flicker noise, MOS transistors also exhibit parasitic noise due to the resistive polygate and substrate resistance. These parasitic noise sources in modern semiconductor technologies should be taken into account, since the polygate resistance R_g and substrate resistance R_b depend mainly on the layout structure of MOS transistors, and the related modeling is quite advanced; also, by using specific layout techniques their noise contribution can be minimized. However, these noise sources are critical in high-frequency applications and can be considered negligible in the case of the detector readout integrated circuits (ICs).

The respective equivalent simplified input noise MOSFET model is shown in Figure 8.2 [18].

This model is based on the fact that the noise performance of any two-port network can be represented by two equivalent noise generators at the input stage of the network [19]. The two equivalent input noise generators are calculated from Figure 8.2 and are shown by Equations 8.4 and 8.5:

$$v_i^2(f) = \frac{\left(i_d^2 + i_f^2\right)}{\left|g_m - j\omega C_{GD}\right|^2} \tag{8.4}$$

$$i_i^2(f) = \left|j\omega\left(C_{GS} + C_{GD}\right)\right|^2 \frac{\left(i_d^2 + i_f^2\right)}{\left|g_m - j\omega C_{GD}\right|^2} \tag{8.5}$$

Since the factor $(g_m/2\pi C_{GD})$ is much higher than the transistor cutoff frequency f_T, the term $j\omega C_{GD}$ can be neglected with respect to g_m for all practical cases of interest. It is also essential to note that these two terms of Equations 8.4 and 8.5 are fully correlated.

High-density semiconductor front-end systems are designed according to low-noise criteria. The noise performance of a detector readout system is generally expressed as the equivalent noise charge (*enc*) and is defined as the ratio of the total *rms* noise at the output of the pulse shaper to the signal amplitude due to one electron charge *q*. The noise contribution of the amplification stage is the dominant source that determines the overall system noise and should, therefore, be optimized. The main noise contributor of the analog part is the CSA input MOSFET, and the noise types associated with this device are 1/*f* and channel thermal noise [12].

The total noise power spectrum at the CSA output due to the thermal and flicker noise sources is given by Equation 8.6 [9,10]:

$$\upsilon^2{}_0(j\omega) = \left|\frac{C_t}{C_f}\right|^2 \upsilon^2{}_{eq} = \left|\frac{C_t}{C_f}\right|^2 \cdot \left(\frac{8}{3}kT\frac{1}{g_m} + \frac{K_f}{C^2{}_{ox}WLf}\right)^2 \tag{8.6}$$

where:

C_t = the total CSA input capacitance
υ_{eq} = the total equivalent noise voltage of the input MOS transistor

The total integrated *rms* noise is given by Equation 8.7:

$$\upsilon^2{}_{total} = \int_0^\infty \left|\upsilon_0(j\omega)\right|^2 \left|H(j\omega)\right|^2 df \tag{8.7}$$

where $H(j\omega)$ is the transfer function consisting of a semi-Gaussian shaper pulse shaper. The *enc*s due to thermal and $1/f$ noise, considering Equations 8.4–8.7 and for an amplifier with capacitive source, are given by Equations 8.8 and 8.9 [9,10]:

$$enc_{th}^2 = \frac{8}{3}kT\frac{1}{g_m}\frac{C_t^2 B(\frac{3}{2},n-\frac{1}{2})n}{q^2 4\pi\tau_s}\left(\frac{n!^2 e^{2n}}{n^{2n}}\right) \tag{8.8}$$

$$enc_{1/f}^2 = \frac{K_f}{C_{ox}^2 WL}\frac{C_t^2}{q^2 2n}\left(\frac{n!^2 e^{2n}}{n^{2n}}\right) \tag{8.9}$$

where:

B = the beta function [9]
q = the electronic charge
τ_s = the peaking time of the shaper
n = the order of the semi-Gaussian shaper

Capacitances C_d, C_f, C_{GS}, and C_{GD} are capacitances of detector, feedback, gate-source, and gate-drain of the input MOSFET, respectively, and C_p is the parasitic capacitance of the interconnection between the detector and the amplifier input, which in this application is considered as negligible. The total input stage capacitance is given by Equation 8.10 [9,18,20,21]:

$$C_t = C_{totalin} = C_d + C_p + C_f + C_{GS} + C_{GD} \tag{8.10}$$

Input MOSFET optimum gatewidths exist for which the respective thermal and flicker *enc*s are minimal. These optimum dimensions are extracted by minimizing the respective *enc*s of Equations 8.8 and 8.9, respectively [9]. The optimal gatewidths are given in Equations 8.11 and 8.12:

$$\frac{\theta enc_{th}}{\theta W} = 0 \Rightarrow W_{th} = \frac{C_d + C_f}{2C_{ox}\alpha L} \tag{8.11}$$

$$\frac{\theta enc_{1/f}}{\theta W} = 0 \Rightarrow W_{1/f} = 3W_{th} \tag{8.12}$$

where α is defined as $\alpha = 1 + (9Xj)/(4L)$ and Xj is the metallurgical junction depth. Equations 8.11 and 8.12 are valid when capacitance C_d is in the picofarad range.

It can easily be determined which noise component (thermal or $1/f$) dominates, by calculating whether the ratio of the respective equivalent noise charges is above or below unity. The noise comparison ratio (*NCR*) depends also on the technology, comes from Equations 8.6 and 8.7, and is given by Equation 8.13 [20]:

$$NCR^2 = \frac{enc_{th}^2}{enc_{1/f}^2} = \frac{8kT2n^2B\left(\frac{3}{2}, n - \frac{1}{2}\right)C_{ox}^2 L^{3/2}}{3\sqrt{2\mu C_{ox}} \cdot 4\pi K_f} \cdot \frac{1}{\tau_s\sqrt{I_{DS}}}\sqrt{W} \tag{8.13}$$

When it is known, for a given technology and for a MOSFET operating in the saturation regime, which noise component, thermal or $1/f$, dominates, the type and optimum dimensions of the preamplifier input MOSFET can be selected, considering that typically p-channel transistors have less $1/f$ noise than their n-channel counterparts, since their majority carriers (holes) are less likely to be trapped due to their lower mobility [18].

It is obvious that CSA input noise is in practice defined by the detector capacitance, the peaking time specification, and the selection of the input MOS type and dimensions (as previously analyzed). A large detector capacitance results in a higher input noise, as it affects the input node biasing. Additionally, the thermal and flicker equivalent noise voltages (*enc*s) are proportional to C_t, whose value is mainly affected by C_d. The peaking time specification determines the operating bandwidth of the readout system. Consequently, it determines the flicker or thermal noise dominance, since the $1/f$ noise is negligible in the frequency region after 10 kHz. Concerning Equations 8.11 and 8.12, these formulas are the result of the optimization methodology, and they are used in calculating *NCR* (Equation 8.13). The dimensions of the input MOSFET are also selected according to these equations ($W_{1/f}$ if it is a p-channel MOSFET [PMOS] or W_{th} if it is an n-channel MOSFET [NMOS]) [18].

8.3 CSA: SHAPER SYSTEM DESIGN

8.3.1 Charge-Sensitive Amplifier Design

The CSA (Figure 8.3) is commonly implemented using a folded cascode structure, built of transistors M1, M2, and M4 [21,22]. This architecture provides a high direct

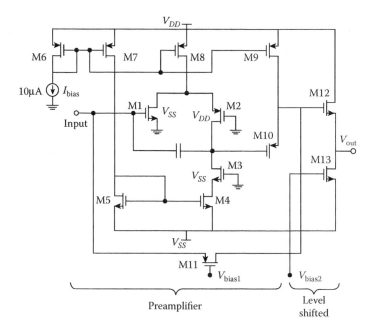

FIGURE 8.3 Charge-sensitive preamplifier using a folded cascode amplifier. (Reprinted from Noulis, T., Siskos, S., Sarrabayrouse, G. and Bary, L., Circuits and systems I: Regular papers, *IEEE Trans. Circuits Syst. Part 1*, 55, 1858 © (2008) IEEE. With permission.)

current (dc) gain and a relatively large operating bandwidth. The folded cascode amplifier has an n-channel input transistor (or drive), but a p-channel transistor is used for the cascode (or common-gate) transistor. A complementary structure can also be configured, always in relation to the noise specs of the application (melecom). This configuration allows the dc level of the output signal to be the same as the dc level of the input signal [22,23]. If the circuit is designed to operate with dc coupling detectors, CSA should be able to supply a current in the range from a few picoamperes up to a few nanoamperes through the feedback loop in order to match the detector leakage current. The open-loop gain of such a configuration is determined by the ratio of currents in the left and the right branch of the folded cascode, the sizes of the input transistor and the load transistor, and the total current I_{d1} in the input transistor [21,22].

Regarding the CSA reset mechanism, the integrated reset device provides a continuous discharge of the feedback capacitance and a dc path for the detector leakage current and defines the dc operating point of the amplifier. A simple resistor would not be a feasible reset device. In fact, in order to make its thermal noise negligible compared with the shot noise associated with the leakage current, its value should not be relatively high (e.g., a 500 MΩ resistor introduces an amount of thermal noise equal to the shot noise of 100 pA leakage current) and cannot be integrated on a small silicon area. An active device (MOS or bipolar) should be used instead [24]. The CSA reset device was configured with a PMOS transistor (M11) biased in the triode region in order to avoid the use of a high-value resistance (Figure 8.3).

Transistors M9 and M10 implement an output buffer for the CSA circuit. The current mirror architecture of transistors M6, M7, and M8 is biasing using the I_{bias} current source. The feedback capacitance C_f is placed between the CSA input node and the gate of the source follower transistor configuration stage for stability reasons. This feedback capacitance C_f is discharged via the input node, which is connected to both C_f and transistor M11. Transistor M3 is in cascade configuration with M4, the MOSFET comprising the current mirror M5, and was placed in the CSA in order to achieve higher output resistance.

In the specific design, the maximum signal swing must be limited for keeping all the devices in the saturation regime, that is, $V_{GS} > V_T$ and $V_{DS} > V_{GS} - V_T$. Therefore, the bias current source I_{bias} has been selected with a specific low value in order to achieve a low power dissipation performance.

The simplified small signal equivalent circuit of the core folded cascode amplifier without the feedback capacitance, the reset device, and transistor M3, which is in cascade configuration with M4, are shown in Figure 8.4.

The open-loop gain $A_V = (V_o/V_i)$ is given by Equation 8.14 [18]:

$$A_V = \frac{-\dfrac{g_{m1}}{g_{ds4}}\left(1 - s\dfrac{C_{gd1}}{g_{m1}}\right)}{\left[1 + sC_{ds1}R_1\left(1 + \dfrac{g_{m1}}{g_{m2}}\right)\right]\left[1 + s\dfrac{C_{p1} + C_{gd1}}{g_{m2}}\right]\left(1 + s\dfrac{C_{p2}}{g_{ds4}}\right)} \tag{8.14}$$

where:

$$C_{p1} = C_{gd1} \tag{8.15}$$

$$C_{p2} = C_{ds1} + C_{gd8} + C_{ds8} + C_{gs2} \tag{8.16}$$

$$C_{p3} = C_{gd2} + C_{gd4} + C_{bd2} + C_{bd4} \tag{8.17}$$

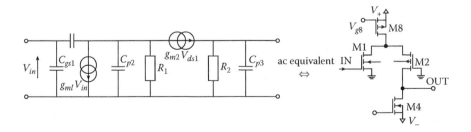

FIGURE 8.4 Small signal equivalent circuit of the open-loop preamplifier. (Reprinted from Noulis, T., Siskos, S., and Sarrabayrouse, G., *IET Circuits Dev. Syst.*, 2, 324–334, 2008. With permission.)

$$R_1 = \frac{1}{g_{ds1} + g_{ds8}} \tag{8.18}$$

$$R_2 = \frac{1}{g_{ds4}} \tag{8.19}$$

The DC gain A_{V0} of the two-port network is given in Equation 8.20:

$$A_{V0} = \frac{-g_{m1}}{g_{ds4}} = -g_{m1}R_2 \tag{8.20}$$

and the dominant pole is

$$P_1 = \frac{-g_{ds4}}{C_{p3}} = -\frac{1}{C_{p3}R_2} \tag{8.21}$$

The PMOS-based equivalent resistor operating as a reset device has a value of

$$R_{eq} = \sqrt{\frac{1}{2I\mu C_{ox}(W/L)}} \tag{8.22}$$

where I is the drain current.

Using the specific reset device implementation, the equivalent resistor can be externally modified by suitably fixing the gate bias voltage of the PMOS-based equivalent resistor. However, since the gate bias voltage is conditioned by the other parameters (the power consumption specification, the process and its allowable supply voltages, the charge–discharge time specification, etc.), the dimensions of the feedback MOS are also suitably selected according to each application [18].

Considering the noise analysis of the folded cascode structure, in Figure 8.5 the basic structure of the folded cascode amplifier (in relation to our preamplifier configuration) is depicted, including all the equivalent MOSFET noise-current sources.

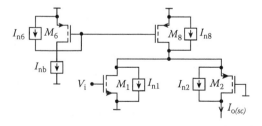

FIGURE 8.5 Folded cascode amplifier equivalent schema structure with noise current sources. (Reprinted from Noulis, T., Siskos, S., and Sarrabayrouse, G., *IET Circuits Dev. Syst.*, 2, 324–334, 2008. With permission.)

The total noise current is equal to (Equation 8.23) [18]

$$I_{o(sc)} = g_{m2}\left(-g_{m1}V_i - I_{n1} - I_{n2} + I_{n8} + g_{m8}\left(I_{nb} - I_{n6}\right)\left(r_{06}\,//g_{m6}{}^{-1}\right)\right)$$

$$\left(r_{01}\,//r_{08}\,//r_{02}\,//g_{m2}{}^{-1}\right) + I_{n2} \cong -g_{m1}V_i - I_{n1} + I_{n8} + g_{m8}\left(\frac{I_{nb} - I_{n6}}{g_{m6}}\right)$$

$$\cong -g_{m1}\left(V_i + V_{ni}\right) \tag{8.23}$$

Performing the approximations $(r_{06}\,//g^{-1}{}_{m6}) \cong g^{-1}{}_{m6}$ and $(r_{01}\,//r_{08}\,//r_{02}\,//g_{m2}{}^{-1}) \cong g_{m2}{}^{-1}$, the equivalent noise voltage is given by Equation 8.24:

$$V_{ni} = g_{m1}\left[-I_{n1} + I_{n8} + \frac{g_{m8}}{g_{m6}}\left(I_{nb} - I_{n6}\right)\right] \tag{8.24}$$

The respective square value is

$$v_{ni}{}^2 = g_{m1}{}^2\left[i_{n1}{}^2 + i_{n8}{}^2 + \left(\frac{g_{m8}}{g_{m6}}\right)^2 \cdot \left(i_{nb}{}^2 + i_{n6}{}^2\right)\right] \tag{8.25}$$

While noise contribution is theoretically generated by M6, M8, and the dc bias current source, in practice their contribution can be neglected when reflected to the amplifier topology input. As is shown by Equations 8.1–8.5, the generated noise current of each MOS transistor separately is dependent on its drain current. Specifically, the equations show that low-noise performance in a MOSFET requires a large value of transconductance (g_m), which, in turn, means that the transistor should have a large W/L ratio and be operated at a high quiescent current level. In practice, as drain current increases, noise performance decreases and power consumption increases. As is obvious from the preamplifier circuit, the current flowing through M8 (which in practice is obtained by the current source and the respective current mirror structure) is "shared" in M1 and M2 of the folded cascode amplifier. Taking into consideration the large dimensions of M1, its noise contribution is the dominant noise source of the CSA, and therefore the noise methodology is focused on its noise optimization. Therefore, the full noise-optimization methodology is focused on the input node conditions and consequently on the input transistor type and dimensions. Other design parameters that can be optimized in terms of noise performance are the peaking time and the order of the shaper, always in relation to the radiation-detection application specifications.

8.3.2 SEMI-GAUSSIAN SHAPING FILTER DESIGN

Semi-Gaussian (S-G) pulse-shaping filters are the most common pulse shapers employed in readout systems. Their use in electronics spectrometry instruments is to measure the energy of charged particles [25], and their purpose is to provide a voltage pulse whose height is proportional to the energy of the detected particle. The theory behind pulse-shaping systems, as well as different realization schemes, can be found in the literature [25,26]. It has been proved that a Gaussian-shaped step response provides optimum signal-to-noise characteristics. However, the ideal S-G shaper is noncasual and cannot be implemented in a physical system. A well-known technique to approximate a delayed Gaussian waveform is to use a CR-RC^n filter [25]. The principle of such a shaper is schematically shown in Figure 8.6.

A high-pass filter (HPF) sets the duration of the pulse by introducing a decay time constant. The low-pass filter (LPF), which follows, increases the rise time to limit the noise bandwidth. Although pulse shapers are often more sophisticated and complicated, the CR-RC^n shaper contains the essential features of all pulse shapers, a lower frequency bound and an upper frequency bound, and it is basically an $(n+1)$ order band pass filter (BPF), where n is the number of integrators. The

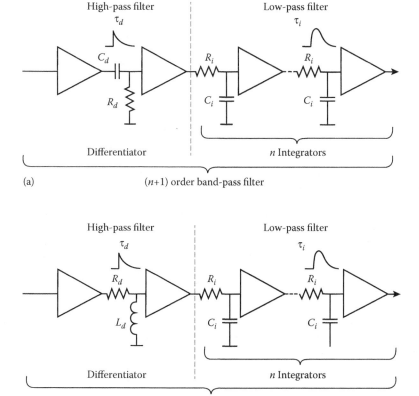

FIGURE 8.6 Principle diagram of a CR-(RC)^n shaping filter: (a) with RC HPF; (b) with RL HPF.

transfer function of an S-G pulse shaper consisting of one CR differentiator and n RC integrators (Figure 8.6) is given by Equation 8.26 [9,10]:

$$H(s)_{\text{shaper}} = \left(\frac{s\tau_d}{1+s\tau_d}\right)\left(\frac{A}{1+s\tau_i}\right)^n = H(s)_{BPF} \tag{8.26}$$

where:

τ_d = time constant of the differentiator
τ_i = time constant of the integrators
A = integrators' dc gain

The number n of the integrators is called shaper order. Peaking time is the time when the shaper output signal reaches peak amplitude and is defined by

$$\tau_s = n\tau_i = nRC \tag{8.27}$$

Increasing the value of n results in a step response, which is closer to an ideal Gaussian but with larger delay. The order n and peaking time τ_s, depending on the application, may or may not be predefined by the design specifications.

The operating bandwidth of an S-G shaper is given by

$$BW = f_i - f_d = \frac{1}{2\pi\tau_i} - \frac{1}{2\pi\tau_d} \tag{8.28}$$

The above combined band-pass frequency behavior enhances the signal-to-noise ratio by separating the "noise sea" from the signal's main Fourier components. The choice of cutoff frequencies defines the shaper's noise performance and consequently the signal-to-noise ratio. In addition, the frequency domain transfer characteristics are strictly linked to shaper time domain behavior or output pulse shape [20]. Pulse shaping has two conflicting objectives. The first is to restrict the bandwidth to match the measurement time. The second is to constrain the pulse so that successive signal pulses can be measured without undershoot or overlap (pileup). Reducing the pulse duration increases the allowable signal rate, but at the expense of electronic noise. In designing the shaper, these conflicting goals should be balanced. Many different considerations lead to the "nontextbook" compromise that optimum shaping depends on the application [27].

Concerning shaper peaking time, in order to achieve a value predefined from the application specifications, the shaper model's passive elements should be suitably selected.

The total CSA–second-order S-G shaper system transfer function using a Laplace representation is

$$H(s)_{\text{total}} = \left(\frac{A_{pr}}{1+s\tau_{pr}}\right)\left(\frac{s\tau_d}{1+s\tau_d}\right)\left(\frac{A_{sh}}{1+s\tau_i}\right)^2 \tag{8.29}$$

where:

A_{pr} = the preamplifier gain

τ_{pr} = its time rise constant

Considering as input signal a Dirac pulse $\delta(t)$, by taking the inverse Laplace transform of the product, the output signal in the time domain is given by Equation 8.30 [21,28,29]:

$$h_{total}(t) = \int_{\sigma-j\omega}^{\sigma+j\omega} H_{total}(s) \cdot e^{st} ds \tag{8.30}$$

Solving the above integral, the output signal of the readout system is

$$h_{total}(t) = A_{pr} \cdot A_{sh} \cdot \left(k_1 e^{-t/\tau_{pr}} + k_2 e^{-t/\tau_d} + k_3 e^{-t/\tau_i} + k_4 t e^{-t/\tau_i} \right) \tag{8.31}$$

where k_1, k_2, k_3, and k_4 are the following constants (Equations 8.32–8.35) [21,28,29]:

$$k_1 = \frac{\dfrac{1}{\tau_{pr}}}{\left(\dfrac{1}{\tau_{pr}}\right)^3 - \left(\dfrac{1}{\tau_{pr}}\right)^2\left(\dfrac{1}{\tau_d} + 2\dfrac{1}{\tau_i}\right) + \dfrac{1}{\tau_{pr}}\dfrac{1}{\tau_i}\left(\dfrac{1}{\tau_i} + 2\dfrac{1}{\tau_d}\right) - \dfrac{1}{\tau_d}\left(\dfrac{1}{\tau_i}\right)^2} \tag{8.32}$$

$$k_2 = \frac{\dfrac{1}{\tau_d}}{\left(\dfrac{1}{\tau_d} - \dfrac{1}{\tau_{pr}}\right) \cdot \left(\dfrac{1}{\tau_i} - \dfrac{1}{\tau_d}\right)^2} \tag{8.33}$$

$$k_3 = \frac{\dfrac{1}{\tau_{pr}} \cdot \dfrac{1}{\tau_d} - \left(\dfrac{1}{\tau_i}\right)^2}{\left(\dfrac{1}{\tau_i} - \dfrac{1}{\tau_{pr}}\right)^2 \cdot \left(\dfrac{1}{\tau_i} - \dfrac{1}{\tau_d}\right)^2} \tag{8.34}$$

$$k_4 = \frac{\left(\dfrac{1}{\tau_{pr}} + \dfrac{1}{\tau_d} - \dfrac{1}{\tau_i}\right)\left(\dfrac{1}{\tau_i}\right)^2 - \dfrac{1}{\tau_{pr}}\dfrac{1}{\tau_d}\dfrac{1}{\tau_i}}{\left(\dfrac{1}{\tau_i} - \dfrac{1}{\tau_{pr}}\right)^2 \cdot \left(\dfrac{1}{\tau_i} - \dfrac{1}{\tau_d}\right)^2} \tag{8.35}$$

Using the above equations, the values of all the shaper model's passive elements are selected.

8.3.2.1 Operational Transconductance Amplifier (OTA)-Based Shaper Approach

The main problem in the design of very-large-scale integration (VLSI) shaping filters for nuclear spectroscopy is the implementation of long shaping times, in the order of microseconds, for which high-value resistors (in the megaohm range), capacitors (in the 100 pF range), or both are demanded. In fact, the practical values in terms of occupied area that can be integrated are in the range of tens of kiloohms for the resistors and in the picofarad range for the capacitors.

Few examples of monolithic shaping structures providing long shaping times can be found in the literature. In all the above topologies, different techniques are used in order to obtain relatively large peaking times within a reasonable occupied area. Specifically, Chase et al. have suggested [30] the use of demultiplying current mirrors to increase the resistor value responsible for setting the filtering time constant of the op-amp-based low-pass filter. De Geronimo et al. have also proposed [31] a versatile architecture using current mirrors to magnify the resistances and still based on voltage-mode op-amps as gain blocks. Finally, Bertuccio et al. [32] adopted a current-mode approach to design a first-order low-pass filter, avoiding the use of operational amplifiers, that results in a very compact topology.

In this chapter, an alternative design technique is described, and an IC semi-Gaussian shaping structure is suggested. The proposed topology, in contrast to the typical shaping structures, is not based on op-amps, which generally demand large-area input transistors and high bias currents, but on operational transconductance amplifiers (OTAs). Extended study of all available OTA-based configurations, in both voltage and current domains, was performed by Noulis et al. [28,29]. All possible implantations were investigated using all the available advanced filter-design techniques.

Also in this chapter, a leapfrog OTA-based architecture is described [21]. Although its operating bandwidth is in the low-frequency region, it is fully integrated and is characterized by low power and low noise performance compatible with the stringent requirements of high-resolution nuclear spectroscopy. However, the main advantage of the specific shaper design is its continuous time-adjustable operating bandwidth, which renders it suitable for a variety of readout applications.

In particular, with respect to Figure 8.6, a two-port passive-element network was designed in order to implement an equivalent fully integrated second-order semi-Gaussian shaping filter. This two-port passive network is shown in Figure 8.7 [21].

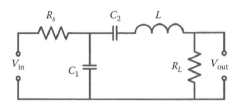

FIGURE 8.7 Equivalent RLC minimum-inductance two-port circuit of a second-order S-G shaper.

The specific passive-element topology has a respective transfer function (Laplace representation) to the typical shaper model. Its Laplace representation transfer function for a second-order S-G shaper is given by Equation 8.36 [21]:

$$H(s) = \frac{\left(\dfrac{1}{LC_1}\right)s}{s^3 + s^2\left(R_S R_L C_1 + L\right)\dfrac{1}{LC_1} + s\left(\dfrac{C_1}{C_2} + R_L + R_S\right)\dfrac{1}{LC_1} + \dfrac{1}{LC_1 C_2}} \qquad (8.36)$$

From the above two-port network, the signal flow graph (SFG) of a second-order S-G shaping filter is extracted (Figure 8.8).

The output signal of the passive network, in relation to the above SFG, is

$$V_{out} = \frac{1}{\dfrac{sL}{R_L} + 1}\left(V_1 - V_{out}\dfrac{1}{sC_2 R_L}\right) \qquad (8.37)$$

Using the extracted SFG and the leapfrog (functional simulation method) design methodology, a second-order shaper is designed. The main advantage of the leapfrog method over other filter design methods, which also provide integrated structures, is the better sensitivity performance and the capability to optimize the dynamic range by properly intervening during the phase of the original passive synthesis [33,34]. In order to implement the shaping filter, operational transconductance amplifiers (OTAs) were selected as the basic building cells. A diagram of an OTA is shown in Figure 8.9.

The OTA is assumed to be an ideal voltage-controlled current source and can be described by

$$I_0 = g_m(V^+ - V^-) \qquad (8.38)$$

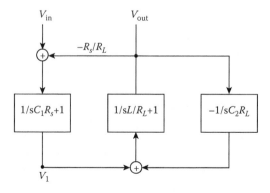

FIGURE 8.8 Signal flow graph of a voltage-mode second-order semi-Gaussian shaper.

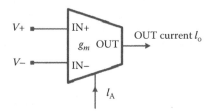

FIGURE 8.9   Diagram of OTA.

where:

I_0 = output current

V^+ and V^- = noninverting and inverting input voltages of the OTA, respectively

Note that g_m (transconductance gain) is a function of the bias current, I_A. The implementation of the S-G shaping filter with OTAs is highly advantageous, since it provides programmable characteristics. In particular, tunability is achieved by replacing the RC and CR sections in the original passive model with active g_m-C sections, where the g_m can be adjusted with an external bias voltage or current. The second-order shaping filter that was designed using the above SFG and the leapfrog method is shown in Figure 8.10.

The capacitor values and OTA transconductances of the above S-G shaper are given in Table 8.1. The shaper configuration was designed in order to provide a peaking time equal to 1.8 μs, which refers to a bandwidth of 260 kHz in the low-frequency region (f_{c1} = 140 Hz). The passive element values of the respective RLC-equivalent two-port network of Figure 8.7 are also given in Table 8.1.

Because of the nonintegrable value of C_2 in the leapfrog shaper, the specific capacitance was substituted with a grounded OTA-C capacitor simulator. The respective OTA architecture is described in Figure 8.11, and its calculated value

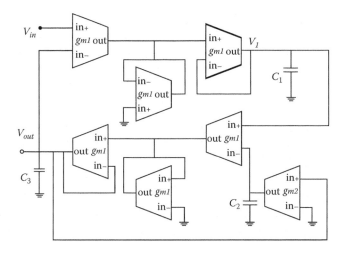

FIGURE 8.10   OTA-based second-order S-G shaper using the leapfrog technique.

TABLE 8.1
IC Shaper and RLC Equivalent
Network Element Values

IC Shaper		RLC Network	
Actives and Passives		**Passives**	
g_{m1}	24 µA/V	Rs	100 Kω
g_{m2}	11.5 µA/V	R_L	100 Kω
g_{m3}	950 nA/V	C_1	10.3 Pf
C_1	13 pF	C_2	7.4 nF
C_2	1.00 nF	L	84 mH
C_3	14 pF		

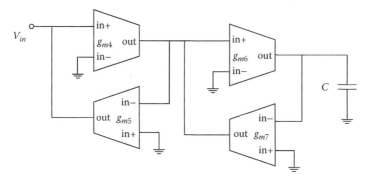

FIGURE 8.11 OTA-based architecture for grounded capacitance simulation. (Reprinted from Noulis, T., Siskos, S., Sarrabayrouse, G. and Bary, L., Circuits and systems I: Regular papers, *IEEE Trans. Circuits Syst. Part 1*, 55, 1857 © (2008) IEEE. With permission.)

is given by Equation 8.39 [33,34]. The values of the transconductances and the capacitor C are $g_{m4}=g_{m5}=g_{m6}=74.4$ µA/V, $g_{m7}=950$ nA/V, and $C=12.8$ pF.

$$C_{eq} = C \frac{g_{m4}g_{m5}}{g_{m6}g_{m7}} \qquad (8.39)$$

8.3.3 Advanced CSA-Shaper Analog Signal Processor Implementation

Using the analysis and methodology presented in Sections 8.2.1 and 8.2.2 on the CSA and shaper design, respectively, an advanced fully integrated radiation-detection analog processing channel was designed and fabricated [29]. The preamplifier–shaper readout front-end system was designed, simulated, and fabricated in a 0.35 µm CMOS process (3M/2P 3.3/5V) by Austria Mikro Systeme (AMS, Unterpremstätten, Austria) for a specific low-energy x-ray silicon strip detector.

The design specifications are given in Table 8.2 [18,21] and Table 8.3 [18,21].

TABLE 8.2
Design Specifications

Detector Diode PIN (Si)

Detector capacitance	$C_d = 2{-}10$ pF
Leakage current	$I_{leak} = 10$ pA
Charge Q collected per event	$Q = 312.5$ ke$^-$
Time t needed for the collection of the 90% of the total charge Q	300 ns

Preamplifier–Shaper

Shaper's order	$n = 2$
Shaper peaking time value	$\tau_s = n\tau_0 = 1.8$ µs
Temperature value	25°C
Power consumption per processing channel	Less than 8 mW

TABLE 8.3
CSA-Level Shifter MOS Transistor Dimensions

MOS Transistors	(W/L) (µm)
M1	310/0.9
M2	100/0.7
M3	25/1.5
M4, M5	2.5/5
M6, M7, M9	2.5/2.5
M8	5/2.5
M10	100/0.7
M11	20/1
M12, M13	5/5

Source: Noulis, T., Deradonis, C., Siskos, S., *VLSI Design J.*, 2007, 71684, 2007.

The preamplifier was designed using the architecture provided in Figure 8.3. Related extended analysis of the design methodology is available in Section 8.2.1. The transistor dimensions and the biases for this configuration were selected according to the design and optimization criteria presented in Section 8.2.1 The transistor dimensions are provided in Tables 8.2 and 8.3, and the biases are provided in this section. The level shifting stage was added as the dc bias level of the signal provided to the shaper to be controlled and the general dynamic range and 1 dB compression point of the filter to be optimized. The feedback capacitance Cf is 550 fF, and it is placed between the input node and the gate of the source follower stage to avoid introduction on the closed loop of the follower stage complex poles and to isolate

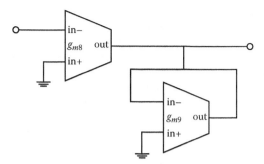

FIGURE 8.12 OTA-based amplification stage. (Reprinted from Noulis, T., Siskos, S., Sarrabayrouse, G. and Bary, L., Circuits and systems I: Regular papers, *IEEE Trans. Circuits Syst. Part 1*, 55, 1859 © (2008) IEEE. With permission.)

the *Cf* from the following stage. The bias current I_{bias} was selected to be 10 µA. The power supplies are VDD = −VSS = 1.65 V, and the reset device bias voltage is fixed to 150 mV. The level shifting bias voltage is equal to −1.18 V [29].

The shaper was implemented with the leapfrog architecture and OTA-based grounded capacitance simulator replacement structure, depicted in Figures 8.10 and 8.11. An inherent drawback of the leapfrog methodology, and consequently of the particular shaper architecture, is that the leapfrog design method provides a gain value of 1/2 [33,34]. In order to cope with this specific loss, an additional gain stage was used after the preamplifier and before the shaper, using an OTA-based structure ($g_m8 = 153$ µA/V, $g_m9 = 11.5$ µA/V) provided in Figure 8.12.

The specific topology gives the capability to program the gain externally by fixing the bias voltage of the two OTA components. In particular, the dc gain of the above OTA structure is $A = V_{out}/V_{in} = g_m8/g_m9$. Considering the S-G OTA-based shaping architecture, capacitor simulator, and amplification topology of Figures 8.10, 8.11, and 8.12, respectively, they were implemented using the CMOS OTA configuration, shown in Figure 8.13.

In the transconductance circuit, a typical CMOS cascode configuration is used, where changing the bias voltage results in approximately equal changes for both the transconductance and the 3 dB frequency. The MOS dimensions of all the OTA circuits are given in Table 8.4.

The PMOS devices in the OTA circuits have their bulk biased in the same voltage to their source (designing separate wells), taking advantage of the specific n-well CMOS process that is used, in order to avoid the body effect and to maximize the signal swing. The power supplies in all the OTA components are also VDD = −VSS = 1.65 V. The bias voltage of each OTA is fixed at −0.85 V in order to achieve the desired transconductance values. The total simplified block diagram of the analog readout ASIC is given in Figure 8.14.

8.3.4 EXPERIMENTAL RESULTS

The fabricated chip in which the prototype x-ray readout ASIC was implanted is shown in Figure 8.15. The measured x-ray IC front-end system output signal is shown

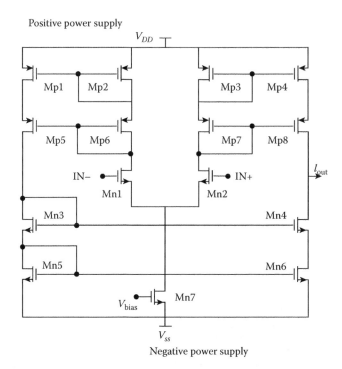

FIGURE 8.13   CMOS operational transconductance amplifier.

TABLE 8.4
OTA MOSFET Dimensions

Transistors	g_{m1}	$g_{m2,9}$	$g_{m3,7}$	$g_{m4,5,6}$	g_{m8}
Mp1, Mp4	25/1	25/2.5	25/35	77/1	80/1
Mp2, Mp3	25/1	25/1	25/1	25/1	26/1
Mp5, Mp8	12/1	12/2.5	12/35	36/1	36/1
Mp6, Mp7	12/1	12/1	12/1	12/1	12/1
Mn1, Mn2	3.7/5	3.7/5	3.7/5	3.7/5	3.7/5
Mn3, Mn4	3.7/5	3.7/12.5	2/100	11/5	11/5
Mn5, Mn6	5/5	5/12.5	3/100	15/5	16/5
Mn7	5/5	5/5	5/5	5/5	5/5

Note: Dimensions in micrometers.

in Figure 8.16. The system provides dc gain equal to 120 dB. However, as will be presented below, this value can be externally modified by suitably fixing the bias voltages and consequently changing the g_m values of the OTA-based amplification topology of Figure 8.12. Regarding the radiation-detection application specifications, the output pulse has a peaking time value of 1.81 μs that shows no undershoot or pileup.

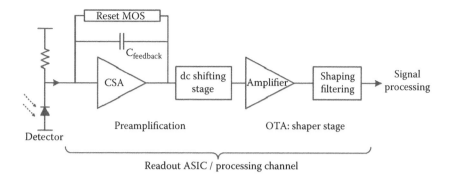

FIGURE 8.14 Block diagram of the IC readout system.

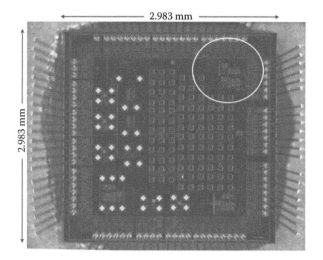

FIGURE 8.15 Microphotograph of the full fabricated microchip (the readout system is circled). (Reprinted from Noulis, T., Siskos, S., Sarrabayrouse, G. and Bary, L., Circuits and systems I: Regular papers, *IEEE Trans. Circuits Syst.* Part 1, 55, 1860 © (2008) IEEE. With permission.)

The power consumption is 1 mW, far lower than the maximum allowable specified value of 8 mW, rendering the system low power [24,35].

The readout ASIC noise performance was also analytically studied. The system enc is 382 e$^-$ for a detector of 2 pF, and noise performance increases with a slope of 21 e$^-$/pF. The enc dependence on the detector capacitance variations is shown in Figure 8.17.

The linearity of the front-end system energy resolution is presented in Figure 8.18. The CMOS readout analog processor achieves an input charge gain–voltage output conversion of 3.31 mV/fC and a linearity of 0.69%.

In terms of the total occupied area, the specific IC system is again advantageous, since it consumes only 0.2017 mm^2 of a total 2.983 × 2.983 mm^2 microchip [36,37].

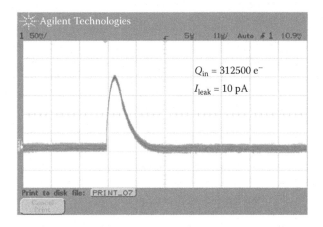

FIGURE 8.16 Readout system output pulse. (Reprinted from Noulis, T., Siskos, S., Sarrabayrouse, G. and Bary, L., Circuits and systems I: Regular papers, *IEEE Trans. Circuits Syst. Part 1*, 55, 1860 © (2008) IEEE. With permission.)

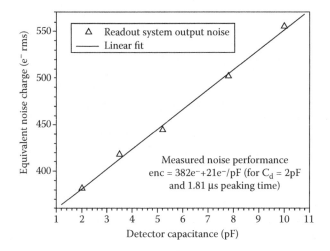

FIGURE 8.17 Readout ASIC enc dependence on detector capacitance. (Reprinted from Noulis, T., Siskos, S., Sarrabayrouse, G. and Bary, L., Circuits and systems I: Regular papers, *IEEE Trans. Circuits Syst. Part 1*, 55, 1860 © (2008) IEEE. With permission.)

Finally, the flexibility of the system for use in a wide range of readout applications taking advantage of the specific shaping stage topology was examined. The system gain is programmable from 118 to 137 dB by changing the OTAs' bias voltage of the gain stage. This gain programmability is demonstrated in Figure 8.19. Furthermore, the peaking time can be similarly externally modified. This is represented in Figure 8.20, in which output signals with peaking times from 1 to 3 μs are provided by suitably fixing the bias voltages of the OTA configurations, which operate in the total shaper architecture as integrators and the OTAs in the capacitance

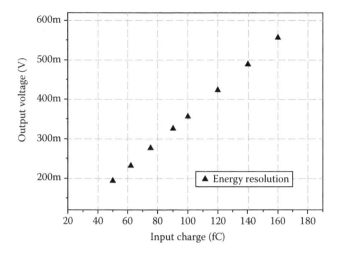

FIGURE 8.18 Readout ASIC measured energy response. (Reprinted from Noulis, T., Siskos, S., Sarrabayrouse, G. and Bary, L., Circuits and systems I: Regular papers, *IEEE Trans. Circuits Syst. Part 1*, 55, 1860 © (2008) IEEE. With permission.)

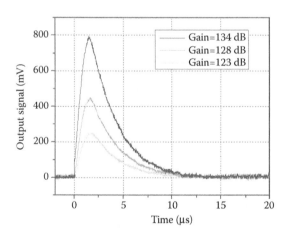

FIGURE 8.19 Gain programmability. (Reprinted from Noulis, T., Siskos, S., Sarrabayrouse, G. and Bary, L., Circuits and systems I: Regular papers, *IEEE Trans. Circuits Syst. Part 1*, 55, 1861 © (2008) IEEE. With permission.)

simulator unit. As the peaking time increases, the shaper upper frequency becomes higher in its band-pass response. This particular OTA-based S-G shaper provides continuously variable peaking time in the range from 950 ns to 3.1 μs. A variation of about 30% in signal amplitude can be detected when moving from the shortest to the longest available peaking time. The output signal undershoot can also be externally adjusted. In Figure 8.21, different output signal undershoots are shown. The undershoot can vary from 0% to 18% of the positive signal amplitude. This results in a respective increment of the low 3 dB frequency, providing a narrower

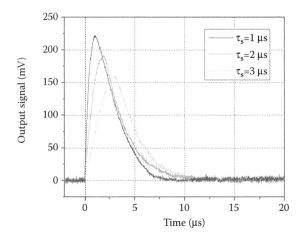

FIGURE 8.20 Adjustable peaking time capability: output signal for different peaking times. (Reprinted from Noulis, T., Siskos, S., Sarrabayrouse, G. and Bary, L., Circuits and systems I: Regular papers, *IEEE Trans. Circuits Syst. Part 1*, 55, 1861 © (2008) IEEE. With permission.)

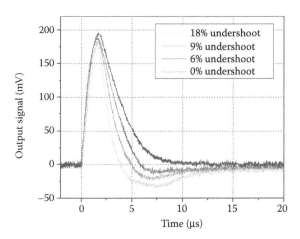

FIGURE 8.21 Variable output signal undershoot capability: different output signal undershoot. (Reprinted from Noulis, T., Siskos, S., Sarrabayrouse, G. and Bary, L., Circuits and systems I: Regular papers, *IEEE Trans. Circuits Syst. Part 1*, 55, 1861 © (2008) IEEE. With permission.)

shaper bandwidth. Consequently, the output noise can also be regulated in relation to each application. Regarding the system, and particularly the S-G shaping filter stability, having only one feedback between its input and output, no stability problems were observed. In addition, no sensitivity problems occurred in relation to the g_m terms, since their value could be fixed externally using the respective bias voltages.

8.4 CURRENT CHALLENGES AND LIMITATIONS

Radiation-detection IC design imposes new challenges and limitations as the technology nodes are shrinking.

In terms of the technology used, CMOS has scaled down to 28 and 20 nm, allowing extremely dense integration and cost minimization for high-volume production. However, in terms of noise optimization, performance, and cost, the process selection should always be defined by the related application. While the noise is becoming lower in submicron technologies, extra noise sources should be taken into account, such as gate-induced noise and channel-reflected noise. In terms of circuit design, advanced noise-optimization methodologies are needed at the level of both circuit architecture and mask design (physical layout), as these noise contributors need to be optimized. While in previous technology nodes these were almost negligible, now, at 28 nm and below, these may contribute up to 50% of the noise, depending always on the application and mostly on the operating circuit/system frequency bandwidth.

In addition, the related transistor noise models, always in relation to the used technology Foundry, are not always adequate to provide an accurate simulation, and the related performance cannot be effectively optimized. Advanced models are needed for the designer to be able to access the trade-offs and to optimize the radiation-detection IC processor.

Furthermore, in terms of radiation hardness, extreme limitations exist. Models for pre- and postradiation conditions are needed, as the performance needs to be accessed in real operation mode. No technology foundry provides this kind of simulation support, and as a result the related performance degradation after irradiation cannot be estimated effectively in all the presilicon design stages and verification. Advanced modeling techniques are required, and device layouts for radiation-hard operation are demanded for the integrated circuits to be able to cover the operation specifications in an irradiated environment. This is a gap in the design flow that constantly renders radiation IC design extremely risky.

REFERENCES

1. E. Beuville, K. Borer, E. Chesi, E. H. M. Heijne, P. Jarron, B. Lisowski and S. Singh, Amplex, a low-noise, low-power analog CMOS signal processor for multi-element silicon particle detectors, *Nuclear Instruments and Methods in Physics Research Section A* 288(1) (1990) 157–167.
2. L. T. Wurtz and W. P. Wheless Jr., Design of a high performance low noise charge preamplifier, *IEEE Transactions on Circuits and Systems I* 40(8) (1993) 541–545.
3. S. Tedja, J. Van der Spiegel and H. H. Williams, A CMOS low-noise and low-power charge sampling integrated circuit for capacitive detector/sensor interfaces, *IEEE Journal of Solid State Circuits* 30(2) (1995) 110–119.
4. Radeka, V., Rehak, P., Rescia, S., Gatti, E., Longoni, A., Sampietro, M., Holl, P., Struder, L. and Kemmer, J. Design of a charge sensitive preamplifier on high resistivity silicon, *IEEE Transactions on Nuclear Science* 35(1) (1988) 155–159.
5. Radeka, V., Rehak, P., Rescia, S., Gatti, E., Longoni, A., Sampietro, M., Bertuccio, G., Holl, P., Struder, L. and Kemmer, J. Implanted silicon JFET on completely depleted high-resistivity devices, *IEEE Electron Device Letters* 10(2) (1989) 91–94.

6. Lund, J. C., Olschner, F., Bennett, P. and Rehn, L. Epitaxial n-channel JFETs integrated on high resistivity silicon for x-ray detectors, *IEEE Transactions on Nuclear Science* 42(4) (1995) 820–823.

7. Lechner, P., Eckbauer, S., Hartmann, R., Krisch, S., Hauff, D., Richter, R., Soltau, H., et al. Silicon drift detectors for high resolution room temperature x-ray spectroscopy, *Nuclear Instruments and Methods in Physics Research Section A* 377 (1996) 346–351.

8. Ratti, L., Manghisoni, M., Re, V. and Speziali, V. Integrated front-end electronics in a detector compatible process: Source-follower and charge-sensitive preamplifier configurations, In: James, R. B. ed., *Hard X-Ray and Gamma-Ray Detector Physics, III Proceedings of SPIE* 4507 (2001) 141–151.

9. Sansen, W. and Chang, Z. Y. Limits of low noise performance of detector readout front ends in CMOS technology, *IEEE Transactions on Circuits and Systems* 37(11) (1990) 1375–1382.

10. Chang, Z. Y. and Sansen, W. Effect of 1/f noise on the resolution of CMOS analog readout systems for microstrip and pixel detectors, *Nuclear Instruments and Methods in Physics Research Section A* 305 (1991) 553–560.

11. Steyaert, M., Chang, Z. Y. and Sansen, W. Low-noise monolithic amplifier design: Bipolar versus CMOS, *Analog Integrated Circuits and Signal Processing* 1(1) (1991) 9–19.

12. Motchenbacher, C. D. and Connelly, J. A. *Low-noise Electronic System Design* (Wiley Interscience, New York, 1998).

13. Xie, D., Cheng, M. and Forbes, L. SPICE models for flicker noise in n-MOSFETs from subthreshold to strong inversion, *IEEE Transactions on Computer-Aided Design* 19(11) (2000) 1293–1303.

14. Noulis, T., Siskos, S. and Sarrabayrouse, G. Comparison between BSIM4.X and HSPICE flicker noise models in NMOS and PMOS transistors in all operating regions, *Microelectronics Reliability* 47 (2007) 1222–1227.

15. Jakobson, C. G., Bloom, I. and Nemirovsky, Y. 1/f noise in CMOS transistors for analog applications from subthreshold to saturation, *Solid-State Electronics* 42(10) (1998) 1807–1817.

16. Nemirovsky, Y., Brouk, I. and Jakobson, C. G. 1/f noise in CMOS transistors for analog applications, *IEEE Transactions on Electron Devices* 48(5) (2001) 921–927.

17. Liu, W. *MOSFET Models for SPICE Simulation including BSIM3V3 and BSIM4* (Wiley, New York, 2001).

18. Noulis, T., Siskos, S. and Sarrabayrouse, G. Noise optimized charge sensitive CMOS amplifier for capacitive radiation detectors, *IET Circuits, Devices & Systems* 2(3) (2008) 324–334.

19. Chang, Z. Y. and Sansen, W. *Low Noise Wide Band Amplifiers in Bipolar and CMOS Technologies* (Kluwer, Norwell, MA, 1991).

20. Noulis, T., Deradonis, C., Siskos, S. and Sarrabayrouse, G., Programmable OTA based CMOS shaping amplifier for x-ray spectroscopy, *Proceedings of 2nd IEEE PRIME* (Otranto, Lecce, Italy, 2006).

21. Noulis, T., Siskos, S., Sarrabayrouse, G. and Bary, L. Advanced low noise X-ray readout ASIC for radiation sensor interfaces. *IEEE Transactions on Circuits and Systems Part 1*, 55(7) (2008) 1854–1862.

22. Vandenbussche, J., Leyn, F., Van der Plas, G., Gielen, G. and Sansen, W. A fully integrated low-power CMOS particle detector front-end for space applications, *IEEE Transactions on Nuclear Science* 45(4) (1998) 2272–2278.

23. Guazzoni, C., Samprieto, M. and Fazzi, A. Detector embedded device for continuous reset of charge amplifiers: Choice between bipolar and MOS transistor, *Nuclear Instruments and Methods in Physics Research Section A* 443 (2000) 447–450.

24. Pedrali-Noy, M., Gruber, G., Krieger, B., Mandelli, E., Meddeler, G., Moses, W. and Rosso, V. PETRIC-A position emission tomography readout integrated circuit, *IEEE Transactions on Nuclear Science* 48(3) (2001) 479–484.

25. Konrad, M. Detector pulse-shaping for high resolution spectroscopy, *IEEE Transactions on Nuclear Science* 15(1) (1968) 268–282.

26. Ohkawa, S., Yoshizawa, M. and Husimi, K. Direct synthesis of the Gaussian filter for nuclear pulse amplifiers, *Nuclear Instruments and Methods in Physics Research Section A* 138(1) (1976) 85–92.

27. Goulding, F. S. and Landis, D. A. Signal processing for semiconductor detectors, *IEEE Transactions on Nuclear Science* 29(3) (1982) 1125–1141.

28. Noulis, T., Deradonis, C., Siskos, S. and Sarrabayrouse, G. Detailed study of particle detectors OTA based CMOS semi-Gaussian shapers, *Nuclear Instruments and Methods in Physics Research Section A* 583 (2007) 469–478.

29. Noulis, T., Deradonis, C. and Siskos, S. Advanced readout system IC current mode semi-Gaussian shapers using CCIIs and OTAs, *VLSI Design Journal* 2007 (2007) 71684.

30. Chase, R. L., Hrisoho, A. and Richer, J. P. 8-channel CMOS preamplifier and shaper with adjustable peaking time and automatic pole-zero cancellation, *Nuclear Instruments and Methods in Physics Research Section A* 409 (1998) 328–331.

31. De Geronimo, G., O'Connor, P. and Grosholz, J. A generation of CMOS readout ASIC's for CZT detectors, *IEEE Transactions on Nuclear Science* 47 (2000) 1857–1867.

32. Bertuccio, G., Gallina, P. and Sampietro, M. 'R-lens filter': An (RC)n current-mode lowpass filter, *Electronics Letters* 35 (1999) 1209–1210.

33. Deliyannis, T., Sun, Y. and Fidler, K. *Continuous-Time Active Filter Design* (CRC Press, Boca Raton, FL, 1999).

34. Sedra, A. S. and Brackett, P. O. *Filter Theory and Design: Active and Passive* (Pitman, London, 1979).

35. Krieger, B., Ewell, K., Ludewigth, B. A., Maier, M. R., Markovic, D., Milgrome, O. and Wang, Y. J. An 8×8 pixel IC for x-ray spectroscopy, *IEEE Transactions on Nuclear Science* 48(3) (2001) 493–498.

36. Limousin, O., Gevin, O., Lugiez, F., Chipaux, R., Delagnes, E., Dirks, B. and Horeau, B. IDeF-X ASIC for Cd(Zn)Te spectro-imaging systems, *IEEE Transactions on Nuclear Science* 52(5) (2005) 1595–1602.

37. Shani, G. *Electronics for Radiation Measurements*, vol. 1. (CRC Press, Boca Raton, FL, 1996), pp. 182–183.

9 Spectral Molecular CT with Photon-Counting Detectors

Michael F. Walsh, Raja Aamir, Raj K. Panta,
Kishore Rajendran, Nigel G. Anderson,
Anthony P. H. Butler, and Phil H. Butler

CONTENTS

9.1 INTRODUCTION

Spectral molecular computed tomography (CT) is a new imaging modality that allows molecular information to be obtained from the energy of x-ray photons. The energy information can be used to find the location of molecules containing highly attenuating atoms. Measuring the quantity and movement of such molecules is a key part of "molecular imaging." In the field of medical imaging, molecular imaging is defined as the visualization, characterization, and measurement of biological processes at the molecular or cellular level in humans and other living systems [1]. By using contrast agents targeted at specific biological processes, spectral molecular CT can achieve molecular imaging. Other applications of spectral molecular CT include soft-tissue imaging, heavy-metal imaging, and imaging of heart disease. This chapter details the use of energy-resolving photon detectors as the key technology for achieving spectral molecular CT.

The attenuation of x-rays through a material varies with energy depending on the physical properties of the material, most importantly the atomic number (Z) of the atoms. Importantly, for each element there is a discontinuity in attenuation at the photon energy corresponding to the binding energy of the electron shells. The most common of these is called the K edge, where the photon energy equals the binding energy of the K-shell electrons. This attenuation change is shown in Figure 9.1 for common CT imaging agents (calcium, iodine, gadolinium, water, and gold). As the binding energy depends on the atomic number, each atom can be separated by measuring the energy location of its K edge. This is used for the material characterization (identification and quantification) of radiographic contrast agents (iodine, gadolinium, gold) and bone (calcium) [2]. Spectral molecular CT has also shown promise in analyzing soft-tissue contrast between fat, muscle tissue, and water [3].

The key technology used for spectral molecular CT is the energy-resolving photon-counting detector. This chapter deals with the Medipix family of photon-processing detectors. The Medipix chips were designed at Conseil européene pour la recherche nucléaire (CERN) and are based on the technology used for measuring high-energy interactions in the large hadron collider (LHC). They are advanced detectors that count individual photon events and categorize them based on their measured energy. The resulting energy bins can be analyzed using material decomposition techniques. Maximum likelihood error (MLE) [4], principal components analysis (PCA) [5,6], self-organizing maps (SOM) [7,8], independent components analysis (ICA) [9], and proprietary techniques have all been used for material decomposition. The results of the use of these techniques allow a color map to be applied to different materials for improved visibility. Figure 9.2 shows how a SOM was used on two radiographs (10, 35 keV) of a mouse containing contrast material in its lungs (barium) and arteries (iodine). The bone was colored blue, and the contrast was colored green. Material

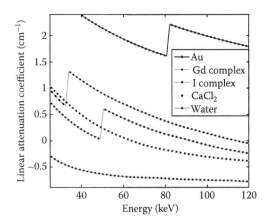

FIGURE 9.1 (**See color insert**) Linear attenuation of $CaCl_2$ (K = 4.03 keV), I complex (K = 33.17 keV), Gd complex (K = 50.24 keV), water (K < 1 keV), and Au (K = 80.72 keV).

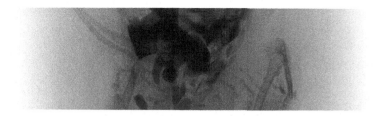

FIGURE 9.2   (See color insert) A spectral radiograph of a mouse, taken with a Medipix3.0 ASIC bonded at 55 μm to 300 μm of silicon. The bone has been colored blue, and the contrast material has been colored green. (From Ronaldson, J.P., et al., *J. Instrum.*, 6, C01056, 2011; Walsh, M. F., et al., *J. Instrum.*, 6, C01095, 2011.)

decomposition algorithms are more effective when applied to 3-D reconstructed images using CT techniques and have been able to distinguish iodine from barium [2]. These materials are only 4 keV apart.

The traditional unit for measuring CT with x-rays is the Hounsfield unit (HU). However, this is a qualitative measure of the attenuation at the location depending on the system parameters, and it is known to vary with energy and material. Since photon-counting detectors have the potential for molecular imaging with quantitative measurements of materials, the HU is no longer appropriate. The alternative units that should be used are the concentration of materials or linear attenuation prior to material analysis.

The Medipix All Resolution System (MARS) small-animal scanners use Medipix technology to obtain 3-D spectral x-ray images. The MARS research group is a collaboration of the University of Canterbury, New Zealand, and the University of Otago, Christchurch, New Zealand. It is a multidisciplinary team of engineers, physicists, mathematicians, biologists, chemists, radiologists, and other medical personnel working on the research and development of spectral molecular CT.

9.2   APPLICATIONS OF SPECTRAL MOLECULAR CT

Material identification and quantification are key applications of spectral molecular CT. In conventional CT, the components of soft tissues are difficult to distinguish from one another, because the average attenuation of different soft-tissue components is very similar when using an energy-integrating detector. One solution to this problem is to increase the attenuation of certain soft-tissue components by introducing non-specific contrast agents into the region of interest, usually by intravenous injection. The most common CT contrast agents are based on iodine. The contrast can be seen in the blood pool and the extracellular space. For instance, tumor deposits in the liver are made conspicuous due to differences in the uptake of the iodine-based contrast injected intravenously, while in the case of deep-venous thrombosis, blood clots from leg veins breaking off to the pulmonary arteries as emboli are made conspicuous by the contrast-laden blood around them. Barium suspensions are another common contrast agent, useful for gastrointestinal tract imaging. In a number of CT examinations, pre-contrast and postcontrast imaging is necessary; for example, for differentiating benign tumors from malignant ones. By using photon-processing detectors such as Medipix, virtual noncontrast images can be calculated with only one postcontrast image.

Some of the initial work on the MARS project involved differentiating iodine contrast from natural soft tissue and identifying calcium in a mouse. One study shows that iodine can be separated using a Medipix2 chip (MXR) with a maximum-likelihood estimation method [4]. The MXR chip has also been used in a MARS-v2 scanner to image a mouse with barium- and iodine-based contrast agents [2]. The Philips research group has created a spectral scanner with a single pixel line of cadmium telluride (CdTe). This has been used to distinguish iodine from gadolinium in a poly(methyl methacrylate) (PMMA) phantom [10]. A chip manufactured by Siemens [11] was able to spectrally separate iodine, gadolinium, and calcium in an acrylic phantom. The advantage of material decomposition thus obtained is further demonstrated in some application-specific results below.

Spectral molecular CT is promising in terms of improved soft-tissue contrast. One application of this is nonalcoholic fatty liver disease (NAFLD). This is the liver component of "metabolic syndrome," a collection of diseases associated with obesity [12], and is present in 17%–33% of American adults [13]. The ability to distinguish fat from liver material would greatly help improve the noninvasive diagnosis of NAFLD. One study shows the ability to distinguish fat from liver in projection images [14]. Another study shows that fat can be separated from water with low-energy imaging using a silicon Medipix3.0 [3].

Gold nanoparticles (AuNP) are showing promise as a contrast agent for molecular imaging. In one study, apolipoprotein E-knockout (E-KO) mice with high-fat diets are compared to wild-type mice, and the system shows that iodine, gold, and calcium are all separable in these mice. The concentration of AuNP in this study is about 4 mg/ml, and there is major concern over toxicity when moving to human spectral molecular CT with AuNP [15]. The imaging of AuNP has also been shown to be helpful in diagnosing cancerous tissue in the neck [16].

One of the leading causes of death today is heart disease. Rupture of atherosclerotic plaque formations are the cause of 70% of heart attacks [17]. The inflammatory cells in atherosclerotic plaque are a major factor in making the plaque unstable and likely to rupture. A macrophage-binding AuNP contrast agent, Au-HDL, has been used to image the inflammatory components of plaque in rabbits [18]. Furthermore, a numerical phantom of an atherosclerotic coronary vessel with gadolinium contrast has been simulated [19]. It was found that the gadolinium and the calcium in the plaque were separable. This was further demonstrated in an experimental setup wherein a Philips spectral scanner was used to scan a PMMA atherosclerotic phantom [20]. The phantom had a simulated calcium plaque and a gadolinium-enhanced lumen. In this study, the calcium and the gadolinium had identical HU attenuation numbers using the conventional detectors but were quite separable using photon counting spectral detectors. Another study shows that low-energy silicon Medipix3.0 chips can be used in small-animal preclinical CT to quantify the fat, calcium, and iron of an excised human atherosclerotic plaque [3].

9.3   PHOTON-COUNTING DETECTORS

Traditional CT detectors are charge-integrating detectors. The signal received from a single pixel from traditional charge-integrating detectors is proportional to the full charge of all the photons. The Medipix detectors, however, are photon-counting detectors, where each photon gives one count, independent of charge. Because the Medipix detectors do not integrate charge, low-energy photons are just as visible as high-energy photons. This means that, compared to traditional charge-integrating detectors, the photon-counting detector has an energy weighting toward the low-energy photons. The ability to count and quantify low-energy photons is essential for soft-tissue contrast.

It has been shown that even without energy-resolving capabilities, photon-counting chips have inherent improvements over conventional CT. For example, a mammographic phantom was scanned with conventional charge-integrating detectors and a Medipix1 [21]. The Medipix1 photon-counting chip had the highest signal-to-noise ratio (SNR), as well as a better modulation transfer function (MTF) than all but one of the detectors. The MTF is an indicator of spatial resolution. This other detector was a similar digital charge-coupled device (CCD) chip with finer pixel resolution. This higher SNR also implies that a lower dose could be applied to achieve an image of similar quality. This has also been demonstrated in the field of dental x-ray imaging [22]. A conventional CCD detector and a Medipix1 chip imaged a tooth with two fillings at different radiation levels. At 40 µGy, the fillings were not visible in the CCD detector, but were clear in the Medipix1. Another study was performed with a custom Siemens spectral x-ray detector [11]. This showed a 10% increase in contrast-to-noise ratio (CNR) in the photon-counting chip, despite having 23% less exposure. This shows that, even without the energy resolution of Medipix chips, the photon-counting approach gives improved CT imaging.

This chapter deals with the Medipix photon-counting detector because it is used by the MARS group. A comparison of photon-counting detectors [23] is shown in Table 9.1. The Medipix3RX has the smallest pixel size of all the chips presented, as it has a focus on imaging small animals. The Medipix3RX is also the only detector with full anti-charge-sharing capabilities. The other chips have less need for anti-charge-sharing, as the larger pixel size makes charge sharing less likely. The choice in pixel size has a trade-off, where increasing the pixel size results in less charge sharing but worse spatial resolution. The pulse pileup [24,25] effects are also based on count rate, which is higher for larger pixel sizes. The ChromAIX and the KTH silicon strip have the best count-rate performance of the listed chips. The Medipix3RX is tied with the KTH silicon strip for the most counters, although the KTH silicon strip may be able to infer more information from the depth of interaction. When the Medipix3RX is in charge summing mode, it only has 4 charge-summing counters, and the GMI CA3 also outperforms it. Of all the listed detectors, the DXMCT-1 and the DXMCT-2 have the best tiling support for covering large areas.

9.3.1 MEDIPIX

The Medipix chips are hybrid technology detectors. They consist of an application-specific integrated circuit (ASIC) connected on a per-pixel level to a reverse-biased sensor layer. The major families of the Medipix are the Medipix1, Medipix2, and Medipix3 families. The Medipix1 was a proof of concept for energy-resolving

TABLE 9.1

Comparison of Photon-Counting Detectors That Have Been or Are Being Developed

Index	Name/ASIC	Operation Mode	Maximum Count Rates (Mcps/pixel)	Pixel Size (μm × μm)	Maximum Count Rates (Mcps/mm²)	No. of Energy Thresholds per Pixel	Tileup Capability	Anticharge Sharing
1	DXMCT-1 [26,27]		5.5	1000 × 1000	5.5	2	2-D	No
2	DXMCT-2 [28]		5.5[a]	500 × 500	22[a]	4	2-D	No
3	Siemens 2010 [29,30]		NA	225 × 225	NA	2 or 4[b]	NA[c]	No
4	ChromAIX [31]		13.5[d]	300 × 300	150[d]	4	1-D	No
5	Hamamatsu [32,33]		1–2	1000 × 1000	1–2	5	1-D	No
6	GMI CA3 [10,20]		1–2	400 × 1000	2–5	6	1-D	No
7	Medipix3RX [34–36]	FM[e]-SPM[f]	0.21[g]	55 × 55	69.4[g]	2	1-D with 2 × N (3-side buttable)	No
		FM[e]-CSM[f]	0.036[g]	55 × 55	11.9[g]	1		Yes
		SM[e]-SPM[f]	0.145[h]	110 × 110	12[h]	8		No
		SM[e]-CSM[f]	0.034[h]	110 × 110	2.8[h]	4 + 4[i]		Yes
8	CIX [37]		3.3	250 × 500	26	1	NA	No
9	Nexis Detector [38,39]		2.0	1000 × 1000	2.0	5	1-D	No

| 10 | MicroDose SI (Silicon strip) [40–42] | 0.056[j] | 50 × 50 | NA | 2 | 1-D | Yes[k] |
| 11 | KTH Silicon strip [43–47] | 2.5 or 7.5[l] | 400 × 500 | 200 or 600[m] | 8[n] | 2-D[o] | No |

Source: Taguchi, K. and Iwanczyk, J.S., *Med. Phys.*, 40, 100901, 2013.

Note: Maximum count rates are measured at full saturation unless otherwise specified. This is not a complete list of all available PCDs, but it provides a comprehensive review of different specifications.

[a] Measured at 15% dead-time loss.

[b] Two per pixel, effectively up to four in chess-pattern mode [29,30].

[c] $2 \times N$ tiling must be possible, as 128 detector rows are formed by detector blocks with 64×64 pixels [29].

[d] Measured in electronics tests without detectors.

[e] Fine pitch mode (FM) and spectroscopic mode (SM).

[f] Single pixel mode (SPM) and charge summing mode (CSM).

[g] Measured with 300 μm thick silicon at 10% dead-time loss.

[h] Measured with 2 mm thick CdTe at 10% dead-time loss.

[i] Four thresholds for summed charge, four for local charge.

[j] Measured at 10% dead-time loss.

[k] Coincident detection and store the counts to one of the counters with no charge summing.

[l] Count rates per layer with 16 layers per pixel. 2.5 Mcps/layer with 0.2% dead-time loss, 7.5 Mcps/layer with 25% loss. The maximum output count rates not measured yet.

[m] Measured with 120 kVp. $200 = 2.5 \times 16/(0.4 \times 0.5)$ with 2% dead-time loss and $600 = 7.5 \times 16/(0.4 \times 0.5)$ with 25% loss. The detection efficiency of 30 mm silicon is 80%.

[n] Each detector pixel has 16 layers along the depth direction, and each layer has 8 energy thresholds. Energy information may also be available from depth of interaction, in which of the 16 layers the interaction took place.

[o] <2% dead area after tiling [48].

photon-counting imaging. It is limited by a larger pixel size (170×170 μm²), and by the fact that it is only able to count positive pulses.

The Medipix2 family consists of the Medipix2 [49], the MXR (Medipix2.1), and the Timepix [50]. From this family onward, the chips have 256×256 pixels at 55×55 μm² for a total size of 14.08×14.08 mm². The pulse-processing technology is capable of inverting the processed pulses, allowing for both positive- or negative-pulse collection depending on the transportation efficiency of the sensor layer.

The Medipix3 family consists of the Medipix3.0 [34], the Medipix3.1 [51], and the Medipix3RX (Medipix3.2) [52]. It is the first Medipix family to have interpixel communication and multiple energy images per exposure. Each 55×55 μm² pixel has two energy counters. Every 2×2 cluster of pixels can be combined to create 128×128 superpixels at 110×110 μm². Each of these superpixels has eight energy counters. This interpixel communication network can also be used to combine a charge that has been shared over multiple pixels, allowing for a more accurately measured spectrum. In order to use the eight-counter mode on the Medipix3 family, the interconnections between the sensor layer and the ASIC must be bonded to only one in every 2×2 pixels.

The sensor layer is a semiconductor doped in a diode arrangement. A high voltage is reverse biased to the sensor, creating a depletion region. When a photon interacts with the sensor, an ionizing electron is ejected. This electron creates electron–hole pairs in the depletion region, and these are separated to the respective electrode under the influence of the electric field strength from the high-voltage bias. This is shown in Figure 9.3. This separation of electron–hole pairs induces a current across the ASIC pixel terminals that is proportional to the energy of the original photon event. This current is compared to a threshold, and, if above, increments a corresponding counter to register the hit.

The choice in sensor layer is important for improving the image quality and decreasing the radiation dose when measuring high-energy photons. Silicon (Si), gallium arsenide (GaAs) [54–56], cadmium telluride (CdTe) [57–59], and cadmium zinc telluride (CZT) are all semiconductors that have been used with the Medipix ASIC. The absorption efficiency depends on the electron density of the atoms in the sensor (equivalent to the effective atomic number (Z)), and the sensor thickness. Si ($Z = 14$) is a widely available sensor, but is only efficient below 30 keV. This means it is not very useful except for imaging small animals and low-energy K edges of contrast agents. GaAs ($Z = 31, 33$), CdTe ($Z = 48, 52$), and CZT ($Z = 48, 30, 52$) are efficient sensors capable of imaging in the human diagnostic energy range up to 120 keV. The availability of nonsilicon sensors is often low. The semiconductor growth and subsequent process of bonding to the ASIC is still an active area of research. This means the spatial quality of the images can be degraded, and acquiring large numbers of these sensors is problematic.

9.3.2 ENERGY RESOLUTION

When performing spectral molecular CT, it is important to be both specific and sensitive. Sensitivity measures the proportion of a material that has been correctly

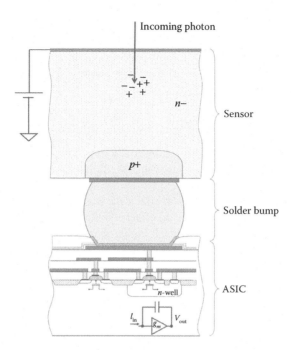

FIGURE 9.3 The side view of a Medipix pixel showing the semiconductor sensor layer and the ASIC (not to scale). Different semiconductor materials can be bump bonded to the ASIC to act as an x-ray detection layer. When a photon is detected, an electron–hole pair is created. The bias voltage applied across the sensor layer drives the charge clouds to the respective electrodes. Within the ASIC, the photon is individually analyzed and can be sorted into multiple energy bins. (From R. Ballabriga and M. Vilass-Cardona Campbell. The design and implementation in $0.13\text{mu}\ \text{m\$}$ CMOS of an algorithm permitting spectroscopic imaging with high spatial resolution for hybrid pixel detectors. PhD Thesis, Ramon Llull University, Barcelona, 2009.)

identified, while specificity measures the proportion of identifications that are correct. Since spectral molecular CT is dependent on energy-resolving photon-counting detectors, the accuracy of the energy measurements is important for both sensitive and specific imaging. The overall accuracy of the energy information is defined by the energy resolution of the chip. The energy resolution is a function of threshold dispersion across pixels, quantization of thresholds, noise on the threshold discriminators, and gain dispersion across pixels. Charge sharing is another source of energy inaccuracy, and is worse in instances of smaller pixels and thicker sensor layers. Pulse pileup also affects the energy accuracy of the detector, and is worse in the case of high count rates.

9.3.3 THRESHOLD DISPERSION AND EQUALIZATION

Due to limitations in the manufacturing process of the Medipix ASIC, each pixel responds to a uniform energy source at a different threshold. This difference in threshold is known as the threshold dispersion of the chip. To improve the threshold dispersion, each pixel has a 3–5-bit adjustment digital to analog converter (DAC) per counter, which allows for the alignment of the thresholds to a set value. The process of selecting the correct adjustment values per pixel and the resulting configuration are known as the threshold equalization process and the equalization configuration, respectively. The term equalization is used to describe both of these, depending on the context. Usually, the equalization is performed using the noise floor of the detector circuitry as a reference energy. Even after equalization, a residual threshold dispersion remains. The equalization does not affect the gain dispersion across the pixels, and this can only be minimized by equalizing close to the energy of interest. The gain dispersion has been measured in a Medipix3RX to be 6.1%–6.5% across different counters [60].

Equalization is important because it affects the energy resolution across the entire detector. The energy resolution also has effects on the spatial quality of the detector. If one pixel is measuring energies much higher than its neighbor, then it will have significantly lower counts. Figure 9.4(a) shows a radiograph of a USB drive with an unequalized quad-Si Medipix3.0 assembly, while Figure 9.4(b) shows the same result after equalization [61].

9.3.4 THE CHARGE-SHARING PROBLEM

When a photon interaction occurs in the sensor layer, it ejects an ionizing electron into the sensor layer. This electron ionizes its neighbors, causing an electron–hole pair charge cloud to form. This diffuses outward as it is driven by the high-voltage electrical field to the pixel terminals. This diffusion, when located near the boundaries between pixels, can cause the charge cloud to be registered in parts over two or more pixels. This effect is called charge sharing.

When charge sharing occurs in a Medipix ASIC, each of the affected pixel terminals generates a pulse. The pulse height analysis circuitry measures the result, and increases the counter if it is above its threshold. Since the charge from the photon events is split, the primary charge is smaller than it otherwise would be, and a lower threshold is needed to measure the event. At thresholds of very low energy (~10 keV,

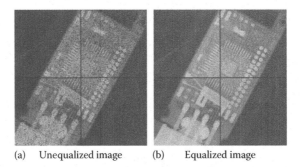

(a)   Unequalized image   (b)   Equalized image

FIGURE 9.4 A comparison of the flat-field corrected images both (a) before and (b) after equalization. (From Walsh, M.F., et al., *Nuclear Science Symposium and Medical Imaging Conference (NSS/MIC), 2011 IEEE*, pp. 1718–1721, 2011.)

depending on x-ray settings), the counts are greatly increased as the lower-intensity pulses are measured. Over a large scale of interactions, this causes the peak of the measured spectrum to be distorted to a lower energy, and causes a high number of counts at the low energy. This high number of low-energy counts is called the charge-sharing tail.

To increase both the sensitivity (the proportion of the object that has been correctly identified) and specificity (the proportion of identifications that are correct) of spectral molecular CT, the charge-sharing problem needs to be addressed. In the Medipix3 family, the interpixel communication allows for charge summing mode (CSM) to be enabled. The basic principal of CSM is to combine the charge from a 2×2 pixel (or 2×2 superpixel) area and assign the combined charge to the pixel with the highest original charge. Figure 9.5 shows the spectrum of molybdenum and indium foil fluorescence, and Am-241 γ-emissions across the whole chip, with and without CSM.

In the Medipix3.0 and Medipix3.1, there is a problem with the charge-allocation algorithm, which causes some pixels to be assigned the charge more times than they should have been [62,63]. This is due to the residual threshold dispersion between pixels, even after equalization. These pixels have a lower threshold, and therefore measure a higher charge than their neighbors. This causes an effective spatial inhomogeneity of counts across the detector and significantly reduces the dynamic range available. The resulting CSM images are of poor quality; however, images without CSM enabled have good spatial quality.

The Medipix3RX has an improved charge-summing algorithm [52]. It requires a non-charge-summing counter to be active. This counter will count photon events as though CSM is not enabled, but will still only assign each photon event once to the pixel with the highest charge. The purpose of this counter is to facilitate the correct assignment of photons in the CSM counters, and it is therefore called the arbitration counter. The other counter has CSM enabled, but only assigns the combined charge to the pixel that counted in the arbitration counter. By using this, the Medipix3RX can obtain images with spectral properties like the CSM measurements in Figure 9.5,

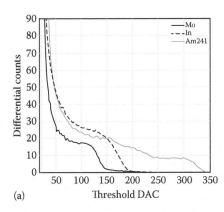 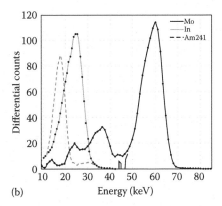

(a)   Threshold DAC   (b)   Energy (keV)

FIGURE 9.5 Molybdenum and indium K_α fluorescence and Am-241 γ-emissions scanned with a Medipix3RX at $110 \times 110 \ \mu m^2$ pixels with (a) CSM disabled and (b) CSM enabled.

but with high spatial quality. The only downside of this technique is that it requires the use of a counter specifically for charge arbitration, and halves the number of energy thresholds available for simultaneous use.

9.4 MARS SCANNERS

In order to acquire spectral x-ray images using the Medipix detectors, the MARS research group has built a series of small-animal CT scanners [64]. These scanners have a rotating gantry around a movable sample bed. The camera can be moved up and down to increase the field of view, and the source and camera can be moved in and out to control magnification. In the MARS-v3 and MARS-v4 scanners, the rotation could only move over one full rotation (~370°). In the MARS-v5 scanners, the gantry uses a slip ring to allow for unlimited rotations. The MARS scanners use Source-Ray 80 or 120 kVp x-ray sources. The scanners use the custom-built MARS camera to interface with the Medipix3 technology and with a python controller interface on a Linux computer.

9.4.1 MARS CAMERA

As part of the development of a full Medipix3.0 CT scanner in 2009, the MARS group created the MARS camera [65]. The MARS camera has a gigabit Ethernet connection for communication and has support for Medipix3.0, Medipix3.1, and Medipix3RX ASICs. The MARS camera can interface with CERN chip-carrier boards, the custom MARS "hexa" carrier boards with tiling of up to six Medipix detectors, or the new MARS fingerboards. The fingerboards are designed to hold one detector and they align side by side, with many fingerboards allowing for unlimited tiling of many MARS cameras. The MARS camera also has a high voltage supply for driving the reverse sensor bias, and Peltier and fan cooling to keep the ASIC temperature stable. Current MARS cameras use CdTe fingerboards in spectroscopic mode and CSM, and can image four energy counters at 7 frames per second.

9.5 CURRENT RESEARCH OF THE MARS TEAM

9.5.1 ATHEROSCLEROSIS AND LIMITATIONS IN CURRENT DIAGNOSTICS

Cardiovascular disease (CVD) is one of the leading causes of death and a major cause of hospitalization in the Western world [66–68]. Fatal events such as myocardial infarctions (heart attacks) and strokes are typically caused by the rupture of vulnerable or unstable atherosclerotic plaques and consequent vascular blockage [69]. Unstable plaques consist of a soft lipid pool covered by a thin layer of fibrous cap, whereas stable plaques typically have a thick fibrous cap [70]. Early detection of plaque vulnerability is critical for preventing the severe downstream effects of heart disease. Atherosclerotic plaques can rupture as a result of the breakdown of the fibrous cap that covers the lipid core. The unstable plaque is prone to inflammatory changes near this thin fibrous cap, which can weaken the fibrous cap and lead to rupturing. Atherosclerosis accounts for ~70% of fatal acute myocardial infarctions or sudden coronary deaths [71].

Atherosclerosis (the progressive narrowing and hardening of arteries) and thrombosis (the rupture of vulnerable plaques) are commonly diagnosed by angiography (invasive coronary imaging) or by intravascular ultrasound. However, the detection of an unstable (vulnerable) atherosclerotic plaque is limited by either a lack of sensitive imaging modalities or by invasive procedures. Positron emission tomography (PET) is sensitive but nonspecific and has limited spatial resolution. It only allows for the identification of the most severely obstructive plaques [72]. It is also limited due to the low uptake of the radioisotope by the atherosclerotic plaque. PET requires radioactive pharmaceuticals to be made shortly before imaging, which requires the expensive addition of an on-site cyclotron. Both bioluminescence and fluorescence molecular-imaging techniques are very sensitive but are not translatable to human imaging due to their poor penetration. It has also been shown that they may not be able to determine the inflammatory status of plaques [73]. Angiography is an invasive procedure and may cause infection, pain, or bleeding at the site of catheter insertion. Catheters are made of plastic and can cause trauma or damage to the blood vessels. The current practice for CVD diagnosis is to perform a preliminary CT scan and then refer patients for either angiography or PET. Although the inflammatory status of the atherosclerotic plaque may be determined by the combination of the above methods, it would be best to have a single imaging modality with improved diagnostic ability and lower risk.

As part of a study on atherosclerosis, excised plaques removed from patients were imaged in a MARS-v4 scanner [70]. The samples were cut into 3 mm segments and imaged at 10, 16, 22, and 28 keV. To confirm the spectral analysis of the images, a histological comparison was done with staining materials. The plaques were then stained with Von Kossa (for calcium detection), Perl's Prussian blue (for iron detection), and Oil-Red O (for lipid detection). Material decomposition was performed on the spectral data using a linear matrix equation [3], although the iron component could not be separated from the calcium. Figure 9.6 shows both the histological samples of one of the atherosclerotic plaques, and the decomposed images of water-like, lipid-like, and calcium-like materials.

This work shows that spectral molecular CT has the potential to noninvasively evaluate the internal structures of vulnerable plaque. Future work involves using high-Z sensors with higher energy thresholds, allowing the separation of the calcium and iron, and the quantification of AuNP bound to biomarkers.

9.5.2 Spectral Imaging for Material Sciences

Not all benefits of photon-counting detectors are limited to spectral molecular CT. Photon-counting detectors can be used to reduce artifacts in material science applications. Material science research aims at developing novel biomaterials that will assist or replace the function of anatomical structures in the human body. The functionality of such materials and assistive devices vary from biomechanical support (e.g., hip replacements) to biocontrol systems (e.g., artificial pacemakers). A common example is the treatment of osteoarthritis of the knee with knee replacements, where

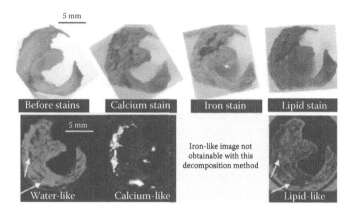

FIGURE 9.6 (**See color insert**) Stained histological images for calcium, iron, and lipids; and spectral molecular CT images of water-like, calcium-like, and lipid-like structures. (Reproduced from Zainon, R., et al., *Eur. Radiol.*, 22, 2581–2588, 2012.)

defective bone and cartilage tissues are replaced by biomaterial components. These components are made of novel materials containing polymers, ceramics, metals, and other composites, and their development is an active area of research. Once they have been implanted into the human body, it is essential to carry out revision in order to identify any adverse reactions, such as implant failure (fracture or loosening), inflammatory responses, infections, and dislocations. Radiological examination plays a vital role in the process of identifying these reactions. In tissue engineering, volumetric information is extracted by imaging ex-vivo or tissue-engineered samples to study biocompatibility or material design aspects. X-ray imaging allows the non-destructive analysis of porous surface materials and tissue ingrowth, thus providing a platform to characterize scaffolds and assistive structures in terms of size, shape, and porosity. Innovative detectors play an important role in improving the quality of examination methods [74].

X-ray tomography becomes challenging when dense-metal implants (e.g., arthroplasty hip prostheses) are imaged. The data acquired become corrupted with dense artifacts in the form of streaks and bands. The main reasons for these are beam hardening and photon starvation [75]. Beam hardening is a physical phenomenon whereby the mean energy of the incident polychromatic x-ray beam undergoes nonlinear changes as it penetrates through highly attenuating objects. Dense materials greatly absorb low-energy photons or soft x-rays and allow high-energy photons or hard x-rays to penetrate without significant attenuation. As a result, the outcoming beam is hardened and sometimes results in empty projections in the data. This leads to incorrect reconstructions and inhomogeneities in metal regions, known as cupping artifacts. Most of these artifacts not only corrupt the metal region, but also the surrounding structural information. Numerous hardware methods and mathematical corrections [76] exist to deal with beam hardening and metal artifacts. Hardware filters are commonly used to preharden the beam, that is, to remove low-energy photons from the beam. However, this process results in poor signal-to-noise ratio, as the soft x-rays provide good contrast information.

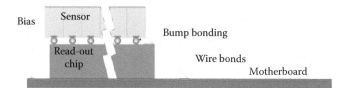

FIGURE 1.1 The hybrid detector consists of two independent parts: a sensor and a read-out chip. These two parts are joined together by bump bonding. The advantage of this technology is that the sensor material can be tailored to the specific application.

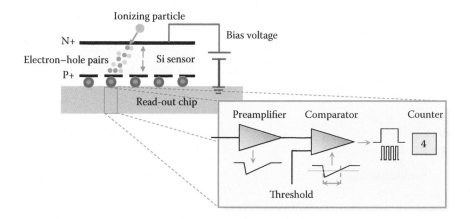

FIGURE 1.3 The schematic illustration of TOT mode. The signal from the electrode is amplified and compared in the comparator. The time the signal is over the comparator level (threshold) represents the energy of the interacting particle. The collected charge is compared in each pixel with the threshold level. If it is lower, the event is not registered and the signal is lost. As a consequence, the charge registered by all pixels is often lower than its original value, which deteriorates the energy resolution.

FIGURE 1.5 The picture taken aboard the ISS shows the REM deployed on a station support computer. The REM is emphasized in a gray box and shown in detail to the right, plugged into the station laptop.

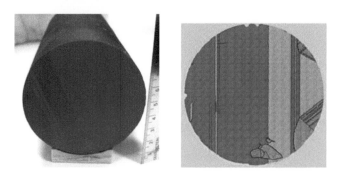

FIGURE 2.3 Cross section of the THM ingot from Figure 2.2 and electron backscatter diffraction (EBSD) orientation map of the adjacent slice revealing twin (red) and grain boundaries (black).

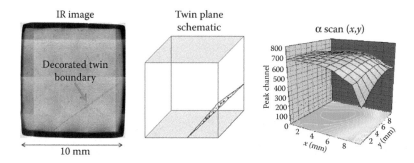

FIGURE 2.4 Effect of a decorated twin plan on charge-collection efficiency. (From Soldner, S.A., Narvett, A.J., Covalt, D.E., and C. Szeles, Characterization of the charge transport uniformity of high pressure grown CdZnTe crystals for large-volume nuclear detector applications, Presented at IEEE 13th International Workshop on Room-Temperature Semiconductor X- and Gamma-ray Detectors, Portland, OR, 2003.)

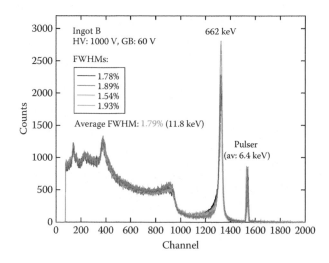

FIGURE 2.12 ^{137}Cs spectra taken with four different CZT detectors in CPG configuration.

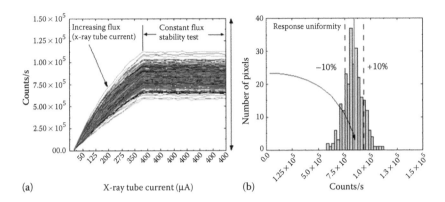

(a) X-ray tube current (μA)   (b) Counts/s

FIGURE 2.14 Test result for a typical 16×16 pixel CdZnTe monolithic detector array. (a) The count rate as a function of the x-ray flux (tube current), which is followed by stability and uniformity tests at a constant maximum tube current. (b) The count distribution at maximum current indicating the response uniformity of the device. (From Szeles, C., Soldner, S.A., Vydrin, S., Graves, J., and Bale, D.S., *IEEE-TNS* 54, 1350, 2007.)

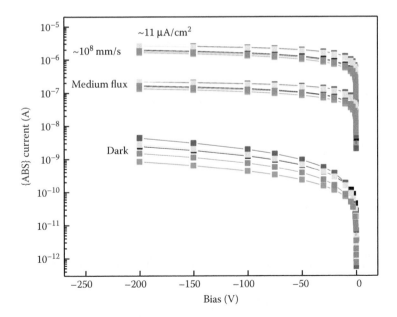

FIGURE 2.20 Bias dependence of dark and x-ray-induced steady-state currents from parallel-plate detectors of different CZT ingots (different colors).

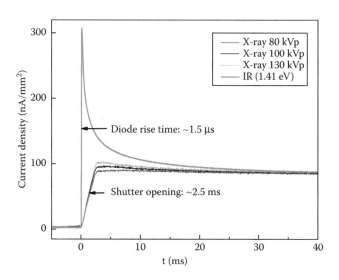

FIGURE 2.23 X-ray vs. IR-induced photocurrent transients.

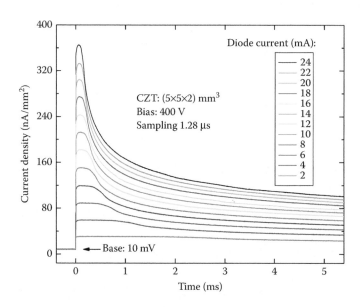

FIGURE 2.24 Example of the IR flux dependence of the photocurrent transients at constant bias.

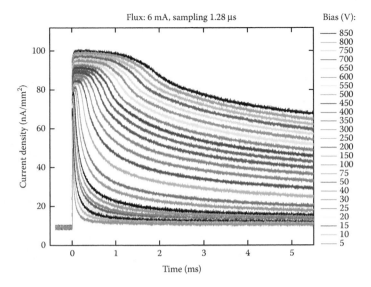

FIGURE 2.25 Example of the bias dependence of the photocurrent transients at constant IR flux.

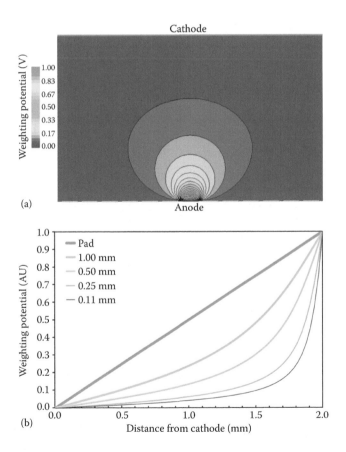

FIGURE 3.1 (a) A simulation of the weighting potential of a 250 μm pitch pixel in a 2 mm thick CdTe detector produced using Sentaurus TCAD. (b) A cross section of the weighting potential for varying pixel pitches in the range 0.11–1.00 mm.

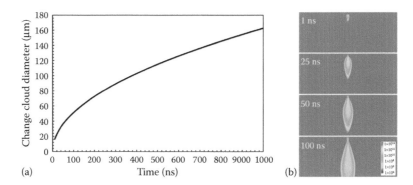

FIGURE 3.2 (a) The variation of charge-cloud diameter with drift time for a CdTe detector under −300 V bias. (b) A TCAD simulation of the dispersion of the electron cloud created by a 60 keV interaction in a 2 mm thick, 5 mm wide CdTe detector under the same bias voltage. The color scale depicts the electron density within the detector.

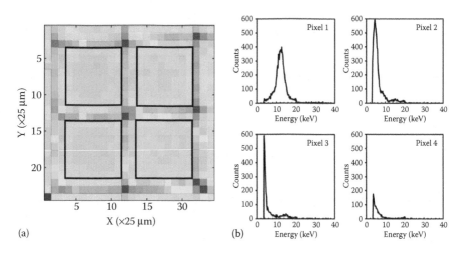

FIGURE 3.3 (a) A map of the measured x-ray intensity on a small pixel CdTe detector measured in steps of 25 μm. Data were taken with a 10 μm × 10 μm beam of 20 keV x-rays at the Diamond Light Source, UK. (b) The x-ray spectra measured in the four pixels when the beam is positioned at X12, Y13. (Adapted from Veale, M.C., Bell, S.J., Seller, P., Wilson, M.D. and Kachkanov, V., *J. Instrumentation*, 7, P07017, 2012.)

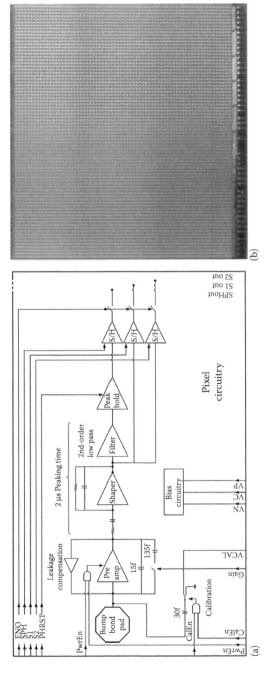

FIGURE 3.6 (a) A block diagram of the HEXITEC ASIC pixel electronics. (b) The 80×80 HEXITEC ASIC: the bottom edge of the ASIC is dedicated to I/O connections. (Adapted from Jones, L.L., Seller, P., Wilson, M.D. and Hardie, A., *Nucl. Inst. Meth. Phys. Res. A*, 604, 34–37, 2009.)

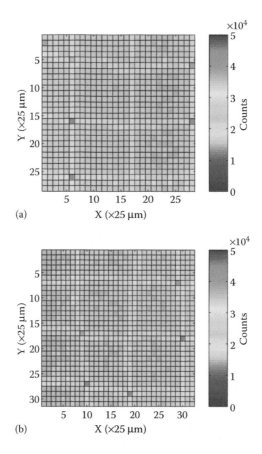

FIGURE 3.14 Maps of the measured x-ray intensity before charge-sharing correction in a good (a) and bad area (b) of a CdZnTe detector. The intensity maps were produced using a 10 μm × 10 μm beam of 20 keV x-rays in steps of 25 μm. (Adapted from Veale, M.C., Bell, S.J., Duarte, D.D., Schneider, A., Seller, P., Wilson, M.D., Kachkanov, V. and Sawhney, K.J.S., *Nucl. Inst. Meth. Phys. Res. A*, 729, 265–272, 2013.)

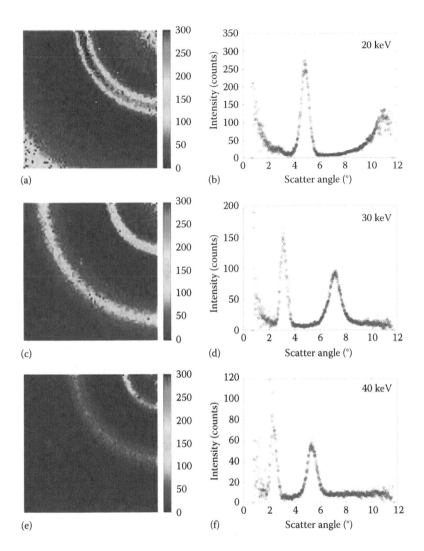

FIGURE 3.15 The x-ray diffraction signatures produced by a polycrystalline caffeine sample. The diffraction patterns and angular resolved signatures observed for photons at (a)–(b) 20 keV, (c)–(d) 30 keV, and (e)–(f) 40 keV. (From O'Flynn, D., Reid, C.B., Christodoulou, C., Wilson, M.D., Veale, M.C., Seller, P. and Speller, R.D., *Proc. SPIE*, 8357, 83570X-4, 2012.)

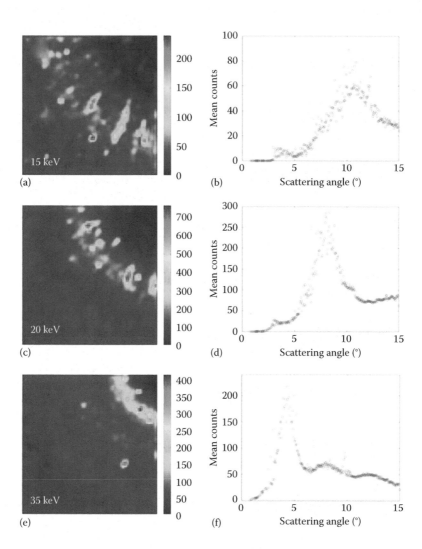

FIGURE 3.16 The x-ray diffraction signatures produced by an explosives simulant containing polycrystalline hexamine and pentaerythritol embedded in a plastic binding material. The diffraction patterns observed for photons at (a) 20 keV, (b) 30 keV, and (c) 40 keV. The angular resolved diffraction signatures detected at each energy are also shown in (d–f). (From O'Flynn, D., Desai, H., Reid, C. B., Christodoulou, C., Wilson, M. D., Veale, M. C., Seller, P., Hills, D., Wong, B., and Speller, R. D., *J. Crime Sci.*, 2, 4, 2013.)

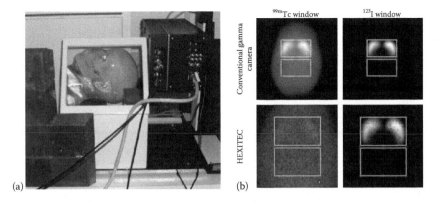

FIGURE 3.19 (a) Dual isotope SPECT imaging of an anthropomorphic brain phantom using the HEXITEC detector. The majority of the phantom is filled with ^{99m}Tc, while the striatal compartments are filled with ^{123}I. (b) A comparison of the imaging performance of a traditional GE Infinia γ camera and the HEXITEC detector system. Images are shown for energy windows around the principal emissions of the ^{99m}Tc and ^{123}I radioisotopes. The use of the HEXITEC detector greatly reduces the cross talk between the two images.

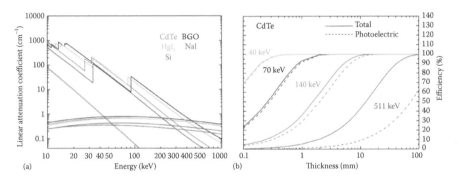

FIGURE 4.1 (a) Linear attenuation coefficients for photoelectric absorption and Compton scattering of CdTe, Si, HgI$_2$, NaI, and BGO. (b) Efficiency of CdTe detectors as a function of detector thickness at various photon energies.

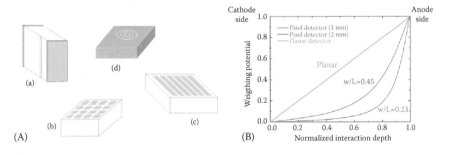

FIGURE 4.5 (A) Single-charge collection electrode configurations widely used in CdTe and CdZnTe detectors: (a) parallel-strip Frisch grid, (b) pixel, (c) strip, and (d) multiple electrodes. (B) Weighting potential of a pixel detector, compared with a planar detector. It is possible to improve the unipolar properties of pixel detectors by reducing the w/L ratio (i.e., pixel size to detector thickness), according to the theory of small pixel effect.

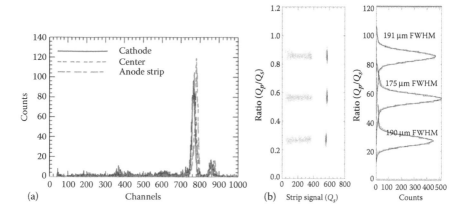

FIGURE 4.10 (a) Spectra of a collimated (0.6 mm spot) ^{57}Co source at three different positions between the collecting electrodes; the variation between the full energies peak and the corresponding energy resolution is within a few percent. (b) On the left, the biparametric distributions of the ratio between the planar electrode signal (Q_p) and the anode collecting-strip signal (Q_s) vs. Q_s for three positions of a 500 keV monochromatic x-ray beam; on the right, the corresponding measured depth resolution for these three different beam positions. The y-axis extension (ratio) is representative of the sensor interelectrode distance. The beam was collimated at 50 μm.

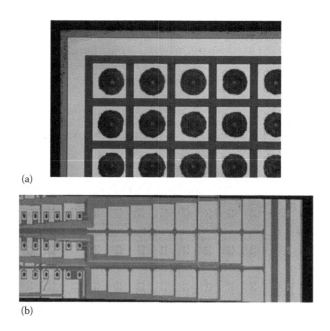

FIGURE 6.3 (a) Hexitec CdTe 250 μm pitch pad geometry with silver glue dots. (b) XFEL LPD redistributed bonding pads 250 × 400 μm² pitch (left) to the 500 μm pitch pixels on a two-layer interconnect silicon detector.

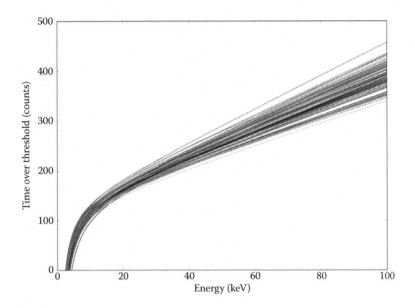

FIGURE 7.7 Calibration functions for several pixels in a Dosepix chip.

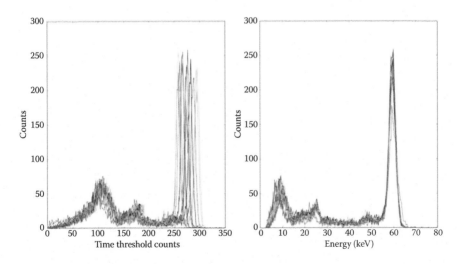

FIGURE 7.8 Measured photon energy spectrum for a [241]Am source for a few pixels in a Dosepix detector before and after calibration.

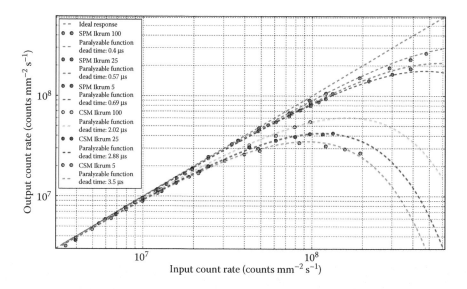

FIGURE 7.11 Count-rate behavior for Medipix3RX in single pixel mode (SPM) and charge summing mode (CSM) for different settings.

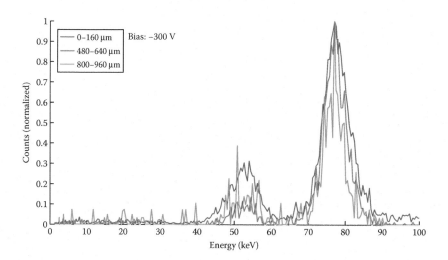

FIGURE 7.17 Time over threshold spectra at different depths in a 1 mm thick CdTe sensor bonded to Timepix. Clusters covering several pixels have been summed together and assigned to the pixel with the highest energy.

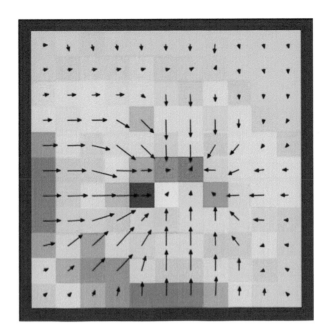

FIGURE 7.19   Arrowplot.

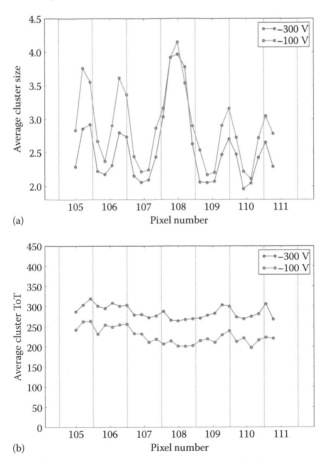

FIGURE 7.20   Line scan in 20 μm steps over a single pixel defect in the CdTe detector with 110 μm pixels. The vertical lines show pixel borders. (a) Cluster size; (b) cluster TOT.

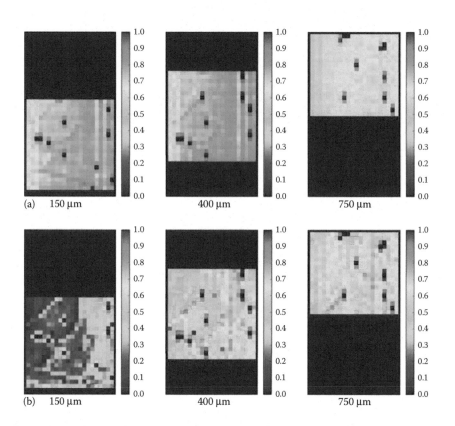

FIGURE 7.21 (a,b) Line defects at different depths in a 1 mm thick CdTe sensor for two different threshold settings: (a) 6 keV and (b) 60 keV.

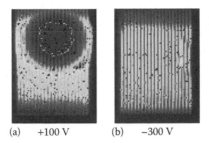

(a)   +100 V   (b)   −300 V

FIGURE 7.23   (a,b) Sum of all scan steps through the circular defect. Shown for both electron and hole collection mode.

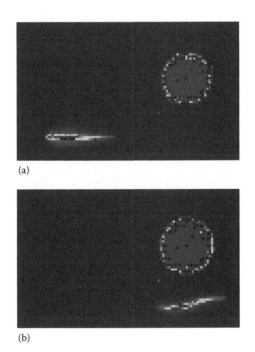

(a)

(b)

FIGURE 7.24   (a,b) A single scan step showing the displacement of the signal.

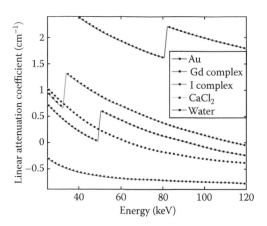

FIGURE 9.1 Linear attenuation of CaCl$_2$ (K = 4.03 keV), I complex (K = 33.17 keV), Gd complex (K = 50.24 keV), water (K < 1 keV), and Au (K = 80.72 keV).

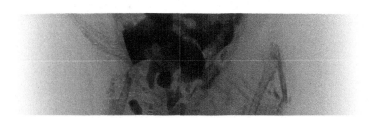

FIGURE 9.2 A spectral radiograph of a mouse, taken with a Medipix3.0 ASIC bonded at 55 μm to 300 μm of silicon. The bone has been colored blue, and the contrast material has been colored green. (From Ronaldson, J.P., et al., *J. Instrum.*, 6, C01056, 2011; Walsh, M. F., et al., *J. Instrum.*, 6, C01095, 2011.)

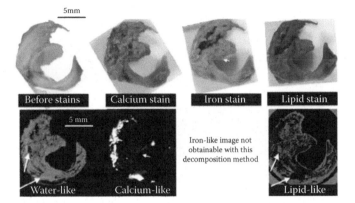

Before stains — Calcium stain — Iron stain — Lipid stain

Water-like — Calcium-like — Iron-like image not obtainable with this decomposition method — Lipid-like

FIGURE 9.6 Stained histological images for calcium, iron, and lipids; and spectral molecular CT images of water-like, calcium-like, and lipid-like structures. (Reproduced from Zainon, R., et al., *Eur. Radiol.*, 22, 2581–2588, 2012.)

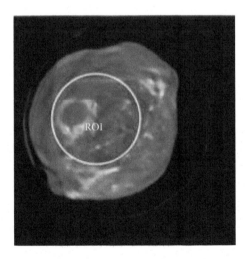

ROI

FIGURE 9.9 Spectral interior reconstruction and material decomposition of a mouse with 0.2 mL or 15 nm Aurovist II AuNP injected in a direct cardiac puncture injection. (From Xu, Q., et al., *Biomed. Eng., IEEE Trans.*, 59, 1711–1719, 2012.)

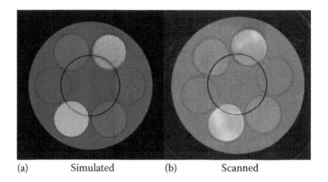

(a)   Simulated   (b)   Scanned

FIGURE 9.10 The interior spectral reconstruction of a phantom containing concentrations of gadolinium and iodine: (a) the simulated data, (b) the data scanned with the real hybrid system. (From Xu, Q., et al., *Biomed. Eng., IEEE Trans.*, 59, 1711–1719, 2012.)

FIGURE 10.4 Photograph showing the experimental hutch of beamline B16 at Diamond Light Source [44]. The incoming beam from the optics hutch is represented by the red line, which impinges on the detector to be characterized (a), once the beam has passed through the CRL lenses (b), which produces the microfocus beam. The sample to be analyzed is typically set in a diffractometer (c). (From Gimenez, E.N., et al., *IEEE Trans. Nucl. Sci.*, 58, 323, 2011.)

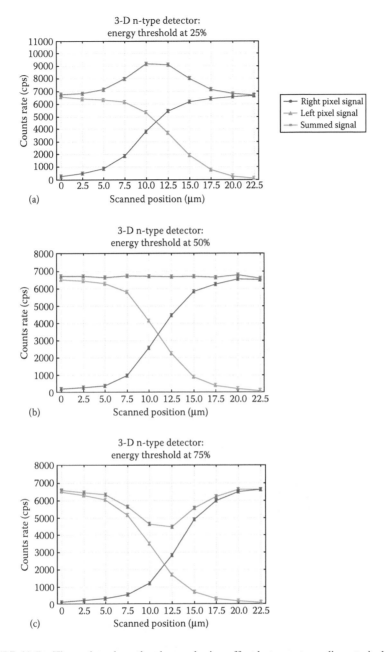

FIGURE 10.5 These plots show the charge-sharing effect between two adjacent pixels as a function of the set discriminator threshold. The red and blue lines show the charge collected by two neighboring pixels when a microfocus scan is performed across them. When the scan moves from the pixel on the left to the one on the right, the charge collection of the left pixel (red) decreases in the boundary region until it reaches negligible values. The beam then moves to the adjacent pixel to the right (blue), whose charge collection increases respectively. The green line shows the summed signal from the left and the right pixels. When the discriminator threshold is set at 25% of the impinging x-ray energy (left image) there are overcounts in the boundary region, and when the discriminator threshold is set at 75% of the impinging x-ray energy (right image) there are count losses. In both cases this is due to the charge sharing between the pixels. When the discriminator threshold is set at half the impinging energy (b), the summed signal is constant, indicating that there are no over- or undercounts due to the charge-sharing effect. (From Gimenez, E.N., et al., *Nucl. Instrum. Methods A*, 633, 114, 2011.)

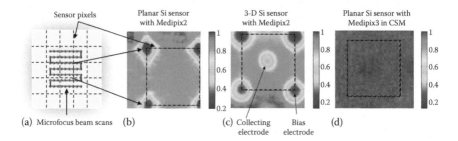

(a) Microfocus beam scans   (b)   (c) Collecting   Bias   (d)
electrode   electrode

FIGURE 10.6 Sketch of a microfocus beam scan on a PAD detector (a) and image of the data acquired from the microfocus beam scan (showing the area of a pixel within the dashed black line) in a planar silicon sensor (b) and a 3-D silicon sensor (c), both with Medipix2 in standard readout; and a planar silicon sensor readout with Medipix3 in charge-summing mode (CSM) (a). For all the sensors, the discriminator threshold was set at half the impinging x-ray energy. Images (b) and (c) show the difference in sensor response due to the charge-sharing effect at the corners for the planar sensor (b), and due to the column electrode and the charge-sharing effect in the 3-D sensor (c). Image (d) shows a uniform sensitive area for the pixel without count losses at the corners as a result of the CSM readout operation mode of the Medipix3 ASIC. (From Gimenez, E.N., et al., *IEEE Trans. Nucl. Sci.*, 58, 323, 2011; MacRaighne, A., et al., *Nucl. Sci. Symp. Conf. Rec.*, 2145, 2009.)

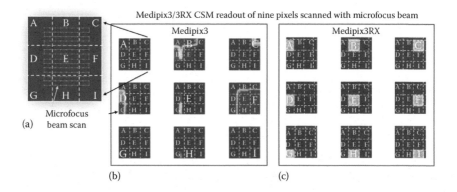

(b)   (c)

FIGURE 10.7 Image (a) shows the microfocus-beam scan path across a sensor region consisting of 9 pixels (from A to I). Image (b) shows the readout from the 9 pixels at different scan-beam positions for the Medipix3 ASIC, while image (c) shows the same from the Medipix3RX ASIC. Both ASICs are operating in charge-summing readout mode (CSM). For the images (b) and (c) the letters identify the pixel region on the sensor, with the large yellow letter indicating the region where the beam impinges along the scan. Image (b) shows the misallocation of events being registered to pixels other than the one they impinged on for the Medipix3. When the microfocus beam impinges on pixel A, events are registered by pixel B, and so on, and results in pixels with different sensitive areas (colored red in the image). The Medipix3RX ASIC implements a redesigned CSM architecture, which corrects the misallocation effect and results in pixels with the same sensitive area as shown in (c). (From Gimenez, E.N., et al., *IEEE Trans. Nucl. Sci.*, 58, 323, 2011.)

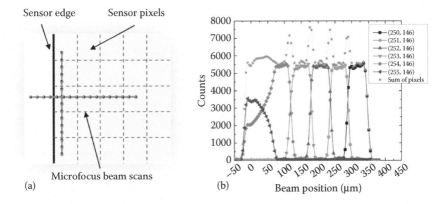

FIGURE 10.9 (a): Sketch of the vertical and horizonal microfocus beam scans used to test edgeless sensors. (b): Plot showing the charge collection of the microfocus scan across the edge of an active edgeless sensor. (From Bates, R., et al., *J Instrum.*, 8, P01018, 2013.)

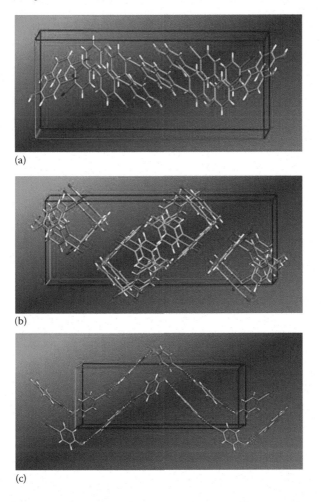

FIGURE 11.2 Simplified views of the 4HCB crystalline unit cell, seen along the crystallographic axes a (green axis), b (red axis), and c (blue axis). (Fraboni, B., Femoni, C., Mencarelli, I., Setti, L., Di Pietro, R., Cavallini, A., and Fraleoni-Morgera, A.: *Adv. Mater.* 2009. 21. 1835. Copyright Wiley-VCH Verlag GmbH & Co. KGaA. Reproduced with permission.)

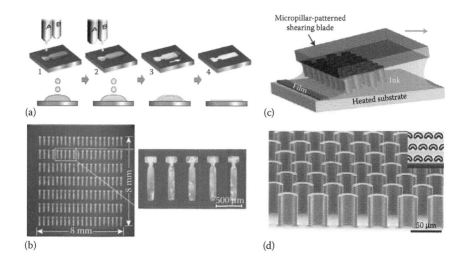

FIGURE 11.4 (a) Solution and (b) overview of a matrix of single crystals obtained with this technique. (c,d) Visual description of the solution-shearing method for originating solution-grown, oriented polycrystalline layers from drops of ink deposited onto substrates (c), and a scanning electron microscopic (SEM) image of the concave-shaped micropillar used for producing the needed shear stress in the deposited solution (d). (Reprinted by permission from Macmillan Publishers Ltd. [*Nature*] (From Minemawari, H., et al., 475, 364), copyright 2011; reprinted by permission from Macmillan Publishers Ltd. [*Nature Materials*] (Diao, Y., et al., 12, 665–671), copyright 2013.)

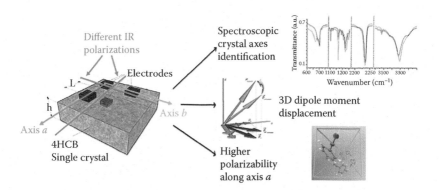

FIGURE 11.7 Sketch of linearly polarized IR measurements carried out over 4HCB single crystals and the information achievable with this technique. (Reproduced with permission from Fraleoni-Morgera, A., Tessarolo, M., Perucchi, A., Baldassarre, L., Lupi, S., Fraboni, B., *J. Phys. Chem. C*, 116, 2563. Copyright [2012] American Chemical Society.)

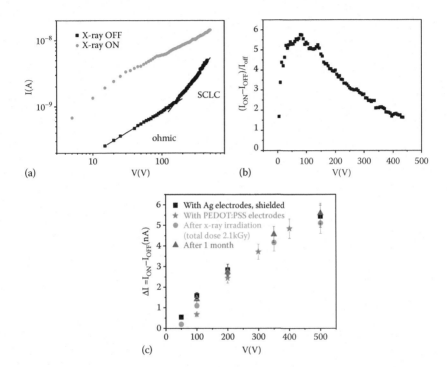

FIGURE 11.8 (a) Current–voltage curves for 4HCB-based devices in the dark (black squares) and under x-ray irradiation from a 35 keV Mo source at 170 mGy s^{-1} (red circles), for the vertical axis. Under x-ray irradiation, no ohmic–SCLC transition is detected along the vertical axis. (b) The maximum response can be achieved at low bias voltage (<100 V). (c) X-ray-induced current variation $\Delta I = (I_{ON} - I_{OFF})$ reported for different bias voltages applied to crystals contacted with Ag electrodes, before (black squares) and after irradiation with a total delivered x-ray dose of 2.1 kGy (red circles). The responses of crystals after an aging period of 1 month (blue triangles) and of crystals contacted with PEDOT:PSS electrodes (green stars) are also reported.

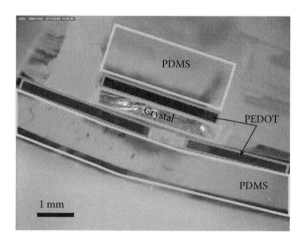

FIGURE 11.9 Optical microscopy image of a cross-sectional view of an optically transparent and bendable all-organic device with a 4HC crystal as the active sensing material. The PDMS substrate, the 4HCB crystal, and the PEDOT:PSS electrodes have been highlighted with color frames.

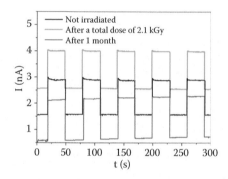

FIGURE 11.10 Electrical response of a 4HCB device to an on-off-switching x-ray beam at a bias voltage V = 100 V. The response is reported for an as-prepared device (black line) after it has been exposed for 3 h to a 170 mGy s^{-1} dose rate (total dose of 2.1 kGy) (red line) and after 1 month of storage in the ambient atmosphere (blue line).

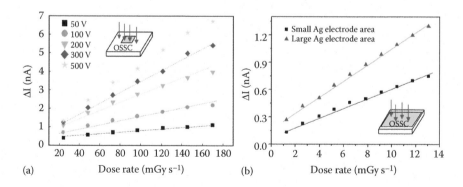

FIGURE 11.11 The x-ray-induced current variation $\Delta I = (I_{ON} - I_{OFF})$ is reported for increasing dose rates and different biases for the vertical axis of a 4HCB crystal on a quartz substrate with small-area (a) and large-area (b) Ag electrodes. (The lines are to guide the eye only.)

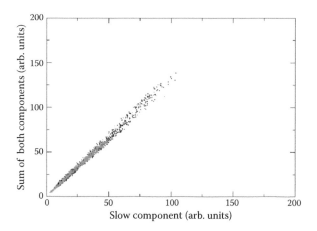

FIGURE 12.5 A spectrum showing a comparison between acquired neutron pulses (which also include coincident γ-rays) from an AmBe source (black) and pulses acquired from a ⁶⁰Co source (red). No discrimination between both sets of pulse data is possible.

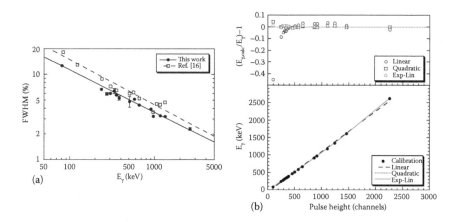

FIGURE 12.6 (a) The measured energy resolutions for γ-ray energies between 81 and 2615 keV. The solid line shows the fit of the power function $E^{-0.4993}$. The dashed line shows the anticipated $E^{1/2}$ behavior. (b) The residuals for the different fits to the calibration curves described. (c) The relationship between the pulse height and γ-ray energy using linear (dashed), quadratic (dotted), and "exponential-linear" (solid) fits. (From Billnert, R., Oberstedt, S., Andreotti, E., Hult, M., Marissens, G., and Oberstedt, A., *Nucl. Instrum. Methods Phys. Res. A*, 647, 94, 2011.)

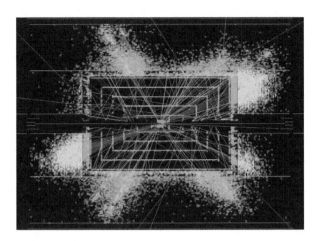

FIGURE 13.2 Simulated WW/ZZ event resulting from a 3 TeV electron collision at the future linear collider CLIC. (Courtesy of C. Grefe.)

FIGURE 13.3 Example of a 511 keV gamma conversion in three crystal pixels of a PET camera, with two successive Compton interactions followed by a photoelectric event. Note that a large fraction of energy escapes the central crystal, where the first interaction took place.

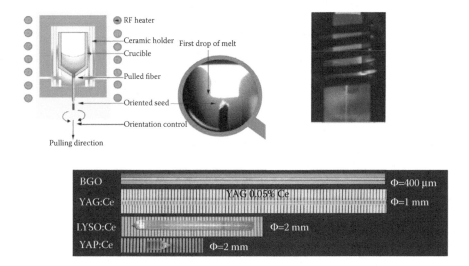

FIGURE 13.4 The micro-pulling-down crystal-growth technology. (Courtesy of Fibercryst.)

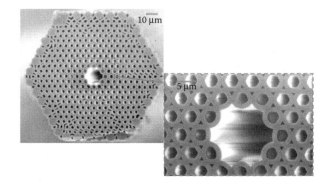

FIGURE 13.5 Photonic crystal fiber.

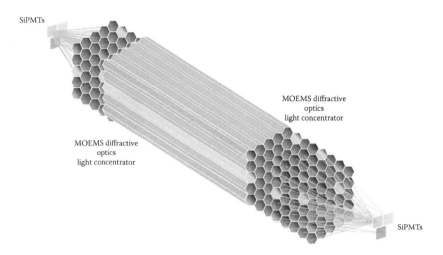

FIGURE 13.6 Concept of a metacable for calorimetry in future linear colliders.

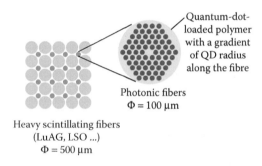

Quantum-dot-loaded polymer with a gradient of QD radius along the fibre

Photonic fibers
$\Phi = 100\ \mu m$

Heavy scintillating fibers
(LuAG, LSO ...)
$\Phi = 500\ \mu m$

FIGURE 13.7 Concept of a metacable for low-energy x- and γ-rays.

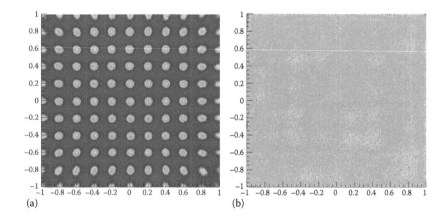

(a)

(b)

FIGURE 14.4 40 mm wide images from a gamma camera with continuous CsI(Na) crystal, obtained with a 11×11 hole mask (a) and a uniform flood field (a), after the first reconstruction steps.

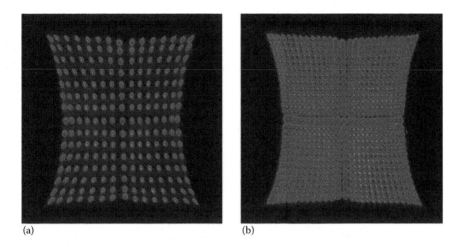

(a)   (b)

FIGURE 14.6   (a,b) Images of calibration masks with 15×15 and 27×27 holes, respectively, taken with a 2×2 PSPMT array coupled to a monolithic CsI(Na) crystal.

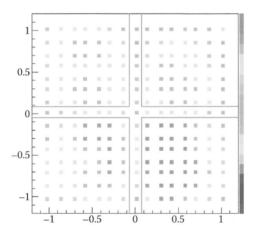

FIGURE 14.7   Color map of photopeak positions in 15×15 reference positions. Points in the central region are systematically shifted toward lower energy values with respect to their direct neighbors.

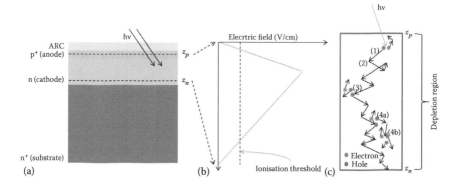

FIGURE 15.7 (a) Simplified cross section of a P-on-N Geiger-mode APD structure. (b) Electric field profile across the photodiode depletion region. (c) A pictorial representation of an avalanche breakdown in a photodiode across the depletion region.

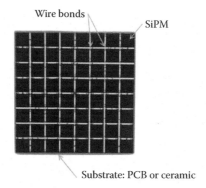

FIGURE 15.20 Epoxy-coated ceramic carrier SiPM array. This example is a 8×8 array of 6×6 mm^2 SiPMs mounted on a PCB substrate.

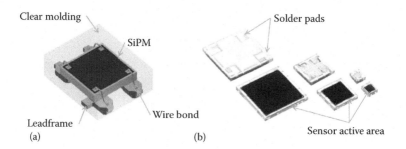

FIGURE 15.22 (a) Clear micro leadframe package drawing. (b) Images of the top and back side of 1×1, 3×3, and 6×6 mm^2 SiPM-packaged sensors.

Mathematical beam-hardening corrections are the most commonly used techniques. Dual-energy correction methods [77] image the atomic number dependent photoelectric component and the density dependent Compton scatter component separately and use them to correct for beam hardening, at the cost of increased exposure.

Photon-counting detectors have the unique advantage of capturing information from multiple energy ranges simultaneously. This can be helpful for correcting beam hardening without complex mathematical models. From experiments reported with dual-energy CT [78], it is understood that high-energy quanta undergo less beam hardening than the low-energy quanta. This information can be discretely extracted from an energy-resolving photon-counting detector without introducing any major external estimations or corrections. Additionally, capturing mid-energy ranges may provide a trade-off between reduced-metal artifacts and reasonable contrast information. An example of spectral beam-hardening reduction using a titanium metal phantom is shown in Figure 9.7. It can be noticed that the cupping artifact in the narrow high-energy bin is negligible when compared to a wide-energy acquisition. The wide-energy bin covers an energy range similar to that of a conventional CT acquisition. Figure 9.8 shows how narrow-energy bins can help minimizing streak artifacts in a titanium metal scaffold.

Spectral imaging is a promising approach for nondestructive evaluation [74] in material science research. The transition from conventional CT to dual-energy modalities has helped reduce metal artifacts. With the use of photon-processing

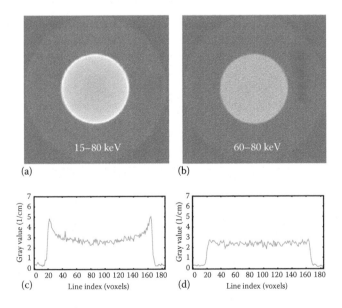

FIGURE 9.7 Titanium metal phantom: 12 mm Ti cylinder in 25 mm perspex. Energy bin image of (a) 15–80 keV and (b) 62–80 keV, while (c) and (d) show the respective intensity profile across the central axis. Note that the high intensity at the edges in (c) is removed from (d). (From Rajendran, K., et al., *J. Instrum.*, 9, P03015, 2014.)

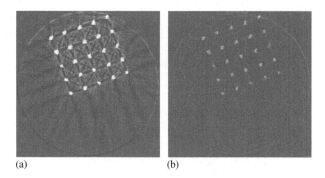

(a)   (b)

FIGURE 9.8 Titanium scaffold: 12 mm cubic lattice structure with porous surface. Energy bin of image (a) 15–80 keV and (b) 62–80 keV. (From Rajendran, K., et al., *J. Instrum.*, 9, P03015, 2014.)

detectors and the ability to capture multienergy information, this has expanded into spectral beam-hardening correction.

9.5.3 NANO-SIZED CONTRAST AGENTS

The advent of nano-sized contrast agents [80] is likely to greatly expand the use of contrast agents, not only in conventional CT but, more importantly, in spectral imaging. Nano-sized iodinated CT contrast agents have been developed with increased circulation time and reduction in adverse effects [81]. The pharmacokinetics, biodistribution, and toxicity of nano-sized iodine-based contrast agents are very different from the conventional water-soluble iodinated dimer contrast agents; this principle also applies to other metal-based nano-sized contrast agents. The size of a nanoparticle and how it is coated strongly influences its function and toxicity [82–84]. Nano-sized materials with a higher atomic number than iodine are more efficient than iodine for use as x-ray contrast agents in human and small animal k-edge imaging. Examples include gold [85–87], tantalum [80], gadolinium [88–91], ytterbium [92], and yttrium [93]. Nano-sized contrast agents can be designed to target very specific applications by modifying their size, shape, and functionality. They can be formulated in differing shapes such as nanospheres, nanorods, nanocapsules, and nanoemulsions; or incorporated into micelles, liposomes, or dendrimers [81]. The nanomaterial may be combined with a pharmaceutical in the core, coated with polyethylene glycol (PEG) to stabilize it and protect from degradation in the blood stream, and then made functional [94] by the attachment of ligands or antibodies to the outside of the core. The resulting macromolecule containing the nanomaterial is thus made multifunctional: the nanomaterial renders it visible and quantifiable on spectral imaging, the ligand or antibody allows it to specifically target a cellular marker, the PEG coating keeps the toxic pharmaceutical away from normal tissue, and the pharmaceutical released only at the target tissue [95] is therapeutic. This combination of imaging-contrast agent and therapeutic agent is known as theranostics [96,97]. By maximizing the payload of the nanocontrast (e.g., gold), and targeting a specific cell marker or cell type, the efficacy of the contrast-agent dose can be enhanced, reducing both cost and toxicity. A single nano-sized contrast agent enables specific imaging of the extracellular space, blood pool, targeted intracellular markers, or specific tissue components [81]. Spectral molecular CT allows for multiple different contrast agents to be imaged at the same time, so that multiple targets can be distinguished, imaged, and measured. Such targets include macrophages [87], monocytes [98], bacteria [99], platelets [100], specific cancers [16], fibroblasts [84], and fibrin [92].

With spectral molecular CT, it is possible to distinguish between two or more contrast agents in use at once. Here are two examples of how multiple contrast agents might be used simultaneously. Strokes and heart attacks are caused by vulnerable atherosclerotic plaque rupturing and blocking a smaller artery downstream. Vulnerable plaque is characterized by a fatty core, bleeding into the plaque, and inflammation. These features are very difficult to assess using current imaging methods. Gold-labeled high-density lipoprotein identifies inflammatory cells [18], while antiplatelet antibodies labeled with gadolinium or iodine could identify the intraplaque hemorrhage [101,102]. Fatty content could be measured with the standard spectral molecular CT technique [3]. These are techniques that will be added to the atherosclerosis work described here. Hypoxic and fibrotic areas in cancers make many cancers hard to treat. Improved molecular imaging of cancer could be achieved by simultaneously assessing the fibrotic content and distribution with one contrast and hypoxic areas with another, measuring drug delivery and distribution within the tumor with a third, and measuring macrophage response with a fourth. This methodology could be used in animal models to streamline the development of anticancer drugs, and then used to provide and monitor individualized treatment of patient-personalized medicine.

The MARS group is not yet actively engaging in multiple nano-sized contrast agents; however, we have successfully imaged mice with gold nanoparticles [103], which can be seen in Figure 9.9.

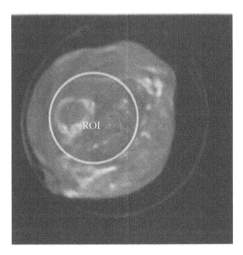

FIGURE 9.9   (See color insert) Spectral interior reconstruction and material decomposition of a mouse with 0.2 mL or 15 nm Aurovist II AuNP injected in a direct cardiac puncture injection. (From Xu, Q., et al., *Biomed. Eng., IEEE Trans.*, 59, 1711–1719, 2012.)

9.5.4 SPECTRAL INTERIOR TOMOGRAPHY

In the field of CT, interior tomography is an emerging field of research for dose reduction. By selecting a narrow field of view (FOV), a small region of interest (ROI) can be imaged. However, the unmeasured volume outside the ROI interferes with the x-ray projection data, leaving the reconstruction as an undersampled mathematical equation. Compressed sensing is a new algorithm that allows for solvable interior reconstructions [104,105]. One requirement for compressed sensing is a rough approximation of the full volume, which can be iteratively updated as the reconstruction continues.

Since spectral molecular CT uses the same principles as CT, it is thought that the same techniques can be used to image a small ROI. For this, a hybrid interior-tomography scanner has been proposed [103]. In this scanner, two x-ray sources are placed orthogonally around the rotation axis. One sends x-rays to a conventional CT detector and provides a low-resolution full-FOV CT image as the rough approximation needed for compressed sensing. The other source sends x-rays to a Medipix detector and provides high-resolution, narrow-FOV, spectral images. This allows for the advantages of spectral molecular CT in the ROI, combined with the dose reduction from interior tomography. The quality of this system has been both simulated and tested with truncated data from Medipix detectors in a MARS scanner. Figure 9.9 shows the material decomposition of a mouse with AuNP injected in a direct cardiac puncture injection, with color being applied based on the spectral data in the ROI. The data was acquired with a CdTe-MXR in a MARS-v4 scanner. The full scan was taken at all energies, and the lowest energy bin was used as the grayscale image across the full FOV. The higher energy bins were truncated and reconstructed using the compressed sensing-based methods.

More recently, this hybrid scanner has been implemented [106]. This uses a single-Si Medipix2 (MXR) assembly, with an Xradia XCT detector (6 mm scintillator plus high-resolution CCD) orthogonally placed along the rotational axis. A contrast-agent phantom was formed to test the functionality of this system. This phantom has 6 capillaries placed in ultrafiltered water. Three of the capillaries have increasing concentrations of iodine-based "Omnipaque 300," and the other three capillaries have increasing concentrations of gadolinium-based "Magnevist." This phantom was both simulated and scanned, with a full FOV reconstruction using the Xradia XCT detector, and a narrow FOV covering the energy range of 10–38 keV in 12 energy bins. Figure 9.10 shows the final reconstructed phantoms in both the simulated and the real data.

These results are early indicators that the methodology of interior tomography works with spectral molecular CT. As the current Medipix chips are small (14.08 mm), this may be an alternative to extensive tiling in enabling the imaging of large patients, such as humans.

9.6 SUMMARY

The MARS research group has developed a small-animal CT scanner for researching and developing spectral molecular CT. The scanners use the Medipix family of

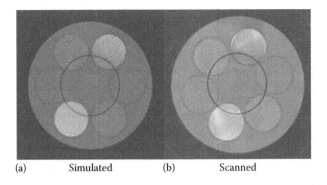

(a)   Simulated   (b)   Scanned

FIGURE 9.10   **(See color insert)** The interior spectral reconstruction of a phantom containing concentrations of gadolinium and iodine: (a) the simulated data, (b) the data scanned with the real hybrid system. (From Xu, Q., et al., *Biomed. Eng., IEEE Trans.*, 59, 1711–1719, 2012.)

chips as the key piece of technology to allow spectral molecular CT. With spectral molecular CT, material separation and decomposition can be performed. The current scanner has already demonstrated many of its uses, including the cardiovascular imaging of atherosclerosis, improvements in the reduction of metal artifacts, the use of targeted contrast agents for molecular imaging, and the separation of materials with interior tomography. The current MARS scanners are only capable of imaging small animals such as mice, but the work now being done is applicable to human imaging. A human-sized scanner is the next major milestone in this development.

REFERENCES

1. D. A. Mankoff. A definition of molecular imaging. *Journal of Nuclear Medicine*, 48(6):18N–21N, 2007.
2. N. G. Anderson, A. P. Butler, N. J. A. Scott, N. J. Cook, J. S. Butzer, N. Schleich, M. Firsching, et al. Spectroscopic (multi-energy) CT distinguishes iodine and barium contrast material in MICE. *European Radiology*, 20(9):2126–2134, 2010.
3. J. P. Ronaldson, R. Zainon, N. J. A. Scott, S. P. Gieseg, A. P. Butler, P. H. Butler, and N. G. Anderson. Toward quantifying the composition of soft tissues by spectral CT with Medipix3. *Medical Physics*, 39(11):6847–6857, 2012.
4. M. Firsching, A. P. Butler, N. Scott, N. G. Anderson, T. Michel, and G. Anton. Contrast agent recognition in small animal CT using the Medipix2 detector. *Nuclear Instruments and Methods in Physics Research Section A: Accelerators, Spectrometers, Detectors and Associated Equipment*, 607(1):179–182, 2009.
5. J. S. Butzer, A. P. H. Butler, P. H. Butler, P. J. Bones, N. Cook, and L. Tlustos. Medipix imaging: Evaluation of datasets with PCA. In *23rd International Conference on Image and Vision Computing New Zealand*, pp. 1–6, 2008.
6. A. R. Kalukin, M. Van Geet, and R. Swennen. Principal components analysis of multienergy x-ray computed tomography of mineral samples. *IEEE Transactions on Nuclear Science*, 47(5):1729–1736, 2000.
7. J. P. Ronaldson, M. Walsh, S. J. Nik, J. Donaldson, R. M. N. Doesburg, D. van Leeuwen, R. Ballabriga, M. N. Clyne, A. P. H. Butler, and P. H. Butler. Characterization of Medipix3 with the MARS readout and software. *Journal of Instrumentation*, 6(01):C01056, 2011.

8. T. Kohonen. The self-organizing map. *Proceedings of the IEEE*, 78(9):1464–1480, 1990.

9. D. G. Kang, Y. Sung, S. S. Kim, S. D. Lee, and J. D. Kim. Multiple object decomposition based on independent component analysis of multi-energy x-ray projections. In *16th IEEE International Conference on Image Processing (ICIP)*, pp. 4173–4176, 2009.

10. J. P. Schlomka, E. Roessl, R. Dorscheid, S. Dill, G. Martens, T. Istel, C. Bumer, et al. Experimental feasibility of multi-energy photon-counting k-edge imaging in pre-clinical computed tomography. *Physics in Medicine and Biology*, 53(15):4031, 2008.

11. P. M. Shikhaliev and S. G. Fritz. Photon counting spectral CT versus conventional CT: Comparative evaluation for breast imaging application. *Physics in Medicine and Biology*, 56(7):1905, 2011.

12. C. G. Lall, A. M. Aisen, N. Bansal, and K. Sandrasegaran. Nonalcoholic fatty liver disease. *American Journal of Roentgenology*, 190(4):993–1002, 2008.

13. G. C. Farrell and C. Z. Larter. Nonalcoholic fatty liver disease: From steatosis to cirrhosis. *Hepatology*, 43(S1):S99–S112, 2006.

14. K. B. Berg, J. M. Carr, M. J. Clark, N. J. Cook, N. G. Anderson, N. J. Scott, A. P. M. Butler, P. H. Butler, and A. P. H. Butler. Pilot study to confirm that fat and liver can be distinguished by spectroscopic tissue response on a Medipix-All-Resolution System-CT (MARS-CT). University of Canterbury. http://ir.canterbury.ac.nz/handle/10092/2908, 2009.

15. J. W. M. Bulte. Science to practice: Can CT be performed for multicolor molecular imaging? *Radiology*, 256(3):675–676, 2010.

16. R. Popovtzer, A. Agrawal, N. A. Kotov, A. Popovtzer, J. Balter, T. E. Carey, and R. Kopelman. Targeted gold nanoparticles enable molecular CT imaging of cancer. *Nano Letters*, 8(12):4593–4596, 2008.

17. M. Naghavi, P. Libby, E. Falk, S. W. Casscells, S. Litovsky, J. Rumberger, J. J. Badimon, et al. From vulnerable plaque to vulnerable patient: A call for new definitions and risk assessment strategies: Part I. *Circulation*, 108(14):1664–1672, 2003.

18. D. P. Cormode, E. Roessl, A. Thran, T. Skajaa, R. E. Gordon, J. P. Schlomka, V. Fuster, et al. Atherosclerotic plaque composition: Analysis with multicolor CT and targeted gold nanoparticles1. *Radiology*, 256(3):774–782, 2010.

19. E. Roessl and R. Proksa. K-edge imaging in x-ray computed tomography using multi-bin photon counting detectors. *Physics in Medicine and Biology*, 52(15):4679, 2007.

20. S. Feuerlein, E. Roessl, R. Proksa, G. Martens, O. Klass, M. Jeltsch, V. Rasche, H. J. Brambs, M. H. K. Hoffmann, and J. P. Schlomka. Multienergy photon-counting k-edge imaging: Potential for improved luminal depiction in vascular imaging. *Radiology*, 249(3):1010–1016, 2008.

21. M. G. Bisogni, D. Bulajic, P. Delogu, M. E. Fantacci, M. Novelli, M. Quattrocchi, V. Rosso, and A. Stefanini. Performances of different digital mammography imaging systems: Evaluation and comparison. *Nuclear Instruments and Methods in Physics Research Section A: Accelerators, Spectrometers, Detectors and Associated Equipment*, 546(12):14–18, 2005.

22. J. Watt, D. W. Davidson, C. Johnston, C. Smith, L. Tlustos, B. Mikulec, K. M. Smith, and M. Rahman. Dose reductions in dental x-ray imaging using Medipix. *Nuclear Instruments and Methods in Physics Research Section A: Accelerators, Spectrometers, Detectors and Associated Equipment*, 513(12):65–69, 2003.

23. K. Taguchi and J. S. Iwanczyk. Vision 20/20: Single photon counting x-ray detectors in medical imaging. *Medical Physics*, 40(10):100901, 2013.

24. K. Taguchi, E. C. Frey, X. Wang, J. S. Iwanczyk, and W. C. Barber. An analytical model of the effects of pulse pileup on the energy spectrum recorded by energy resolved photon counting x-ray detectors. *Medical Physics*, 37(8):3957, 2010.

25. K. Taguchi, M. Zhang, E. C. Frey, X. Wang, J. S. Iwanczyk, E. Nygard, N. E. Hartsough, B. M. W. Tsui, and W. C. Barber. Modeling the performance of a photon counting x-ray detector for CT: Energy response and pulse pileup effects. *Medical Physics*, 38(2):1089, 2011.

26. J. S. Iwanczyk, E. Nygrd, O. Meirav, J. Arenson, W. C. Barber, N. E. Hartsough, N. Malakhov, and J. C. Wessel. Photon counting energy dispersive detector arrays for x-ray imaging. *IEEE Transactions on Nuclear Science*, 56(3):535–542, 2009.

27. W. C. Barber, E. Nygard, J. S. Iwanczyk, M. Zhang, E. C. Frey, B. M. W. Tsui, J. C. Wessel, N. Malakhov, G. Wawrzyniak, and N. E. Hartsough. Characterization of a novel photon counting detector for clinical CT: Count rate, energy resolution, and noise performance. *Proceedings of the SPIE Medical Imaging Conference*, 7258:725824, 2009.

28. W. C. Barber, J. C. Wessel, E. Nygard, N. Malakhov, G. Wawrzyniak, N. E. Hartsough, T. Gandhi, A. Arodzero, M. Q. Damron, D. Moraes, P. Jarron, P. Weilhammer, and J. S. Wanczyk. High flux x-ray imaging with CdZnTe arrays. In *IEEE Nuclear Science Symposium and Medical Imaging Conference*, IEEE, Anaheim, CA, 2012.

29. S. Kappler, F. Glasser, S. Janssen, E. Kraft, and M. Reinwand. A research prototype system for quantum-counting clinical ct. *Proceedings of the SPIE Medical Imaging Conference*, 7622:76221Z, 2010.

30. S. Kappler, T. Hannemann, E. Kraft, B. Kreisler, D. Niederloehner, K. Stierstorfer, and T. Flohr. First results from a hybrid prototype ct scanner for exploring benefits of quantum-counting in clinical CT. *Proceedings of the SPIE Medical Imaging Conference*, 8313:83130Z, 2012.

31. R. Steadman, C. Herrmann, O. Mülhens, D. G. Maeding, J. Colley, T. Firlit, R. Luhta, M. Chappo, B. Harwood, and D. Kosty. Chromaix: A high-rate energy-resolving photon-counting asic for spectral computed tomography. *Proceedings of the SPIE Medical Imaging Conference*, 7622:762220–1, 2010.

32. Y. Tomita, Y. Shirayanagi, S. Matsui, T. Aoki, and Y. Hatanaka. X-ray color scanner with multiple energy discrimination capability. In *Optics & Photonics 2005*, pp. 59220A–59220A. International Society for Optics and Photonics, 2005.

33. Y. Tomita, Y. Shirayanagi, S. Matsui, M. Misawa, H. Takahashi, T. Aoki, and Y. Hatanaka. X-ray color scanner with multiple energy differentiate capability. *IEEE Nuclear Science Symposium Conference Record*, 6:3733–3737, 2004.

34. R. Ballabriga, M. Campbell, E. H. M. Heijne, X. Llopart, and L. Tlustos. The Medipix3 prototype, a pixel readout chip working in single photon counting mode with improved spectrometric performance. *IEEE Transactions on Nuclear Science*, 54(5):1824–1829, 2007.

35. E. N. Gimenez. Characterization of Medipix3 with synchrotron radiation. *IEEE Transactions on Nuclear Science*, 58(1, 2):323–332, 2011.

36. T. Koenig, E. Hamann, S. Procz, R. Ballabriga, A. Cecilia, M. Zuber, X. Llopart, M. Campbell, A. Fauler, and T. Baumbach. Charge summing in spectroscopic x-ray detectors with high-Z sensors. *IEEE Transactions on Nuclear Science*, 60(6):4713–4718, 2013.

37. E. Kraft, P. Fischer, M. Karagounis, M. Koch, H. Krueger, I. Peric, N. Wermes, C. Herrmann, A. Nascetti, and M. Overdick. Counting and integrating readout for direct conversion x-ray imaging: Concept, realization and first prototype measurements. *IEEE Transactions on Nuclear Science*, 54(2):383–390, 2007.

38. F. Rupcich and T. Gilat-Schmidt. Experimental study of optimal energy weighting in energy-resolved CT using a CZT detector. *Proceedings of the SPIE Medical Imaging Conference*, 8668:86681X, 2013.

39. V. B. Cajipe, R. F. Calderwood, M. Clajus, S. Hayakawa, R. Jayaraman, T. O. Tumer, B. Grattan, and O. Yossifor. Multi-energy x-ray imaging with linear CZTpixel arrays and integrated electronics. *IEEE Nuclear Science Symposium Conference Record*, 7:4548–4551, 2004.

40. E. Fredenberg, M. Hemmendorff, B. Cederström, M. Åslund, and M. Danielsson. Contrast-enhanced spectral mammography with a photon-counting detector. *Medical Physics*, 37:2017, 2010.

41. E. Fredenberg, M. Lundqvist, B. Cederström, M. Åslund, and M. Danielsson. Energy resolution of a photon-counting silicon strip detector. *Nuclear Instruments and Methods in Physics Research Section A: Accelerators, Spectrometers, Detectors and Associated Equipment*, 613(1):156–162, 2010.

42. M. Åslund, B. Cederström, M. Lundqvist, and M. Danielsson. Physical characterization of a scanning photon counting digital mammography system based on si-strip detectors. *Medical Physics*, 34:1918, 2007.

43. C. Xu, M. Danielsson, S. Karlsson, C. Svensson, and H Bornefalk. Performance characterization of a silicon strip detector for spectral computed tomography utilizing a laser testing system. *Proceedings of the SPIE Medical Imaging Conference*, 7961:79610S, 2011.

44. C. Xu, M. Danielsson, S. Karlsson, C. Svensson, and H. Bornefalk. Preliminary evaluation of a silicon strip detector for photon-counting spectral CT. *Nuclear Instruments and Methods in Physics Research Section A: Accelerators, Spectrometers, Detectors and Associated Equipment*, 677:45–51, 2012.

45. C. Xu, H. Chen, M. Persson, S. Karlsson, M. Danielsson, C. Svensson, and H. Bornefalk. Energy resolution of a segmented silicon strip detector for photon-counting spectral CT. *Nuclear Instruments and Methods in Physics Research Section A: Accelerators, Spectrometers, Detectors and Associated Equipment*, 715:11–17, 2013.

46. C. Xu, M. Danielsson, and H. Bornefalk. Evaluation of energy loss and charge sharing in cadmium telluride detectors for photon-counting computed tomography. *Nuclear Science, IEEE Transactions on*, 58(3):614–625, 2011.

47. C. Xu, M. Persson, H. Chen, S. Karlsson, M. Danielsson, C. Svensson, and H. Bornefalk. Evaluation of a second-generation ultra-fast energy-resolved ASIC for photon-counting spectral CT. *IEEE Transactions on Nuclear Science*, 60(1):437–45, 2013.

48. H. Bornefalk and M. Danielsson. Photon-counting spectral computed tomography using silicon strip detectors: a feasibility study. *Physics in Medicine and Biology*, 55(7):1999, 2010.

49. X. Llopart, M. Campbell, R. Dinapoli, D. San Segundo, and E. Pernigotti. Medipix2: A 64-k pixel readout chip with 55-m square elements working in single photon counting mode. *IEEE Transactions on Nuclear Science*, 49(5):2279–2283, 2002.

50. X. Llopart, R. Ballabriga, M. Campbell, L. Tlustos, and W. Wong. Timepix, a 65k programmable pixel readout chip for arrival time, energy and/or photon counting measurements. *Nuclear Instruments and Methods in Physics Research Section A: Accelerators, Spectrometers, Detectors and Associated Equipment*, 581(1):485–494, 2007.

51. M. F. Walsh, S. J. Nik, S. Procz, M. Pichotka, S. T. Bell, C. J. Bateman, R. M. N. Doesburg, et al. Spectral CT data acquisition with Medipix3.1. *Journal of Instrumentation*, 8(10):P10012, 2013.

52. R. Ballabriga, J. Alozy, G. Blaj, M. Campbell, M. Fiederle, E. Frojdh, E. H. M. Heijne, X. Llopart, M. Pichotka, S. Procz, L. Tlustos, and L. Wong. The Medipix3RX: A high resolution, zero dead-time pixel detector readout chip allowing spectroscopic imaging. *Journal of Instrumentation*, 8(02):C02016, 2013.

53. R. Ballabriga and M. Vilass-Cardona Campbell. The design and implementation in $0.13mu m$ CMOS of an algorithm permitting spectroscopic imaging with high spatial resolution for hybrid pixel detectors. PhD Thesis, Ramon Llull University, Barcelona, 2009.

54. L. Tlustos, M. Campbell, C. Frjdh, P. Kostamo, and S. Nenonen. Characterisation of an epitaxial GaAs/Medipix2 detector using fluorescence photons. *Nuclear Instruments and Methods in Physics Research Section A: Accelerators, Spectrometers, Detectors and Associated Equipment*, 591(1):42–45, June 2008.

55. L. Tlustos, G. Shelkov, and O. P. Tolbanov. Characterisation of a GaAs(Cr) Medipix2 hybrid pixel detector. *Nuclear Instruments and Methods in Physics Research Section A: Accelerators, Spectrometers, Detectors and Associated Equipment*, 633, Supplement 1(0):S103–S107, 2011.

56. A. Zwerger, A. Fauler, M. Fiederle, and K. Jakobs. Medipix2: Processing and measurements of GaAs pixel detectors. *Nuclear Instruments and Methods in Physics Research Section A: Accelerators, Spectrometers, Detectors and Associated Equipment*, 576(1):23–26, 2007.

57. R. Aamir, M. F. Walsh, S. P. Lansley, R. M. Doesburg, R. Zainon, N. J. A. De Ruiter, P. H. Butler, and A. P. H. Butler. Characterization of CdTe x-ray sensor layer on Medipix detector chips. *Materials Science Forum*, 700:170–173, 2011.

58. R. Aamir, N. G. Anderson, A. P. H. Butler, P. H. Butler, S. P. Lansley, R. M. Doesburg, M. Walsh, and J. L. Mohr. Characterization of Si and CdTe sensor layers in Medipix assemblies using a microfocus x-ray source. In *Nuclear Science Symposium and Medical Imaging Conference (NSS/MIC)*, October, pp. 4766–4769. IEEE, 2011.

59. M. Chmeissani, C. Frojdh, O. Gal, X. Llopart, J. Ludwig, M. Maiorino, E. Manach, et al. First experimental tests with a CdTe photon counting pixel detector hybridized with a Medipix2 readout chip. *IEEE Transactions on Nuclear Science*, 51(5), 2004.

60. R. K. Panta, M. Walsh, S. Bell, N. Anderson, A. Butler, and P. Butler. Energy calibration of the pixels of spectral x-ray detectors. *IEEE Transactions on Medical Imaging* [Epub ahead of print] 2014.

61. M. F. Walsh, R. M. N. Doesburg, J. L. Mohr, R. Ballabriga, A. P. H. Butler, and P. H. Butler. Improving and characterising the threshold equalisation process for multi-chip Medipix3 cameras in single pixel mode. In *Nuclear Science Symposium and Medical Imaging Conference (NSS/MIC)*, October, pp. 1718–1721. IEEE, 2011.

62. D. Pennicard, R. Ballabriga, X. Llopart, M. Campbell, and H. Graafsma. Simulations of charge summing and threshold dispersion effects in Medipix3. *Nuclear Instruments and Methods in Physics Research Section A: Accelerators, Spectrometers, Detectors and Associated Equipment*, 636(1):74–81, 2011.

63. E. N. Gimenez, R. Ballabriga, M. Campbell, I. Horswell, X. Llopart, J. Marchal, K. J. S. Sawhney, N. Tartoni, and D. Turecek. Study of charge-sharing in MEDIPIX3 using a micro-focused synchrotron beam. *Journal of Instrumentation*, 6(01):C01031, 2011.

64. M. F. Walsh, A. M. T. Opie, J. P. Ronaldson, R. M. N. Doesburg, S. J. Nik, J. L. Mohr, R. Ballabriga, A. P. H. Butler, and P. H. Butler. First CT using Medipix3 and the MARS-CT-3 spectral scanner. *Journal of Instrumentation*, 6(01):C01095, 2011.

65. R. M. N. Doesburg, M. N. Clyne, D. van Leeuwen, N. J. Cook, P. H. Butler, and A. P. H. Butler. Fast ethernet readout for Medipix arrays with MARS-CT. 2009.

66. Heart Foundation. General heart statistics for New Zealand. Heart Foundation, 2014. http://www.heartfoundation.org.nz/know-the-facts/statistics.

67. M. Crooke. New Zealand cardiovascular guidelines: Best practice evidence-based guideline: The assessment and management of cardiovascular risk December 2003. *Clinical Biochemist Reviews*, 28(1):19, 2007.

68. American Heart Association. Atherosclerosis. American Heart Association, 2014. http://www.heart.org/HEARTORG/Conditions/Cholesterol/WhyCholesterolMatters/Atherosclerosis_UCM_305564_Article.jsp.

69. V. L. Roger, A. S. Go, D. M. Lloyd-Jones, E. J. Benjamin, J. D. Berry, W. B. Borden, D. M. Bravata, S. Dai, E. S. Ford, and C. S. Fox. Heart disease and stroke statistics 2012 update: A report from the American Heart Association. *Circulation*, 125(1):e2–e220, 2012.

70. R. Zainon, J. P. Ronaldson, T. Janmale, N. J. Scott, T. M. Buckenham, A. P. H. Butler, P. H. Butler, et al. Spectral CT of carotid atherosclerotic plaque: comparison with histology. *European Radiology*, 22(12):2581–2588, 2012.

71. M. E. Lobatto, V. Fuster, Z. A. Fayad, and W. J. M. Mulder. Perspectives and opportunities for nanomedicine in the management of atherosclerosis. *Nature Reviews Drug Discovery*, 10(11):835–852, 2011.

72. D. Vancraeynest, A. Pasquet, V. Roelants, B. L. Gerber, and J. L. J. Vanoverschelde. Imaging the vulnerable plaque. *Journal of the American College of Cardiology*, 57(20):1961–1979, 2011.

73. P. Baturin, Y. Alivov, and S. Molloi. Spectral CT imaging of vulnerable plaque with two independent biomarkers. *Physics in Medicine and Biology*, 57(13):4117, 2012.

74. S. Procz, K. A. Wartig, A. Fauler, A. Zwerger, J. Luebke, R. Ballabriga, G. Blaj, M. Campbell, M. Mix, and M. Fiederle. Medipix3 CT for material sciences. *Journal of Instrumentation*, 8(01):C01025, 2013.

75. M. J. Lee, S. Kim, S. A. Lee, H. T. Song, Y. M. Huh, D. H. Kim, S. H. Han, and J. S. Suh. Overcoming artifacts from metallic orthopedic implants at high-field-strength MR imaging and multi-detector CT. *Radiographics*, 27(3):791–803, 2007.

76. E. van de Casteele, D. van Dyck, J. Sijbers, and E. Raman. A model-based correction method for beam hardening artefacts in x-ray microtomography. *Journal of X-ray Science and Technology*, 12(1):43–57, 2004.

77. R. E. Alvarez and A. Macovski. Energy-selective reconstructions in x-ray computerised tomography. *Physics in Medicine and Biology*, 21(5):733–744, 1976.

78. F. Bamberg, A. Dierks, K. Nikolaou, M. F. Reiser, C. R. Becker, and T. R. C. Johnson. Metal artifact reduction by dual energy computed tomography using monoenergetic extrapolation. *European Radiology*, 21(7):1424–1429, 2011.

79. K. Rajendran, M. F. Walsh, N. J. A. de Ruiter, A. I. Chernoglazov, R. K. Panta, A. P. H. Butler, P. H. Butler, et al. Reducing beam hardening effects and metal artefacts in spectral CT using Medipix3RX. *Journal of Instrumentation*, 9(3):P03015, 2014.

80. N. Lee, S. H. Choi, and T. Hyeon. Nano-sized CT contrast agents. *Advanced Materials*, 25(19):2641–2660, 2013.

81. F. Hallouard, N. Anton, P. Choquet, A. Constantinesco, and T. Vandamme. Iodinated blood pool contrast media for preclinical x-ray imaging applications a review. *Biomaterials*, 31(24):6249–6268, 2010.

82. C. Xu, G. A. Tung, and S. Sun. Size and concentration effect of gold nanoparticles on x-ray attenuation as measured on computed tomography. *Chemistry of Materials*, 20(13):4167–4169, 2008.

83. T. Mironava, M. Hadjiargyrou, M. Simon, V. Jurukovski, and M. H. Rafailovich. Gold nanoparticles cellular toxicity and recovery: Effect of size, concentration and exposure time. *Nanotoxicology*, 4(1):120–137, 2010.

84. N. Khlebtsov and L. Dykman. Biodistribution and toxicity of engineered gold nanoparticles: a review of *in vitro* and *in vivo* studies. *Chemical Society Reviews*, 40:1647–1671, 2011.

85. S. Elzey, D. H. Tsai, S. A. Rabb, L. L. Yu, M. R. Winchester, and V. A. Hackley. Quantification of ligand packing density on gold nanoparticles using ICP-OES. *Analytical and Bioanalytical Chemistry*, 403(1):145–149, 2012.

86. C. C. Chien, H. H. Chen, S. F. Lai, Y. Hwu, C. Petibois, C. S. Yang, Y. Chu, and G. Margaritondo. X-ray imaging of tumor growth in live mice by detecting gold-nanoparticle-loaded cells. *Scientific Reports*, 2:article 610, 2013.

87. E. Roessl, D. Cormode, B. Brendel, K. J. Engel, G. Martens, A. Thran, Z. Fayad, and R. Proksa. Preclinical spectral computed tomography of gold nano-particles. *Nuclear Instruments and Methods in Physics Research Section A: Accelerators, Spectrometers, Detectors and Associated Equipment*, 648(Suppl. 1):S259–S264, 2011.

88. J. Vymazal, E. Spuentrup, G. Cardenas-Molina, A. J. Wiethoff, M. G. Hartmann, P. Caravan, and E. C. Parsons Jr. Thrombus imaging with fibrin-specific gadolinium-based Mr contrast agent EP-2104R: Results of a phase II clinical study of feasibility. *Investigative Radiology*, 44(11):697–704, 2009.

89. T. Koenig, J. Schulze, M. Zuber, K. Rink, J. Butzer, E. Hamann, A. Cecilia, et al. Imaging properties of small-pixel spectroscopic x-ray detectors based on cadmium telluride sensors. *Physics in Medicine and Biology*, 57(21):6743–6759, 2012.

90. C. S. Broberg, S. S. Chugh, C. Conklin, D. J. Sahn, and M. Jerosch-Herold. Quantification of diffuse myocardial fibrosis and its association with myocardial dysfunction in congenital heart disease. *Circulation: Cardiovascular Imaging*, 3(6):727–734, 2010.

91. J. Cai, T. S. Hatsukami, M. S. Ferguson, W. S. Kerwin, T. Saam, B. Chu, N. Takaya, N. L. Polissar, and C. Yuan. *In vivo* quantitative measurement of intact fibrous cap and lipid-rich necrotic core size in atherosclerotic carotid plaque comparison of high-resolution, contrast-enhanced magnetic resonance imaging and histology. *Circulation*, 112(22):3437–3444, 2005.

92. D. Pan, C. O. Schirra, A. Senpan, A. H. Schmieder, A. J. Stacy, E. Roessl, A. Thran, S. A. Wickline, R. Proska, and G. M. Lanza. An early investigation of ytterbium nanocolloids for selective and quantitative multicolor spectral CT imaging. *ACS Nano*, 6(4):3364–3370, 2012.

93. L. J. Higgins and M. G. Pomper. The evolution of imaging in cancer: Current state and future challenges. *Seminars in Oncology*, 38(1):3–15, 2011.

94. N. Erathodiyil and J. Y. Ying. Functionalization of inorganic nanoparticles for bioimaging applications. *Accounts of Chemical Research*, 44(10):925–935, 2011.

95. S. Taurin, H. Nehoff, and K. Greish. Anticancer nanomedicine and tumor vascular permeability; where is the missing link? *Journal of Controlled Release*, 164(3):265–275, 2012.

96. L. S. Wang, M. C. Chuang, and J. A. Ho. Nanotheranostics—a review of recent publications. *International Journal of Nanomedicine*, 7:4679, 2012.

97. K. Y. Choi, G. Liu, S. Lee, and X. Chen. Theranostic nanoplatforms for simultaneous cancer imaging and therapy: current approaches and future perspectives. *Nanoscale*, 4:330–342, 2012.

98. P. G. Camici, O. E. Rimoldi, O. Gaemperli, and P. Libby. Non-invasive anatomic and functional imaging of vascular inflammation and unstable plaque. *European Heart Journal*, 33(11):1309–1317, 2012.

99. H. T. Ta, S. Prabhu, E. Leitner, F. Jia, D. Von Elverfeldt, K. E. Jackson, T. Heidt, A. K. N. Nair, H. Pearce, and C. von Zur Muhlen. Enzymatic single-chain antibody tagging a universal approach to targeted molecular imaging and cell homing in cardiovascular disease. *Circulation Research*, 109(4):365–373, 2011.

100. D. von Elverfeldt, C. von zur Muhlen, K. Wiens, I. Neudorfer, A. Zirlik, M. Meissner, P. Tilly, A. L. Charles, C. Bode, and K. Peter. In vivo detection of activated platelets allows characterizing rupture of atherosclerotic plaques with molecular magnetic resonance imaging in mice. *PloS one*, 7(9):e45008, 2012.

101. C. E. Hagemeyer and K. Peter. Targeting the platelet integrin GPIIb/IIIa. *Current Pharmaceutical Design*, 16(37):4119–4133, 2010.

102. N. G. Anderson and A. P. H. Butler. Clinical applications of spectral molecular imaging: potential and challenges. *Contrast Media and Molecular Imaging*, 9(1):3–12, 2014.

103. Q. Xu, H. Yu, J. Bennett, P. He, R. Zainon, R. Doesburg, A. Opie, et al. Image reconstruction for hybrid true-color micro-CT. *IEEE Transactions on Biomedical Engineering*, 59(6):1711–1719, 2012.

104. Q. Xu, X. Mou, G. Wang, J. Sieren, E. A. Hoffman, and H. Yu. Statistical interior tomography. *IEEE Transactions on Medical Imaging*, 30(5):1116–1128, 2011.

105. H. Yu and G. Wang. Compressed sensing based interior tomography. *Physics in Medicine and Biology*, 54:2791–2805, 2009.

106. J. R. Bennett, A. M. T. Opie, Q. Xu, H. Yu, M. Walsh, A. Butler, P. Butler, G. Cao, A. Mohs, and G. Wang. Hybrid spectral micro-CT: System design, implementation, and preliminary results. *IEEE Transactions on Biomedical Engineering*, 61(2):246–253, 2013.

10 Characterization of Photon-Counting Detectors with Synchrotron Radiation

Eva N. Gimenez

CONTENTS

10.1 INTRODUCTION

Hybrid single-photon-counting semiconductor detectors exhibit characteristics such as large dynamic range, fast frame rate, and negligible noise above threshold. Together with the capability of achieving large areas via a modular approach, a small pixel size as a result of advances in size reduction through the use of the latest developments in complementary metal–oxide–semiconductor (CMOS) technology [1] makes these types of detectors of increasing interest for x-ray applications ranging from medical imaging [2,3] to astrophysics [3], high-energy physics [4], and synchrotron applications [5,6].

In particular, these detectors are of interest for synchrotron applications that make use of imaging techniques, such as crystallography and ptychography, which require very large dynamic range. Hybrid photon-counting detectors can also be used to take advantage of the features of synchrotron radiation to research new technological advances that are directly related to these detectors. Such research can lead to the enhanced performance of hybrid photon-counting detectors for x-ray applications in synchrotron facilities and also for applications in other fields.

The next sections introduce hybrid photon-counting semiconductor detectors and synchrotron radiation. This is followed by a description of the link between synchrotrons and hybrid photon-counting detectors. Firstly, an overview of their use for synchrotron applications will be presented. Secondly, the latest research developments in this detector technology will be discussed, as well as how synchrotron beams can be used to characterize and evaluate their performance.

10.2   HYBRID PHOTON-COUNTING SEMICONDUCTOR DETECTORS

Hybrid semiconductor photon-counting detectors consist of monolithic semiconductor photodiodes, with either a strip (1-D) or an array (2-D) configuration (the latter is also known as pixel array detectors or PAD). This is bump bonded to front-end electronics consisting of CMOS-based application-specific readout chips (ASICs) (Figure 10.1). Silicon is the semiconductor material most commonly used as a sensor in hybrid photon-counting detectors. Its standard thickness of 0.3–0.5 mm is suitable for detecting x-ray photons in the range of 6–15 keV at room temperatures and down to 1.75 keV for chilled in-vacuum detector environments [7]. Recently, first assemblies using higher-atomic-number materials such as 0.75–1 mm thick cadmium telluride CdTe [8], gallium arsenide (GaAs) [9,10], and germanium (Ge) [11] have been developed and can extend the use of this technology to detect photons

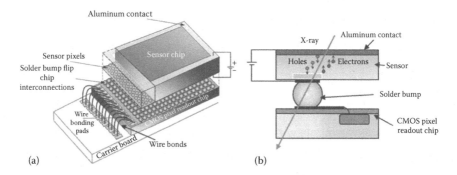

FIGURE 10.1   Sketch of a hybrid PAD photon-counting detector (a) and zoomed image corresponding to a pixel within the detector (b). (From ADVACAM. Perspective of the sensor interconnectivity to the TSV readout ASIC: Realization of a large area photon counting camera. http://indico.cern.ch/event/209454/session/12/contribution/49/material/slides/0.pdf.)

of a few hundred keV. However, module sizes of these high-atomic-number materials are currently limited by the relatively small wafer size [12], whereas silicon-detector modules of larger size are obtained from wafers of 6 inches. At the lower end of the energy range, the photon interacts with these semiconductor materials, releasing energy mainly via the photoelectric effect, and additionally via Compton and Rayleigh scattering at higher energies [13]. This generates a charge cloud of electrons and holes proportional to the total x-ray photon energy absorbed in the sensor.

The semiconductor is operated fully depleted, and thus the electron and hole charge clouds travel under the effect of the applied electric field toward the anode and cathode, respectively (Figure 10.1b). The transport of the charge carriers that form the charge cloud is affected by the diffusion and drift velocity. The diffusion of carriers is related to their random movement and the charge-cloud distribution spread, and depends on the temperature, the carrier mobility, and the carrier concentration. The drift velocity is related to the acceleration caused by the electric field, and thus is proportional to the electric field applied and the carrier mobility [14]. Once the charge arrives at the electrode plate, it is collected and integrated by a charge-sensitive preamplifier in the front-end ASIC, producing a voltage step-like function with a long exponential decay time. It is then filtered by a shaping amplifier into a voltage pulse, the height of which is proportional to the charge released by the x-ray photon. The pulse height is then compared in the discriminator to a preset value acting as a threshold, and each pulse larger than the threshold increments a counter by one. The counter depth defines the dynamic range of the detector. In this way, the signal is digitized, with each count corresponding to a single photon of energy above the threshold. The energy threshold is used to suppress undesired effects such as electronic noise and background fluorescence, or reduce effects such as charge sharing between adjacent pixels [15].

10.2.1 CHALLENGES IN PADS

The challenge of increasing the performance of these detectors lies in achieving large areas without dead spaces between modules, reducing the gap between chips within the same module, suppressing the charge-sharing effect, and developing detectors for the higher-energy x-ray range.

A drawback of present hybrid photon-counting detectors results from the dead space of several millimeters created between modules when designing large-area detectors (Figure 10.2). This dead space is caused by the space required for wire bonds at the edge of the ASICs in order to transmit the data (Figure 10.1a). One of the leading technological developments that can eliminate this gap is the use of through-silicon vias (TSV) [16,17]. TSVs consist of laser-drilled holes in the chip, which, after being filled with a conducting material, will allow vertical interconnection access (VIA) between ASICs. By this means it will be possible to stack layers, achieving a compact three-dimensional (3-D) configuration. Another research field that will contribute to reducing the gap between modules (and between chips within the same module) is the development of edgeless sensors. Here, the number of guard rings is reduced and the guard-ring separation distance is decreased from hundreds

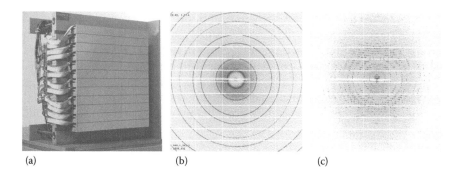

(a)   (b)   (c)

FIGURE 10.2   A Pilatus 6M hybrid photon-counting detector of 424×435 mm² area, consisting of 5×12 modules (a), a silicon powder diffraction image (b), and a single crystal diffraction from the protein thaumatin (c) taken with this detector. (Courtesy of BL I04 of Diamond Light Source.) The dead space due to the module tiling used to construct large-area detectors can be observed in all three figures: in the detector assembly and in the acquired data images. ((a) From A. Bergamaschi et al., Experience and results from the 6 megapixel PILATUS system. *Proceedings of the 16th International Workshop on Vertex Detectors*, Lake Placid, NY, 2007. http://pos.sissa.it/archive/conferences/057/049/Vertex%202007_049.pdf.)

to tens of micrometers [16,18]. An alternative approach to decrease the dead areas between modules is to tilt the modules to cover the gap areas [19].

A further undesired effect arises as a consequence of a charge being shared between adjacent sensor pixels or strips when x-rays impinge on the pixel edge. This results in count losses as the charge cloud is split between 2 and 4 pixels (depending on whether the photon impinged on the edge or in the corner). Such charge clouds generate a pulse with a height inferior to the energy threshold, which consequently does not contribute to the counter. This effect increases the smaller the strip or pixel pitch size and the thicker the sensor, strongly affecting the pixel array sensor configuration.

The approaches proposed to eliminate the charge-sharing effect can be grouped into three types, depending on whether the detector mechanical design changes and which detector component (the sensor or the front-end ASIC) is investigated. An example of the first approach is to tilt the detectors with respect to the direction of the impinging beam [21]. The development of 3-D sensors is an example of the second approach. Here, planar electrodes are replaced with columnar electrodes that go through the bulk of the material either completely [22] or partially [23] (Figure 10.3). The idea behind this approach is that the charge cloud does not need to travel the entire distance of the bulk thickness in order to be collected by the electrodes; it therefore undergoes less diffusion, and consequently less charge is shared with adjacent pixels. A further benefit of this electrode geometry is that it requires less voltage in order to deplete the sensor. The third approach implements features in the front-end ASIC that try to correct the charge-sharing effect. This approach has been boosted by the continuous reduction in CMOS feature size, which enables increased component density and thus enhanced ASIC functionality in the same footprint. This is exemplified by the Medipix3 ASIC [1], developed within the Medipix collaboration [24], which has implemented an arbitration unit and a comparator that eliminates the charge-sharing effect when operating the chip in charge-summing mode (CSM).

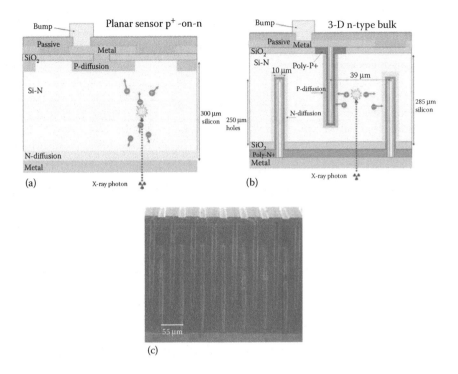

FIGURE 10.3   Sketch of a silicon sensor with planar electrodes (a) and with 3-D electrodes (b); and a real image of the 3-D double-sided silicon sensor (c). ((a) and (b) from Pellegrini, et al., *Nucl. Instrum. Methods A*, 592, 38, 2008; (c) courtesy of CNM-CSIC.)

Most of these developments have made use of synchrotron x-ray beams to characterize and subsequently improve these detector technologies. Section 10.3 discusses how synchrotron beams have been used successfully to characterize hybrid photon-counting detectors, and will introduce the latest developments, such as edgeless and 3-D sensors, and the CSM mode of operation implemented in Medipix3 ASICs.

10.3 SYNCHROTRON RADIATION

Synchrotron radiation (SR) consists of high-brilliance x-ray photons that are emitted when relativistic charged particles (electrons in synchrotron facilities) change their trajectory due to the action of magnetic fields. Predicted by Ivanenko and Pomeranchuk, SR was first observed by Elder at General Electric in a 70 MeV synchrotron in 1947 [25]. At that time, SR was considered a parasitic by-product, since the charged particle loses energy through this mechanism. However, since then, SR has evolved from being an undesired effect to becoming an essential tool used to help understand the structure and nature of materials. Purpose built synchrotron facilities have been constructed for material science research, and SR has increased in energy and brilliance over the years. Modern synchrotrons consist of a linear accelerator, a synchrotron booster, and a storage ring. The storage ring consists of many straight sections formed into a loop, with each corner comprising either a bending magnet or an insertion device. The magnetic field generated in these devices affects the electron's trajectory, which then loses energy through photon emission. This emission is known as synchrotron radiation.

SR provides a high brilliance and an intense polarized beam of photons collimated in the direction of flight with an angular spread of about 1 mrad and a pulse-length time structure down to 100 ps [26]. The photon beam emerges tangentially to the storage ring into an *optics hutch*, where several stages of lenses, mirrors, slits, and collimators are used to steer the beam and select other beam features, such as its monochromatic energy (using a silicon crystal monochromator, a grating monochromator, or a multilayer mirror) and its size. Following the optics stage, the beam arrives at the *experimental hutch*, where the sample to be studied is located, together with the detector used to detect the photons after they have impinged on the sample, as well as the rest of the equipment necessary for the experiment (e.g., a sample stage and a diffractometer). The combination of optical hutch and experimental hutch is known as a *beamline*. Around the synchrotron storage-ring perimeter, there can be as many beamlines as there are bending magnets or insertions devices, typically 20–40 beamlines in modern facilities [27]. In a single synchrotron facility, many techniques can be applied, as each beamline specifies the features of the beam setup in the optical hutch depending on the focus of the research being undertaken.

In SR facilities, material information is gathered by studying the interaction of x-ray photons with matter via diverse imaging and spectroscopic techniques. X-ray diffraction processes provide structural information. In crystallography, the material structure is inferred from the (single-crystal or powder) x-ray diffraction pattern of the atoms using Bragg's law (Equation 10.1), after an x-ray beam with a wavelength (λ) similar to the sample's lattice size (d) impinges on the material [28]. This is the technique that revealed the DNA double-helix structure [29].

$$\text{Bragg's law:} \quad 2d \sin \theta = n\lambda \tag{10.1}$$

Material composition can be studied through different spectroscopic techniques (x-ray fluorescence [XRF], x-ray absorption [XAS], and x-ray emission [XES]) by analyzing the energy of the released photons after scanning the sample through an energy range. The elemental constituents of the material can be identified by detecting the characteristic excitation lines (fluorescence, absorption, and emitted photons) of the material [26].

For each of these techniques, and depending on the aim of the study, detectors with specific designs and requirements are necessary. Imaging applications require good spatial resolution, which implies detector modules with small strip/pixel size, tiled in some cases to form detectors with a large area. Spectroscopic applications lean toward good energy resolution, which requires detectors with a large number of carriers generated per photon interaction (event) and fast charge collection [30]. Additionally, for synchrotron applications, detectors for both techniques require fast-frame rates, good signal-to-noise ratio, large dynamic range, radiation hardness, negligible dead time, and low noise. Spectroscopic detectors are beyond the scope of this chapter (but additional information can be found in [26]), since the focus is on imaging detectors such as hybrid photon-counting semiconductor detectors.

10.3.1 HYBRID PHOTON-COUNTING DETECTORS
FOR SYNCHROTRON APPLICATIONS

Hybrid photon-counting detectors are particularly useful for single-crystal and powder-diffraction techniques in synchrotron applications [31]. Diffraction occurs when an x-ray beam with a wavelength similar to the distance between atoms impinges on the sample. If the sample is a single crystal, this produces a pattern of spots of different intensities (Figure 10.2c), whose distance is related to the separation of structures within the material, given by Bragg's law (Equation 10.1). Single-crystal diffraction is widely used in protein crystallography with beam energies between 8 and 12 keV, and provides information about the molecular and atomic distribution within the protein (to within a few nanometers) (Equation 10.2). If the sample consists of a number of crystals randomly orientated, the diffraction produces a pattern of cones (Figure 10.2 b) whose distance is related to the planes in the lattice. This technique is known as powder diffraction.

$$E \text{ (keV)} = 1.24 / \lambda \text{ (nm)} \tag{10.2}$$

Hybrid photon-counting semiconductor detectors deliver the performance required for x-ray diffraction applications in SR facilities because of their inherent characteristics, such as high dynamic range, fast frame rate, negligible noise and dead time, and large detection area. The high-dynamic-range counters and fast frame rate are both necessary to cope with the high brilliance delivered by modern SR facilities. The ability to record both strong and weak signals simultaneously is enhanced by the high dynamic range and also by having a detector with negligible noise. The fast frame rate is useful for acquiring data before the sample undergoes the effects of radiation damage caused by excess exposure to a high-brilliance beam. Radiation damage can also be reduced by using detectors with negligible dead time. Large-area detectors are of interest for this technique because the further the detector is from the sample, the larger the signal-to-scatter background ratio. Finally, the smaller the pixel size, the better the spatial resolution that can be achieved, and good spatial resolution is required in crystallography in order to determine the distance between the diffraction patterns [32].

Due to their characteristic suitability for x-ray diffraction applications, hybrid photon-counting detectors are becoming a standard component of many synchrotron beamlines around the world. Systems commercially available, such as Pilatus commercialized by DECTRIS [5,33,34] and XPAD detectors commercialized by imXPAD [6,35], which feature a pixel size of $172 \times 172 \ \mu m^2$ and $130 \times 130 \ \mu m^2$, respectively, cover these applications. However, other synchrotron applications, such as coherent diffraction, require detectors to have pixel-pitch sizes smaller than 100 μm in order to take advantage of this x-ray technique. The Medipix ASIC range of detectors developed at CERN [1,24,36], which have a $55 \times 55 \ \mu m^2$ pixel size, fulfill these requirements, as does the EIGER ASIC [37] developed by PSI and commercialized by DECTRIS, which features 75×75 μm^2 pixels. One of the first detectors developed within the Medipix family of ASICs for synchrotron applications was the Medipix2-based MAXIPIX detector developed by ESRF [36,38]. Large-area Medipix3-based detectors are under development to cover these and other applications, such as the EXCALIBUR [39] developed by Diamond Light Source, the LAMBDA [40] developed by DESY, or the HEXA [41] developed by ANKA-KIT. All these Medipix-based detector developments are conducted by the synchrotrons and associate institutes within the Medipix collaboration. However, decreasing the pixel pitch increases the charge sharing between pixels, leading to a loss of count efficiency at the borders and corners of the pixels. The different developments aimed at reducing this effect are discussed in Section 10.4.2.

Detectors are mainly used as measurement instruments to conduct research in synchrotron facilities. However, they themselves can also be the focus of research in synchrotron facilities, as features of SR beams can be useful tools for characterizing detectors. This will be discussed in Section 10.4.

10.4 CHARACTERIZING HYBRID PHOTON-COUNTING DETECTORS WITH SYNCHROTRON RADIATION

10.4.1 SYNCHROTRON BEAM CONFIGURATION

Synchrotron beams are very versatile and vary from beamline to beamline, but general features such as their monochromaticity and high flux are essential characteristics of detectors. The monochromaticity of the delivered synchrotron beam can be used to evaluate the linearity response of the detector, and its high flux can provide information on detector dead time. The radiation hardness of the detector can be obtained by directly exposing it to the *white/pink beam* (i.e., the beam that comes directly from the bending magnet without being processed in the optical hutch). Energy resolution at different energies and image resolution can be also studied.

The beam can also be reduced in size to the order of few microns using a compound refractive lens (CRL) [42]. A CRL consists of a number of lenses (between tens and hundreds depending on the energy used and the distances to the focal point), made of beryllium and of parabolic shape, which are stacked in line in a He-filled chamber [43]. This microfocused beam tool, also referred to as *pencil beam*, has proved of real benefit in studying how charge is generated and collected in hybrid photon-counting semiconductor detectors. Figure 10.4 shows the CRL setup in order to characterize the Medipix3 detector using microfocus scans.

As a result of the small size of the beam (down to the order of 1–5 μm) compared to the detector pixel size (larger than 55 μm), the location where the charge impinges on the detector can be controlled with high precision, and thus effects such as the charge sharing between adjacent pixels or the reduction of the guard ring in edgeless sensors can be investigated. This is performed by scanning the detector surface with the pencil beam in the regions of interest, such as boundaries between pixels, at the limit of the sensors, or where material defects are observed. The use of SR, and in

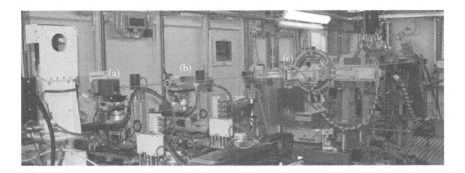

FIGURE 10.4 **(See color insert)** Photograph showing the experimental hutch of beamline B16 at Diamond Light Source [44]. The incoming beam from the optics hutch is represented by the red line, which impinges on the detector to be characterized (a), once the beam has passed through the CRL lenses (b), which produces the microfocus beam. The sample to be analyzed is typically set in a diffractometer (c). (From Gimenez, E.N., et al., *IEEE Trans. Nucl. Sci.*, 58, 323, 2011.)

particular the microfocused beam, has proved to be a very useful tool in the research and development of new ASIC technologies and sensor materials. This is analyzed in further detail in the next sections.

10.4.2 CHARGE SHARING BETWEEN PIXELS

The spatial resolution of PAD detectors can be improved by reducing the pixel size. However, this leads to an increase in the number of events whose generated charge cloud is split between adjacent pixels, that is, along the sides and at the corners of pixels. This effect is not negligible. It can lead to over- or undercounting of events, depending on the setting of the discriminator threshold level (Figure 10.5), and is known as the *charge-sharing effect* [45]. When a monochromatic beam of energy (E) impinges on the boundary region between pixels, the charge cloud generated by a unique photon splits between two adjacent pixels. (1) If the discriminator threshold level is set to less than half of the beam energy ($E_{th} < E/2$), the collected charge per pixel can be larger than the threshold level and thus triggers two counts per event (one in each pixel). This process is referred to as double counts. Similarly, if the region where the photon interacts with the pixel is at the corners of the pixel, this can lead to triple and quadruple counts per event. (2) If the discriminator threshold level is set to values higher than half the beam energy ($E_{th} > E/2$), the charge cloud split between two adjacent pixels may not be enough to exceed the threshold level in any of the pixels, and thus no event is recorded, leading to count losses [21,46]. In the latter case, the higher the discriminator threshold is set, the larger the area that would not be sensitive to photons. This consequently reduces the effective sensitive area of the detector [47]. Figure 10.5 shows the effect of over- and undercounts in the border region of two adjacent pixels depending on the value set in the discriminator threshold. Different approaches have been considered to reduce or eliminate the charge-sharing effect in PAD detectors.

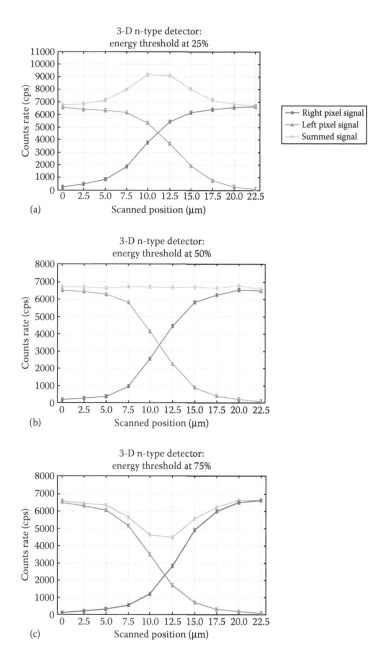

FIGURE 10.5 **(See color insert)** These plots show the charge-sharing effect between two adjacent pixels as a function of the set discriminator threshold. The red and blue lines show the charge collected by two neighboring pixels when a microfocus scan is performed across them. When the scan moves from the pixel on the left to the one on the right, the charge collection of the left pixel (red) decreases in the boundary region until it reaches negligible values. The beam then moves to the adjacent pixel to the right (blue),

10.4.2.1 Modifying the Setup

The first approach is to set the discriminator threshold level to half of the impinging energy ($E_{th} = E/2$). This method is only effective for monochromatic beams, and reduces the over- and under counts between the sides of 2 pixels, but not in the corners. In the corner region, the charge cloud generated by the event of energy (E) can be split between the four adjacent pixels (i.e., E/4) and, therefore, will not be enough in any of the pixels to exceed the half-energy threshold needed (i.e., E/2). Consequently, it will not trigger a count [46]. Therefore, in this region there will still be count losses due to the charge-sharing effect. An additional factor that has an effect on this process is the trimming of the pixels in the detector, also referred to as pixel-threshold equalization [42,45,68]. The pixel-trimming process aims at leveling all the pixel responses to the same impinging energy. Ideally, this response would be a delta-shape function. However, in practice, the pixel response to the same impinging energy describes a Gaussian distribution due to mismatches between electronic components, and to the discrete number of trim steps available to trim the response of each pixel (which are limited by the logic bit depth) [49]. Therefore, even when setting the discriminator threshold to half the beam energy, the response to the same impinging energy can vary slightly between adjacent pixels despite being trimmed, and double counts or loss of counts may still occur in this region.

A second approach to reducing the charge-sharing effect is to tilt the detector with respect to the impinging beam [19,21]. With this detector setup, the sensitive area where x-ray interactions can generate a charge cloud that can split between neighbors is reduced. However, tilting the detector results in attenuation of the beam intensity, and additional procedures are required to calibrate the image.

10.4.2.2 Modifying the Sensor

An alternative approach is to use 3-D silicon sensors instead of planar sensors. A 3-D silicon sensor has columnar electrodes passing completely or partially through the bulk of the material (Figure 10.3). This is in contrast to a standard planar electrode, which lies close to the surface layer of the material. The purpose of columnar electrodes is to reduce the distance the charge cloud has to travel to be collected, and by this means reduces the charge shared between pixels. Consequently, this sensor design decreases the charge collection time, which benefits faster acquisitions and reduces the voltage needed to deplete the sensor. This technology emerged within the framework of high-energy physics to track particles, where shorter charge-cloud collection times were required in order to increase the number of tracked particles in

---

FIGURE 10.5 (Continued) whose charge collection increases respectively. The green line shows the summed signal from the left and the right pixels. When the discriminator threshold is set at 25% of the impinging x-ray energy (left image) there are overcounts in the boundary region, and when the discriminator threshold is set at 75% of the impinging x-ray energy (right image) there are count losses. In both cases this is due to the charge sharing between the pixels. When the discriminator threshold is set at half the impinging energy (b), the summed signal is constant, indicating that there are no over- or undercounts due to the charge-sharing effect. (From Gimenez, E.N., et al., *Nucl. Instrum. Methods A*, 633, 114, 2011.)

high-flux conditions. The 3-D sensor design shows an improvement in reducing the charge-sharing effect compared to the planar electrode configuration, but the main drawback of this approach is that the columns are 3-D structures within the sensor (Figure 10.3) [48,50]. This reduces the amount of sensor material employed to stop the x-ray and convert its energy into carriers. Consequently there is a loss of sensitivity where the 3-D electrodes are located (Figure 10.6).

10.4.2.3 Modifying the ASIC

The approach followed by some photon-counting ASIC developers is to implement architectural functionalities that can reduce or eliminate the charge-sharing effect, resulting in the detector always recording one event per impinging x-ray. This is the case with the Medipix3 ASIC, [1] developed within the Medipix collaboration at Conseil Européene pour la Recherche Nucléaire (CERN). Medipix3 has implemented two new readout modes in addition to the standard readout mode (also referred to as *single pixel mode* [SPM]). The first new mode aims at eliminating charge-sharing events, and is referred to as *charge summing mode* (CSM); the second, aimed at providing spectroscopic information, is known as *color mode* (CM).

The Medipix3 chip is part of the Medipix family of ASICs and consists of 256×256 pixels of 55 μm pixel pitch. This is one of the smallest-pixel-size ASICs dedicated to photon-counting semiconductors detectors presently available in 2014 (refer to Section 10.3.1). However, the small pixel size results in a larger charge-sharing effect for the Medipix3 ASIC compared to other dedicated ASICs. Therefore, the implementation of functionality that eliminated the charge-sharing effect was more relevant to this ASIC than to others. This step forward was made possible due to advances in the CMOS fabrication technology. The predecessor to the Medipix3 chip, the Medipix2 [36], was implemented in 0.25 μm CMOS technology. Medipix3

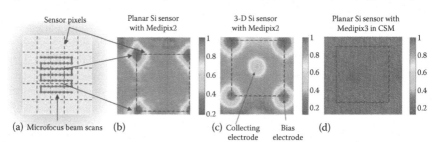

FIGURE 10.6 **(See color insert)** Sketch of a microfocus beam scan on a PAD detector (a) and image of the data acquired from the microfocus beam scan (showing the area of a pixel within the dashed black line) in a planar silicon sensor (b) and a 3-D silicon sensor (c), both with Medipix2 in standard readout; and a planar silicon sensor readout with Medipix3 in charge summing mode (CSM) (a). For all the sensors, the discriminator threshold was set at half the impinging x-ray energy. Images (b) and (c) show the difference in sensor response due to the charge-sharing effect at the corners for the planar sensor (b), and due to the column electrode and the charge-sharing effect in the 3-D sensor (c). Image (d) shows a uniform sensitive area for the pixel without count losses at the corners as a result of the CSM readout operation mode of the Medipix3 ASIC. (From Gimenez, E.N., et al., *IEEE Trans. Nucl. Sci.*, 58, 323, 2011; MacRaighne, A., et al., *Nucl. Sci. Symp. Conf. Rec.*, 2145, 2009.)

is designed in 8-metal 0.13 μm CMOS technology. The feature size reduction releases space available to implement additional functionality per pixel, and for implementing the new operating readout modes CSM and CM.

The great benefit of characterizing detectors with a synchrotron beam was demonstrated during the development of the Medipix3 ASIC. The Medipix3 ASIC underwent several stages of redesign after initial characterization with a synchrotron beam was used to understand the ASIC performance [42,70]. This led to an improved final version of the ASIC: the Medipix3RX [53]. Among the features that changed between the first and last versions of the Medipix3 ASIC are the CSM operation mode, the pixel-trimming circuit, and the continuous read-write feature.

10.4.2.3.1 Synchrotron Beam Tests for ASIC Development

The first released version of the Medipix3 ASIC had a new front-end architecture compared to the Medipix2 chip. The new front-end architecture implemented the charge summing mode (CSM) of operation and was aimed at eliminating the charge-sharing effect that particularly affected highly segmented PAD semiconductor detectors. CSM grouped the pixels in clusters of 4 and added together the signal produced by the collected charge in each of the 4 pixels at a summing node. Each summing node reconstructed the pulse from the charge generated by the impinging x-ray at each group of the 4 pixels. The total reconstructed pulse at each summing node is compared to the discriminator threshold. In the case where the reconstructed pulses of two adjacent summing nodes are above the threshold simultaneously, an arbitration circuit selects the one with the largest pulse as the hit, that is, the one that remains above the discriminator level for longest. Therefore, if the Medipix3 ASIC is operating in CSM, when the charge cloud generated from a single event splits between adjacent pixels, the total pulse can be reconstructed from each cluster of 4 pixels triggering a single count per impinging event. This eliminates the over- and undercounts recorded due to the charge-sharing effect.

Synchrotron beam tests were performed to characterize the first production of Medipix3-based silicon photon-counting detectors [42]. The detector consisted of a 300 μm thick pixel array silicon sensor, bump bonded to a Medipix3 ASIC. The tests showed that when the detector was operated in CSM readout mode, the ASIC presented large pixel dispersion due to an unexpected mismatch between transistors, which could not be fixed by the trimming algorithm. When microfocus pencil-beam surface scans were performed on clusters of 9 adjacent pixels, it was shown that the CSM readout counted each event only once and eliminated the charge-shared events (Figure 10.6 d). However, this procedure also showed that events were being misallocated, that is, events impinging on the surface of one pixel were being registered by the neighboring pixel. This implied that the total pixel-sensitive area was inhomogeneous, that is, for some pixels, the sensitive area was larger than the pixel pitch of 55 μm, while for others it was smaller (Figure 10.7).

The effect observed was a consequence of the arbitration process not having the same threshold level against which all pulses could be compared [53]. The arbitration circuit was sensitive to the threshold discriminator dispersion between summing nodes; that is, the time that the signal remained above the threshold was reduced or increased depending on whether the threshold was at a higher or lower level,

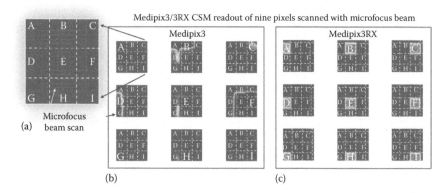

FIGURE 10.7 (**See color insert**) Image (a) shows the microfocus-beam scan path across a sensor region consisting of 9 pixels (from A to I). Image (b) shows the readout from the 9 pixels at different scan-beam positions for the Medipix3 ASIC, while image (c) shows the same from the Medipix3RX ASIC. Both ASICs are operating in charge-summing readout mode (CSM). For the images (b) and (c) the letters identify the pixel region on the sensor, with the large yellow letter indicating the region where the beam impinges along the scan. Image (b) shows the misallocation of events being registered to pixels other than the one they impinged on for the Medipix3. When the microfocus beam impinges on pixel A, events are registered by pixel B, and so on, and results in pixels with different sensitive areas (colored red in the image). The Medipix3RX ASIC implements a redesigned CSM architecture, which corrects the misallocation effect and results in pixels with the same sensitive area as shown in (c). (From Gimenez, E.N., et al., *IEEE Trans. Nucl. Sci.*, 58, 323, 2011.)

respectively. This weakness of the charge summing mode architectural design generated an event misallocation effect that was additionally magnified by the pixel dispersion observed in CSM due to the mismatch between transistors. Results obtained from synchrotron-beam tests of the first Medipix3 ASIC version led to a redesign of the CSM architecture in the front-end electronics. This was implemented in the upgraded version of the Medipix3 ASIC, named Medipix3RX.

The new CSM architecture in the Medipix3RX ASIC requires two discriminator thresholds rather than just one, as used in the first version [52]. One of the thresholds has similarities to the threshold in Medipix3, as the charge collected in adjacent pixels is still added at the summing nodes in order to reconstruct the pulse. This discriminator threshold is used to compare the reconstructed pulses between the summing nodes, and the hit pulse value will correspond to the largest of the reconstructed pulses at the summing nodes. The difference lies in the fact that this threshold is not used to allocate the count. Count allocation is performed using the second threshold, which compares the charge collected at each individual neighboring pixel and allocates the hit to the pixel that recorded the highest charge value. Finally, if the reconstructed pulse in one of the summing nodes exceeds the first threshold, and the hit is allocated to a pixel adjacent to that summing node, then the event is recorded in the counter. Additionally, the new Medipix3RX ASIC design uses more controlled production, which has reduced the mismatch between transistors that had previously led to increased pixel dispersion in the first version of the Medipix3 chip.

Synchrotron beam tests were used to characterize the Medipix3RX ASIC, following the same procedures used with its predecessor Medipix3. This allowed easy comparison. When operating in CSM readout mode, the results showed that the pixel dispersion had reduced significantly to levels similar to that observed when the detector was operated in standard readout mode. Furthermore, the implemented CSM functionality eliminated the charge-sharing effect and at the same time corrected the event misallocation (Figure 10.7b) observed in the first version of the Medipix3 ASIC.

10.4.3 EGDELESS SENSORS

One of the challenges in photon-counting detectors is to increase their active area when tiling modules in order to create large-area detectors. There are two technological developments that contribute to meeting this goal. The first is to reduce the dead area at the borders of the sensor, a development defined as *edgeless sensor*. The second is to stack different layers and interconnect them using TSV (see Section 10.2.1). The latter will replace wire bonds reducing the intermodule space required. The present approach to reducing the effect of the space required for wire bonds is to group all wire-bond connections at one side of the ASIC. This enables the development of modules consisting of two rows and several columns (e.g., from 2 columns in quad silicon detectors to the 8 columns of the Excalibur modules). However, tiling these modules still leaves gaps of several millimeters (Figure 10.2). TSV technology can be tested electronically, whereas edgeless-sensor technology can be evaluated using an SR beam, specifically a microfocused beam.

In standard sensors, guard rings are located around the periphery of the sensor, surrounding the strip or pixel structure (Figure 10.8b). Their aim is to avoid the contribution of surface current to the electronic noise, and charge loss at the boundaries of the sensor when a photon impinges on the edge, while at the same time keeping the electric field for charge collection homogeneous. While standard guard rings vary in width from 200 to 500 μm, the dead area at the periphery of the sensitive structure can increase by 1 mm or more once secure space is left for sensor dicing. The dead areas at the sensor periphery need to be reduced in order to tile modules closer together. This implies the reduction or elimination of the guard-ring structure.

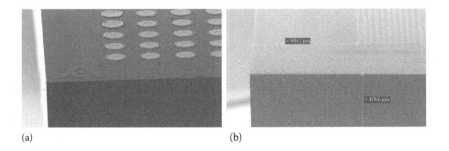

(a)   (b)

FIGURE 10.8 Silicon PAD sensor images with the same bump-bond opening size and spacing showing the difference in dead space between an "active edge" edgeless sensor without guard rings (a) and a standard sensor with guard rings (b). (Courtesy of Advacam Oy.)

These so-called edgeless sensors aim to eliminate the guard-ring structure without affecting the charge collection; that is, to avoid charge losses at the borders that affect the detector sensitivity in those regions, which would normally contribute to higher leakage current from the sensor. The differences in the area surrounding the pixel-sensitive structure are shown in Figure 10.8. Figure 10.8a shows a PAD active edge (edgeless) silicon sensor, while Figure 10.8b shows the PAD sensor with the same bump-bond opening size and spacing but with a standard guard ring at the edges of the sensitive structure.

There are several techniques used to develop edgeless sensors, such as active edge [54], slim edge [55], scribe-cleave and passivate (SCP) [56], and current terminating rings (CTR) [57]. They differ in the design and process followed to reduce the dead area without charge losses. These techniques can be tested using SR beam where a microfocus beam scans along and across the edges of the sensor (Figure 10.9). Evaluation of the charge-collection process (the amount of charge collected) is performed (Figure 10.9b), and simultaneously, the leakage current is monitored. This provides information for evaluating the quality of these technologies by analyzing charge collection and leakage current.

The implementation of these technologies will bring advantages not only to imaging techniques in synchrotron application, but also to other fields of research, such as high-energy physics, because of the increase in sensitive area.

10.4.4 NEW SENSOR DESIGNS AND MATERIALS

The development of new sensors is another research area that can take advantage of the use of the synchrotron beam for both characterization and evaluation of their eligibility for use in applications that differ from those that led to their development. The SR beam, particularly the microfocused beam, was used to characterize and understand the charge-collection process in recent sensors, such as the 3-D

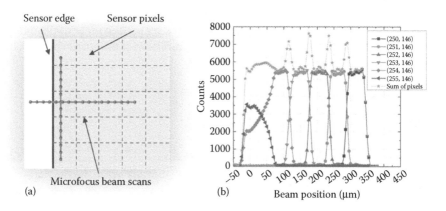

FIGURE 10.9 **(See color insert)** (a) Sketch of the vertical and horizonal microfocus beam scans used to test edgeless sensors. (b) Plot showing the charge collection of the microfocus scan across the edge of an active edgeless sensor. (From Bates, R., et al., *J Instrum.*, 8, P01018, 2013.)

silicon sensors mentioned in Sections 10.2.1 and 10.4.2.2. These sensors emerged as a solution to the requirements of high-energy physics applications, such as the Large Hadron Collider and Atlas upgrade at CERN [57,59]. Additionally, their design could have been a possible solution to the charge-sharing effect observed in PAD detectors. Tests with the SR beam were essential to evaluate their performance in reducing this effect and their suitability for x-ray imaging applications.

A further example of the use of SR is to characterize high-atomic-number PAD sensors, such as CdTe [60–63], CZT [64], or GaAs [10], bump bonded to a hybrid photon-counting detector. High-atomic-number materials are of interest because of their high x-ray stopping power and their effectiveness in detecting x-rays in the region starting from around 20 keV and going up to a few hundred keV. Each of these materials present their own challenges. In general, they are difficult to grow in large areas [12] and result in poor image quality due to various factors. Such factors include defects due to inclusions, strains, or dislocations of the material, nonideal charge transport, ageing due to x-ray exposure, and polarization effects. These effects depend strongly on whether holes or electrons are collected, the temperature of operation, and whether the electrode type is Schottky or Ohmic [15]. This is certainly the case for CdTe sensors.

Testing these materials with an SR beam can help in understanding the mechanisms affecting charge collection, and to evaluate if they can be used for applications that require good image resolution. For example, a microfocused beam tilted with respect to the sensors was used to study the charge collection close to these defects [65]. The rationale was to impinge a CdTe sensor at a tilted angle, where a defect had been previously located. Tilting the detector with respect to the beam provides information on the depth of interaction (DOI). Information on their spectroscopic performance can also be studied [63,66].

10.4.5 GLOBAL DETECTOR RESPONSE

In the previous sections, the suitability of the microfocus SR beam to study charge collection, evaluate the charge-sharing effect, and characterize edgeless and high-atomic-number sensors has been discussed. There are further SR beam configurations that are of interest in efforts to characterize the performance of hybrid photon-counting detectors [21,42,67–70]. Among the detector features that can be tested with an SR beam are linearity, dead time, radiation hardness, homogeneous response, and quantum efficiency. These features can also be tested with an x-ray tube or radioactive sources, but the SR beam has significant advantages.

Two of the advantages of the SR beam are monochromaticity and high brilliance. The SR beam monochromaticity spans a large range of energies (usually dependent on the beamline design), and the energy can easily be tuned by changing the mirror angles within the monochromator. For some monochromator configurations (silicon double-crystal mirror (DCM) or double multilayer mirror (DMM)), radiation with three times the energy of the main beam (known as the *third harmonic*) can be observed, but its flux is several orders of magnitude less than the main beam (also known as the *first harmonic*). Other monochromator configurations (DCM together with DMM) eliminate the third harmonic contribution altogether. A monochromatic beam that can span a broad energy range is useful for characterizing the linearity response of the detector [42,67]. In this case, an unfocused beam configuration is preferred in order to simultaneously cover a wider area of the detector.

Another useful feature of the SR beam is its high brilliance. This can also be increased or decreased by changing the monochromator setting. A high-brilliance beam together with selectable widths of aluminum filters are used to acquire information about the detector dead time. The structure of the SR beam in the synchrotron storage ring needs to be taken into consideration for this test [71]. The SR beam is not continuous but follows a distribution of discrete bunches of electrons a few picoseconds long, with bunch spacing of a few nanoseconds. Additionally, the synchrotron may run in a nonuniform mode where not all bunches are filled [42,67].

Synchrotron radiation can be used to test the radiation hardness of a detector, where the current intensity of the beam is known and the exposure time is controlled. The type of radiation hardness test can vary depending on the SR beam configuration. One option is to irradiate the detector with a direct beam straight from the insertion device (known as white or pink beam) [67,70]. Since in this case the beam is very intense, a fast shutter is required in order to evaluate the radiation sensitivity of the detector. When working with this beam configuration, caution should be taken to protect all the electronics inside the experimental hutch from the direct beam, especially the electronics associated with the detector. Another option is to use the microfocus beam to irradiate specific areas of the detector front-end electronics in order to check if they are radiation sensitive [70]. Evaluation of the radiation damage can be performed by monitoring the leakage current at the moment of irradiation, and afterwards by retrimming the thresholds or by annealing the detector.

The homogeneity of the pixel detector response can be tested by using the k-fluorescence emission of materials such as copper (8 keV), Ge (9.8 keV), and platinum (67 keV)[25]. In this case, the SR beam impinges on a sheet of fluorescence material with energy above the binding energy of the electrons in the k shell of the material. Photons of the material are emitted via the photoelectric effect. There are two possible configurations. First, the fluorescence material foil is set in transmission mode perpendicular to the beam, and the detector is aligned with the fluorescence material and the beam. This setup requires the use of a beam stop to avoid the direct beam impinging on the detector. Second, the fluorescence sheet is set in reflective mode, that is, with its surface at a 45° angle to the beam, and the detector surface is turned 90° with respect to the beam, effectively in an L-shaped configuration [42]. The advantage of the latter configuration is that no beam stop is required and the whole detector surface is flood illuminated by the fluorescence photons (no beam stop shadow). The detector needs to be located a minimum distance from the fluorescence material in order to be illuminated homogeneously, creating what is known as a flat-field image [19]. A good-quality flat-field image requires a large number of counts and provides information about the intrinsic differences between pixels, which can be corrected by calculating the flat-field coefficients. Additionally, analysis of scans increasing the discriminator threshold provides useful information on the detector pixel dispersion [42].

10.5 SUMMARY

Hybrid photon-counting semiconductor detectors are of interest to a broad range of scientific fields that require the detection of x-ray photons; this is due to their characteristics such as high dynamic range, fast frame rate, negligible noise, and low dead time. Additionally, this technology has in recent years lent itself to extensive use in synchrotron applications, due to the ability to produce large-area detectors by tiling modules and reducing the pixel pitch. However, there are still challenges ahead, such as reducing the intermodule gap, eliminating charge sharing, and developing high-atomic-number detectors. These are areas of current research interest where progress is ongoing.

The versatility of the SR beam has proved to be extremely beneficial in evaluating the latest advances in the sensors, ASICs, and stacking processes involved in detectors using these sensors. In particular, microfocus beam configurations have proved a very useful technique into helping evaluate new technological developments by controlling the location of charge generation and analyzing the process of charge collection. These developments will lead to enhanced hybrid photon-counting detector performance and will improve their suitability for a broader range of x-ray applications in a number of different research fields.

REFERENCES

1. R. Ballabriga, The design and implementation in 0.13 um CMOS of an algorithm permitting spectroscopic imaging with high spatial resolution for hybrid pixel detectors (2010), PhD. Thesis, Universitat Ramon Llull, Barcelona, Spain.
2. K. Taguchi and J.S. Iwanczyk, Vision 20/20: Single photon counting x-ray detectors in medical imaging, *Med. Phys.*, 40(10) (2013) 10090.
3. S. Del Sordo, et al., Progress in the development of CdTe and CdZnTe semiconductor radiation detectors for astrophysical and medical applications, *Sensors*, 9 (2009) 3491.
4. S. Eisenhardt, et al., Production and tests of hybrid photon detectors for the LHCb RICH detectors, *Nucl. Instrum. Methods A*, 595 (2008) 142–145.
5. B. Henrich, et al., PILATUS: A single photon counting pixel detector for X-ray applications, *Nucl. Instrum. Methods A*, 607 (2009) 247.
6. P. Pangaud, et al., XPAD3-S: A fast hybrid pixel readout chip for X-ray synchrotron facilities, *Nucl. Instrum. Methods A*, 591(1) (2008) 159–162.
7. J. Wernecke, et al., Characterization of an in-vacuum PILATUS 1M detector, *J. Synchrotron Rad.*, 21 (2014) 529–536.
8. L. Jones, et al., HEXITEC ASIC—A pixellated readout chip for CZT detectors, *Nucl. Instrum. Methods A*, 604(1–2) (2009) 34–37.
9. C. Schwarz, et al., X-ray imaging using a hybrid photon counting GaAs pixel detector, *Nucl. Phys. Proc. Suppl.*, 78 (1999) 491.
10. E. Hamman, Characterization of high resistivity GaAs as sensor material for photon counting semiconductor pixel detectors (2013), PhD. Thesis, Albert-Ludwigs-Universität, Freiburg, Germany.
11. D. Pennicard, et al., Development of high-Z sensors for pixel array detectors, 19th International Workshop on Vertex Detectors, PoS(VERTEX 2010)027. Scotland.
12. Semiconductor Wafer Inc. 2~6 Group wafer (2011), Semiconductor Wafer Inc. http://www.semiwafer.com/products/26groupwafer.htm (accessed in 2014).
13. M.J. Berger, et al., XCOM: Photon cross sections database. National Institute of Standards and Technology. http://www.nist.gov/pml/data/xcom/.

14. A. Owens, *Compound Semiconductor Radiation Detectors* (2012) CRC Press, Boca Raton.
15. R. Steadman, et al. ChromAIX: A high-rate energy-resolving photon-counting ASIC for 2 spectral computed tomography, *SPIE Med. Imag.*, (2010) 762220.
16. J. Kalliopuska. Perspective of the sensor interconnectivity to the TSV readout ASIC: Realization of a large area photon counting camera. Advacam (2013). http://indico.cern.ch/event/209454/session/12/contribution/49/material/slides/0.pdf.
17. Allvia. TSV and interposer technologies at Allvia. Allvia (2010). http://www.ipc.gatech.edu/workshop/2010/6.pdf (accessed in 2014).
18. G. Pellegrini, et al., Edgeless detectors fabricated by dry etching process, *Nucl. Instrum. Methods A*, 563 (2006) 70.
19. C. Le Bourlot, et al., Synchrotron X-ray diffraction experiments with a prototype hybrid pixel detector, *J Appl. Crystallogr.*, 45 (2012) 38–47.
20. A. Bergamaschi, et al., Experience and results from the 6 megapixel PILATUS system, In *Proceedings of the 16th International Workshop on Vertex Detectors,* Lake Placid, NY (2007). http://pos.sissa.it/archive/conferences/057/049/Vertex%202007_049.pdf.
21. C. Brönnimann, et al., Synchrotron beam test with a photon-counting pixel detector, *J. Synchrotron Radiat.*, 7(5) (2000) 301.
22. C.J. Kenney, et al., Active-edge planar radiation sensors, *Nucl. Instrum. Methods A*, 565(1) (2006) 272.
23. G. Pellegrini, et al., First double-sided 3-D detectors fabricated at CNM-IMB, *Nucl. Instrum. Methods A*, 592 (2008) 38.
24. Medipix. CERN. http://medipix.web.cern.ch/medipix/ (accessed in 2014).
25. A.L. Robinson, X-ray data booklet: Section 2.2: History of synchrotron radiation. http://xdb.lbl.gov/Section2/Sec_2-2.html (accessed in 2014).
26. H. Saisho and Y. Gohshi (eds), *Applications of Synchrotron Radiation to Materials Analysis*, Analytical Spectroscopy Library, Vol. 7 (1996), Elsevier, Amsterdam.
27. lightsources.org, Lightsources of the world, lightsources.org (2013) http://www.lightsources.org/regions (accessed in 2014).
28. J. Drenth, *Principles of Protein X-Ray Crystallography*, (2007) 3rd edn, Springer, Berlin.
29. J.D. Watson and F. Crick, Molecular structure of deoxypentose nucleic acids, *Nature*, 171 (1953) 738.
30. G.F. Knoll, *Radiation Detection and Measurement*, (2000) 3rd edn, Wiley, New York.
31. B.B. He, *Two-Dimensional X-Ray Diffraction*, (2009) Wiley, New York.
32. S.G. Angelo, Development of a mixed-mode pixel array detector for macromolecular crystallography, *Nucl. Sci. Symp. Conf. Rec. IEEE*, 7 (2004) 4667.
33. C. Brönnimann, et al., The Pilatus 1M detector, *J. Synchrotron Rad.*, 13 (2006) 120.
34. Dectris. https://www.dectris.com/ (accessed in 2014).
35. imXPAD. http://www.imxpad.com/ (accessed in 2014).
36. X. Llopart, et al., Medipix2, a 64k pixel read out chip with 55 µm square elements working in single photon counting mode, *IEEE Trans. Nucl. Sci.*, 49 (2002) 2279.
37. R. Dinapoli, EIGER: Next generation single photon counting detector for X-ray applications, *Nucl. Instrum. Methods A*, 650(11) (2011) 79.
38. C. Ponchut, et al., Photon-counting X-ray imaging at kilohertz frame rates, *Nucl. Instrum. Methods A*, 576 (2007) 109.
39. J. Marchal, et al., EXCALIBUR: A small-pixel photon counting area detector for coherent X-ray diffraction—Front-end design, fabrication and characterisation, *J. Phys. Conf. Ser.*, 425 (2013) 062003.
40. D. Pennicard, et al. Development of LAMBDA: Large area Medipix-based detector array, *J. Instrum.*, 6 (2011) C11009.

41. T. Koening, et al., On the energy response function of a CdTe Medipix2 Hexa detector, *Nucl. Instrum. Methods A*, 648 (2011) 265.
42. E.N. Gimenez, et al., Characterization of Medipix3 with synchrotron radiation, *IEEE Trans. Nucl. Sci.*, 58 (2011) 323.
43. L. Alianelli, et al., Ray-tracing simulation of parabolic compound refractive lenses, *Spectrochim. Acta B*, 62 (2007) 593.
44. K.J.S. Sawhney, et al., A test beamline on Diamond Light Source, *AIP Conf. Proc.*, 1234 (2010) 387.
45. L. Rossi, P. Fischer, T. Rohe, N. Wermes, *Pixel Detectors: From Fundamentals to Applications*, (2006) Springer, Berlin.
46. E.N. Gimenez, et al., Study of charge-sharing in MEDIPIX3 using a micro-focused synchrotron beam, *J Instrum.*, 6 (2011) C01031.
47. J. Marchal, Theoretical analysis of the effect of charge-sharing on the detective quantum efficiency of single-photon counting segmented silicon detectors, *J Instrum.*, 5 (2010) P01004.
48. E.N. Gimenez, et al., 3D Medipix2 detector characterization with a micro-focused X-ray beam, *Nucl. Instrum. Methods A*, 633 (2011) 114.
49. J. Uher, et al., Equalization of Medipix2 imaging detector energy thresholds using measurement of polychromatic X-ray beam attenuation, *J. Instrum.*, 6 (2011) C11012.
50. A. Mac Raighne, et al., Precision scans of the pixel cell response of double sided 3D pixel detectors to pion and X-ray beams, *J. Instrum.*, 6 (2011) P05002.
51. A. MacRaighne, et al., Synchrotron tests of 3D Medipix2 and TimePix X-ray detectors, *Nucl. Sci. Symp. Conf. Rec.*, (2009) 2145-2150.
52. R. Ballabriga, et al., The Medipix3RX: A high resolution, zero dead-time pixel detector readout chip allowing spectroscopic imaging, *J. Instrum.*, 8 (2013) C02016.
53. D. Pennicard, et al., Simulations of charge summing and threshold dispersion effects in Medipix3, *Nucl. Instrum. Methods A*, 636 (2011) 74.
54. C.J. Kenney, et al., Active-edge planar radiation sensors, *Nucl. Instrum. Methods A*, 565 (2006) 272.
55. G. Pellegrini, et al., Recent results on 3D double sided detectors with slim edges, *Nucl. Instrum. Methods A*, 731 (2013) 198.
56. V. Fadeyev, et al., Scribe-cleave-passivate (SCP) slim edge technology for silicon sensors, *Nucl. Instrum. Methods A*, 731 (2013) 260.
57. E. Noschis, et al., Final size planar edgeless silicon detectors for the TOTEM experiment, *Nucl. Instrum. Methods A*, 563 (2006) 41.
58. R. Bates, et al., Characterisation of edgeless technologies for pixellated and strip silicon detectors with a micro-focused X-ray beam, *J. Instrum.*, 8 (2013) P01018.
59. S. Altenheiner, et al., Planar slim-edge pixel sensors for the ATLAS upgrades, *J Instrum.*, 7 (2012) C02051.
60. D. Greiffenberg, et al., Characterization of Medipix2 assemblies with CdTe sensor using synchrotron radiation, *Nucl. Sci. Symp. Conf. Rec.*, (2008) 287–290.
61. A. Cecilia, et al., Investigation of crystallographic and detection properties of CdTe at the ANKA synchrotron light source, *J Instrum.*, 6 (2011) P10016.
62. M. Ruat, et al., Characterization of a pixelated CdTe X-ray detector using the Timepix photon-counting readout chip, *Nucl. Sci. Symp. Conf. Rec.*, (2012) 2392.
63. D. Maneuski, et al., Imaging and spectroscopic performance studies of pixellated CdTe Timepix detector, *J. Instrum.*, 7 (2012) C01038.
64. S. Midgley, et al., Hybrid pixel detector development for medical radiography, *Nucl. Instrum. Methods A*, 573 (2007) 129.
65. E. Frojdh, et al., Probing defects in a small pixellated CdTe sensor using an inclined mono energetic X-ray micro beam, *IEEE Trans. Nucl. Sci.*, 60(4) (2013) 2864.

66. R. Bellazzini, et al., Chromatic X-ray imaging with a fine pitch CdTe sensor coupled to a large area photon counting pixel ASIC, *J. Instrum.*, 8 (2013) C02028.

67. C. Ponchut, et al., Evaluation of a photon-counting hybrid pixel detector array with a synchrotron X-ray source, *Nucl. Instrum. Methods A*, 484 (2002) 396–406.

68. P. Kraft, et al., Characterization and calibration of PILATUS detectors, *IEEE Trans. Nucl. Sci.*, 56 (2009) 758.

69. K. Medjoubi, et al., Performance and applications of the CdTe- and Si-XPAD3 photon counting 2D detector, *J. Instrum.*, 6 (2011) C01080.

70. E.N. Gimenez, et al., Evaluation of the radiation hardness and charge summing mode of a Medipix3-based detector with synchrotron radiation, *Nucl. Sci. Symp. Conf. Rec. IEEE*, (2010) 1976-1980.

71. J.E. Bateman, The effect of beam time structure on counting detectors in SRS experiments, *J. Synchrotron Radiat.*, 7 (2000) 307.

11 Organic Semiconducting Single Crystals as Novel Room-Temperature, Low-Cost Solid-State Direct X-Ray Detectors

Beatrice Fraboni and Alessandro Fraleoni-Morgera

CONTENTS

11.1 INTRODUCTION

The detection of ionizing radiation is an important task for a number of technologically and socially relevant activities, ranging from environmental monitoring to industrial, security, and health applications. As such, a large variety of detectors for ionizing radiations have been developed in the past, based either on gas-filled containers (such as ionization chambers or Geiger counters), on solid or liquid materials capable of producing visible photons upon exposure to the radiation (scintillators), or on gaseous, liquid, or solid materials (or a combination of them) capable of visualizing the track of ionizing particles (cloud chambers, bubble chambers, spark chambers, silicon detectors, etc.). Ionizing radiation detectors may also be categorized as being "direct" or "indirect" detectors. A "direct detector" is one in which the incoming ionizing radiation is transduced by a sensor directly into an electrical signal. In an "indirect detector," the transduction occurs in a two-step process, the first step being performed by a first sensor (for example, a scintillator) and the second step being performed by a second sensor (for example, a photodiode). Indirect detectors are therefore constituted of two (or more) sensors, complemented by a signal amplifying/processing unit. This approach induces a loss of information in the process, especially at low radiation doses, and, most importantly, a rather complex and fragile device structure. State-of-the-art direct detectors are made with inorganic semiconductors, such as silicon, cadmium telluride, and gallium nitride, in which the incoming radiation directly generates electron–hole pairs that constitute the collected output electrical signal. However, some of these materials (e.g., Si, Ge) have less than ideal bandgap widths and show high dark currents, making them unattractive candidates for room-temperature operation, while others contain a large number of defects and electrically active impurities that affect the recombination and collection processes of the photogenerated carriers, as well as detector stability under irradiation and polarized operation. In general, to achieve a high radiation-detection efficiency it is necessary to have materials that exhibit a very low dark current ($<10^{-7}$ A), and thus possess a high resistivity ($>10^9$ Ω·cm). Large-area x-ray-detecting panels for digital imaging have been developed for medical applications, but are mostly based on the "indirect detection" approach, with amorphous silicon (or other inorganic semiconductors) coated by a scintillator material. The complexity of these devices and the high fabrication cost of detector-grade inorganic semiconductors limit their application in other fields. The main drawbacks of such large-area detectors are the poor conversion efficiency and signal-to-noise ratio typical of indirect detecting systems, plus the high cost, not taking into account the high complexity of operation (requiring trained operators). Therefore, alternative materials and technologies enabling the fabrication of direct detectors on large areas and at acceptable prices have to be sought.

Organic materials have recently attracted major attention because they demonstrated, in the case of other electronic/optoelectronic devices (e.g., transistors and light-emitting devices), room-temperature operation, low environmental impact, low fabrication costs, and the possibility of fabricating transparent and flexible devices, allowing unprecedented and integrated device functions and architectures [1–7].

As detectors of ionizing radiation, organic semiconductors have so far been mainly proposed in the indirect-conversion approach, either as scintillators [8,9] or as photodiodes [10–12], which detect visible photons coming from a scintillator and convert them into an electrical signal. A few examples of organic materials used for the direct detection of radiation have been recently reported in the literature, referring to semiconducting or conducting polymer thin films, or charge-transfer-conducting organic crystals. Interestingly, it has been shown that thin films based on semiconducting polymers can withstand high x-ray doses without significant material degradation [13]. However, to obtain this result, the polymers must be properly encapsulated to protect them from environmental oxygen and water; otherwise, significant degradation of the device occurs rapidly [14]. It has to be noted that effective encapsulation of semiconducting polymer-based devices leads to markedly increased production costs and complexity of device geometry. A few thin-film, organic-based direct detectors have been reported recently, but they rely on the presence in the detector of metallic electrodes or intentionally introduced metallic or semimetallic nanoparticles, which act as sources of secondary photoelectrons, to produce reasonable photoconversion efficiencies [13,15–21].

We recently showed that organic semiconducting single crystals (OSSCs) can be used as effective direct x-ray detectors. In particular, devices based on solution-grown

OSSCs (from two different molecules: 4-hydroxycyanobenzene [4HCB] and 1,8-naphthaleneimide [NTI]) (Figure 11.1) have been fabricated and operated in air, under ambient light, and at room temperature, at voltages as low as a few tens of volts, delivering highly reproducible performances and a stable linear response to the x-ray dose rate, with notable radiation hardness and resistance to aging [22].

The role of high-atomic number components (e.g., metals) in the device response has been elucidated, evidencing the intrinsic response of the organic crystals to x-rays, which allowed the fabrication of well-performing all-organic and optically transparent devices.

It is important here to remember how the use of single crystals has been fundamental to the development of semiconductor microelectronics and solid-state science. Whether based on inorganic [23,24] or organic [25,26] materials, the devices that show the highest performance rely on single-crystal interfaces, due to their nearly perfect translational symmetry and exceptionally high chemical purity. Moreover, organic single crystals show band-like transport behavior [27], and top-performing rubrene single crystals (vapor grown, only a few microns thick) have been recently reported to reach hole mobilities of 40 cm^2 V^{-1} s^{-1} [28] and electron mobilities up to 11 cm^2 V^{-1} s^{-1}, measured in a field-effect transistor configuration, which allows the current flowing through the crystal to be controlled and amplified [29–32]. The same crystals, when sufficiently thin (i.e., less than 3 µm), are reportedly conformable/flexible [33].

The observed performance indicates that OSSCs are very promising candidates for a novel generation of low-cost, room-temperature x-ray detectors, which will allow unprecedented applications to be developed in the radiation-detector research

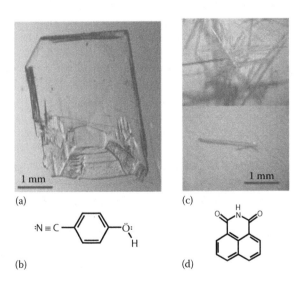

(a)

(b)

(c)

(d)

FIGURE 11.1 Single crystals and molecular structures of 4HCB (a,b) and NTI (c,d). (From Fraboni, B. Ciavatti, A. Merlo, F., Pasquini, L., Cavallini, A., Quaranta, A., Bonfiglio, A., and Fraleoni-Morgera, A.: *Adv. Mater.* 2012. 24. 2289. Copyright Wiley-VCH Verlag GmbH & Co. KGaA. Reproduced with permission.)

field, mostly complementary to the high-performance ones presently based on inorganic semiconductors.

11.2 ORGANIC SEMICONDUCTING SINGLE-CRYSTAL GROWTH

OSSCs are usually grown using chemical vapor deposition (CVD), which involves the use of multizone heated tubes, in which a vapor of molecules is carried by a convenient inert gas onto cold walls. The molecules condense onto these walls, and tight control over the temperatures of the system is necessary to encourage them to form a good crystal [29,34]. However, this method is rather sensitive to system perturbations (mainly temperature and carrier gas flow variations), and it is time and energy intensive. In addition, it is well documented that even small amounts of impurities can have a dramatic impact on the OSSCs' transport properties [35,36], and the CVD method is not particularly effective in getting rid of these impurities, unless repeated and slow cleaning/deposition cycles are performed. Finally, due to the nature of this growth process, it is difficult to grow very large crystals. Crystal growth from the melt could represent an interesting alternative to CVD. Starting from a melt that can be contained in an arbitrary vessel (provided, of course, that the latter is able to withstand the melting conditions), this process does not have tight constraints in terms of final crystal size. Moreover, consistent know-how on this technique is available from inorganic materials for electronic applications, in which crystallization from the melt has become a scientific and technological field of its own [37]. Nonetheless, although in some cases crystallization of OSSCs from the melt has proved to be successful [38], stability problems (especially due to enhanced photooxidation rates at temperatures approaching melting temperature) often limit the possibility of taking advantage of this crystallization technique for organic materials.

These problems may be overcome using growth from solution. This approach has high versatility, with the capability to deliver very large (up to several cubic centimeters) and pure crystals, low energy requirements (i.e., no need for dramatic heating, cooling, or vacuum), and ease of implementation. This latter point is worthy of particular attention, since it implies extremely low costs, especially when areas as large as square meters are projected to be covered with single crystals (as would be necessary, for example, for flexible, large-area ionizing-radiation detectors aimed at monitoring landfills or seaports). Moreover, solution growth allows good control over many crystal properties, including size and even the developed crystallographic phase [39]. A partial drawback of solution growth is its relative slowness, due to the small amounts of matter that are involved in the process, especially in dilute solutions. In fact, for applications such as nonlinear optics, where crystals having centimeter-long sides are needed, at least for fundamental studies, several days (often weeks) of growth are needed to obtain a crystal of satisfactory size [40,41]. On the other hand, many applications, such as organic electronics, do not require such large crystals, and a suitable crystal size may be as low as the nanometer or micrometer range [42], though most common sizes are around hundreds of microns—a few millimeters [26,28,43]. Solution growth has also been used to produce organic (not necessarily semiconducting) single crystals used as scintillators in ionizing-radiation detectors [44,45], demonstrating that this technique delivers crystals able to usefully interact with ionizing radiation. In addition, solution-grown OSSCs have repeatedly demonstrated good electronic quality [43,46].

In this regard, it was recently reported that single crystals based on solution-grown 4HCB (Figure 11.1a and b) are able to detect x-rays via direct generation of electrical current on x-ray irradiation [22]. 4HCB crystals have rather complex crystal unit cell structures, which we have characterized by x-ray diffraction (XRD) methods (Figure 11.2), and which result from enhanced hydrogen-bond formation between the –OH and –CN groups of the molecule [47,48]. Despite this structural complexity, they present marked reproducibility of their electronic properties (mobilities, traps density, etc.) over tens of crystals having different sizes, deriving from different

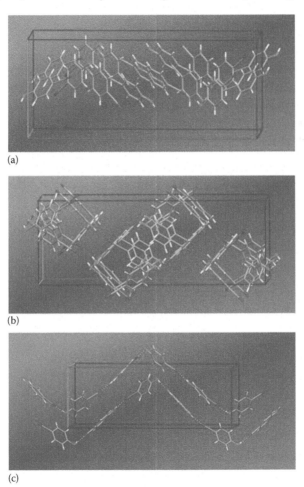

(a)

(b)

(c)

FIGURE 11.2 (**See color insert**) Simplified views of the 4HCB crystalline unit cell, seen along the crystallographic axes a (green axis), b (red axis), and c (blue axis). (From Fraboni, B., Femoni, C., Mencarelli, I., Setti, L., Di Pietro, R., Cavallini, A., and Fraleoni-Morgera, A.: *Adv. Mater.* 2009. 21. 1835. Copyright Wiley-VCH Verlag GmbH & Co. KGaA. Reproduced with permission.)

batches, and even grown in different conditions (in terms of temperature, initial solution concentration, etc.) [49,50].

A further point of advantage of 4HCB crystals is that they can be easily grown from widely available solvents/nonsolvents, such as ethylic ether or petroleum ether, and their size can be controlled by working on parameters such as the solvent/nonsolvent volume ratio or the 4HCB concentration in the starting solutions [51]. The single crystals obtained in this way are very robust to physical manipulation (they can be easily moved and positioned on substrates, electrode arrays, sample holders, etc.) and to environmental conditions (normal atmosphere, light, room temperature). The possibility of tuning the crystal size with simple changes of the growth recipe is of particular importance in view of a thorough understanding of their functional properties.

Another possibility offered by the versatility of the solution growth technique is to obtain crystals directly grown onto substrates constituted by a thin dielectric layer (SiOx) thermally grown on doped Si, so to constitute a native field-effect-transistor (FET) substrate. The crystals obtained in this way were tested under synchrotron-originated polarized IR spectroscopy as bare crystals or as the active channels (after source and drain upper electrodes were fabricated on the crystals, as shown in Figure 11.3) in operating FETs, revealing interesting and in some ways surprising crystal polarization effects due to gating, which apparently develops a charge transport layer either with marked metallic behavior or much thicker (on the order of several hundreds of microns) than currently assumed to be the case for these systems [52].

The possibility of growing OSSCs from solution does not only offer exciting chances of performing fundamental studies, but, even more interestingly, paves the way for promising applications of these materials in practical devices.

In fact, it has been recently shown that it is possible to fabricate high-quality OSSCs via inkjet printing [53] or shear-aided crystallization [54] (Figure 11.4). These intriguing technological possibilities highlight the potential for large-area fabrication

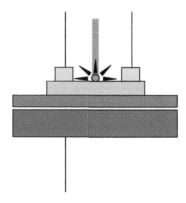

FIGURE 11.3 Layout of a 4HCB single-crystal-based device probed under FT-IR linearly polarized light. (Reproduced with permission from Fraleoni-Morgera, A., Tessarolo, M., Perucchi, A., Baldassarre, L., Lupi, S., Fraboni, B., *J. Phys. Chem. C*, 116, 2563. Copyright [2012] American Chemical Society.)

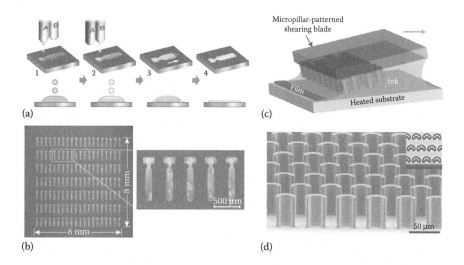

FIGURE 11.4 **(See color insert)** (a) Solution and (b) overview of a matrix of single crystals obtained with this technique. (c,d) Visual description of the solution-shearing method for originating solution-grown, oriented polycrystalline layers from drops of ink deposited onto substrates (c), and a scanning electron microscopic (SEM) image of the concave-shaped micropillar used for producing the needed shear stress in the deposited solution (d). (Reprinted by permission from Macmillan Publishers Ltd. [*Nature*] (From Minemawari, H., et al., 475, 364), copyright 2011; reprinted by permission from Macmillan Publishers Ltd. [*Nature Materials*] (Diao, Y., et al., 12, 665–671), copyright 2013.)

of arrays of OSSCs-based devices. In particular, this is true for large-area, composite devices (which represent by far the largest class of electronic products today, and the most desired goal for the whole field of organic optoelectronics). In this frame, the accurate and reproducible placement of OSSCs onto electrode matrices or in precise positions onto a given substrate still represents a challenge far from being solved. Crystal growth from solution, thanks to its unique cocktail of low costs, rapid production times (when small crystals are considered), low energy requirements, and manufacturing versatility, could well represent a viable fabrication pathway for solving this problem, opening the way for a truly widespread organic electronics.

11.3 ORGANIC SEMICONDUCTING SINGLE-CRYSTAL PROPERTIES

In order to achieve the goal of low-cost organic electronics produced on a large scale [55–60], besides the need to find a convenient processing technique, a number of issues related to the charge transport in organic materials have to be clarified. When compared with inorganic semiconductors such as silicon or III–V compounds, it seems clear that technology based on organic semiconductors would vastly benefit from a much more systematic, basic understanding of the electronic properties of these materials. In particular, a clear and well-defined comprehension of their electronic behavior (including a complete description of the fundamental electronic states of the charge carriers) has still to be gained and related to their structural properties. In this respect, OSSCs provide a unique example of intrinsic electronic transport in organic materials, thanks to their long-range molecular order and to the limited charge-carrier trapping and hopping phenomena present in thin films, usually due to grain boundaries, interfaces, and structural imperfections. A useful investigation tool in this sense may be found in organic field-effect transistors (OFETs), which can provide precious information on the nature of the charge-transport phenomena in organic materials. Indeed, single-crystal organic transistors, in which the active channel is a single crystal, show the highest mobilities in organic materials, of several tens of $cm^2 V^{-1} s^{-1}$ [29,35,36,61], since a key parameter contributing to the mobility is the degree of molecular packing and the long-range order. Any process that modifies the molecular packing in the OSSC layer can affect the orbital overlap and therefore the transfer integral, which then affects charge-carrier mobilities. Several reviews have already discussed the chemical structure–molecular packing relationships for a variety of small-molecule OSSCs [62,63]. The methods employed for varying the molecular packing in a controlled way use chemical modifications of the OSSC or specific surface–OSSC interactions, such as the use of interfacial layers in the device-fabrication process or OSSC postprocessing conditions [64–67].

Nonetheless, the low symmetry of organic molecules leads to anisotropically packed crystal structures, which affects their transport properties (for example, the direction of the strongest p-orbital overlap usually coincides with the direction of the highest carrier mobility [4,28–30,35,68]). On the one hand, this asymmetry introduces difficulties for clearly understanding the transport behavior of the organic crystal, but on the other hand it offers the possibility of investigating the correlation between the three-dimensional molecular stacking order of OSSCs and their recently discovered anisotropic electronic transport properties. In particular, in recent years different mobilities have been measured along the three dimensions for micrometer- or nanometer-sized organic single crystals, either vapor deposited [4,28–30,35,68] or solution grown [43,51,53,50,69].

By means of space-charge-limited current (SCLC) [29,47,50,70] and by FET analyses [47,49], it is possible to carry out a complete characterization of several key electronic transport parameters of OSSCs.

We have recently reported on the electrical characterization of solution-grown, macroscopic 4HCB single crystals, and we will focus our attention on this specific crystal as a paradigm for OSSCs' properties and applications.

In particular, we have determined (i) the majority carrier mobility, (ii) the density of states (DOS) distribution of the intrinsic electronic states, (iii) their concentration, and (iv) their energy levels in the bandgap. We provide evidence that all these properties are clearly anisotropic along the three crystal dimensions [49,50]. This anisotropicity of the electronic properties is due to the mentioned complex lattice structure of the 4HCB crystal (Figure 11.2), which was tentatively related to the transport properties by comparing the XRD-derived crystal structure with the three-dimensional anisotropic charge mobilities [47].

Figure 11.5 shows a sketch of the electrical contacts on the crystals, fabricated in alignment with the three main crystal axes and used to carry out the electrical characterization analyses aimed at correlating the crystal's transport properties with its structural molecular packing.

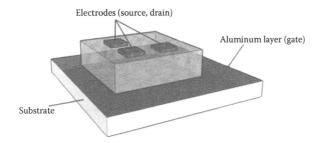

FIGURE 11.5 Layout of a 4HCB single crystal to be probed for anisotropic electronic properties. (From B. Fraboni, C. Femoni, I. Mencarelli, L. Setti, R. Di Pietro, A. Cavallini, A. Fraleoni-Morgera, *Adv. Mater.*, 21, 1835, 2009; B. Fraboni, R. Di Pietro, A. Castaldini, A. Cavallini, A. Fraleoni-Morgera, L. Setti, I. Mencarelli, C. Femoni, *Org. Electr.*, 9, 974, 2008.)

SCLC analyses carried out at room temperature allow the crystal's electrical transport properties to be assessed and the charge-carrier mobility and the DOS distribution along the three axes to be determined [47,49,50] (Figure 11.6a). Photocurrent spectroscopy analyses allowed the effective electrical bandgap of 4HCB crystals to be determined, estimated at about 4 eV (Figure 11.6b).

Indeed, as shown in Figure 11.6a, the transport properties of 4HCB crystals reflect their anisotropic packing, and a clear difference in mobility values is found along the three axes. Charge-carrier mobilities of 0.1 $cm^2 V \cdot s^{-1}$ can be easily and routinely obtained along the a axis of 4HCB solution-grown crystals, with very reproducible and stable results [50].

While the anisotropic electrical response of OSSCs is extremely interesting from the point of view of fundamental studies, it poses the problem of properly aligning the crystals and the electrodes in order to select and control the appropriate carrier-transport properties needed for specific device applications. Of course, proper determination of the crystallographic orientation of OSSCs is easily obtained via XRD,

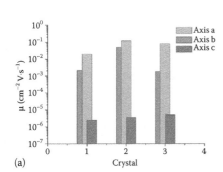

(a)

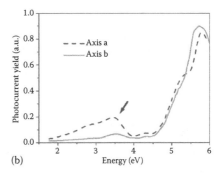

(b)

FIGURE 11.6 (a) Charge-mobility values determined from SCLC-based measurements in three different 4HCB single crystals along the three crystal axes, showing a remarkable reproducibility in each case. (b) Photocurrent yields along the two crystallographic axes a and b for a 4HCB single crystal. ((a) Reprinted from *Org. Electr.*, 9, Fraboni, B., et al., 974. Copyright [2008], with permission from Elsevier; (b) reprinted from *J. Cryst. Growth*, 312, Fraleoni-Morgera, A., et al., 3466, Copyright [2010], with permission from Elsevier.)

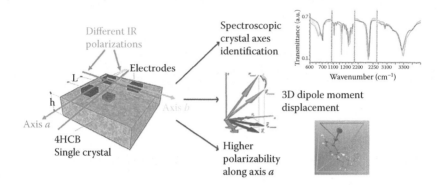

FIGURE 11.7 (See color insert) Sketch of linearly polarized IR measurements carried out over 4HCB single crystals and the information achievable with this technique. (Reproduced with permission from Fraleoni-Morgera, A., Tessarolo, M., Perucchi, A., Baldassarre, L., Lupi, S., Fraboni, B., *J. Phys. Chem. C*, 116, 2563. Copyright [2012] American Chemical Society.)

but this method has some limitations when applied to large arrays of crystals (as, for example, those shown in Figure 11.4b).

An effective way of assessing crystal orientation, more practical than XRD analyses for large numbers of crystals, is linearly polarized infrared spectroscopy (LP-IR). This technique has been demonstrated to be able to identify the crystallographic orientation due to the peculiar response of anisotropic crystals to linearly polarized radiation, allowing the crystal orientation to be clearly identified with no need for more intrusive and time-demanding XRD. In addition, LP-IR can be used to extract extremely useful fundamental information on the electrostatic intermolecular forces occurring within the crystal and possible molecular polarization phenomena, helping to understand the relations between the inner crystal structure and its electronic properties (Figure 11.7) [48,52].

11.4 X-RAY ELECTRICAL PHOTORESPONSE

The low dark current typical of organic semiconducting single crystals, their relatively good mobility, and the possibility of tuning their volume up to cubic millimeters with solution-growth techniques open up the possibility of using them as solid-state radiation detectors for x-rays.

We report here the main results observed on x-ray irradiation of 4HCB OSSCs. As shown in Figure 11.8a, a strong modification is induced in the current measured as a function of applied voltage along the vertical axis when a 4HCB crystal is exposed to an x-ray beam (35 keV, 170 mG s^{-1}). A similar effect is visible along the planar axis (not reported here) [22].

Interestingly, the normalized photocurrent ($I_{ON}-I_{OFF}$)/I_{OFF} versus V curve presents a maximum at rather low voltages for both axes, suggesting that practical devices may be operated at voltages as low as a few tens of volts, hence with low power requirements (Figure 11.8b). The photoresponse has been measured on 4HCB crystals contacted

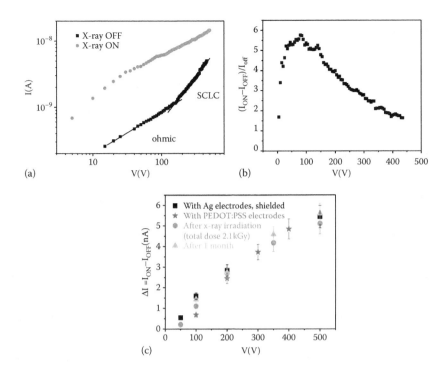

FIGURE 11.8 **(See color insert)** (a) Current–voltage curves for 4HCB-based devices in the dark (black squares) and under x-ray irradiation from a 35 keV Mo source at 170 mGy s⁻¹ (red circles), for the vertical axis. Under x-ray irradiation, no ohmic–SCLC transition is detected along the vertical axis. (b) The maximum response can be achieved at low bias voltage (<100 V). (c) X-ray-induced current variation $\Delta I = (I_{ON} - I_{OFF})$ reported for different bias voltages applied to crystals contacted with Ag electrodes, before (black squares) and after irradiation with a total delivered x-ray dose of 2.1 kGy (red circles). The responses of crystals after an aging period of 1 month (blue triangles) and of crystals contacted with PEDOT:PSS electrodes (green stars) are also reported.

with electrodes fabricated with different materials—a metal (Ag) and a conducting polymer (poly(ethylenedioxythiophene):poly(styrenesulfonate) [PEDOT:PSS])—to assess their charge-collecting properties and the crystal intrinsic response. The results, shown in Figure 11.8c for the vertical and planar axes, clearly indicate that highly reproducible and consistent results can be obtained with different types of electrodes.

Thanks to these results, all-organic devices have been fabricated using a low-Z organic elastomer as a substrate (poly(dimethylsiloxane) [PDMS]) and a transparent and flexible conducting polymer (PEDOT:PSS) as electrode material. Such devices are composed of proven biocompatible materials and were found to be optically transparent and reasonably conformable (Figure 11.9), suggesting possible applications in bioelectronics.

The radiation hardness of these crystals is also quite remarkable. 4HCB-based devices were continuously exposed for 3 h to a 170 mGy s⁻¹ dose rate (total dose of 2.1 kGy), and then tested under an x-ray beam that was switched on and off; after this

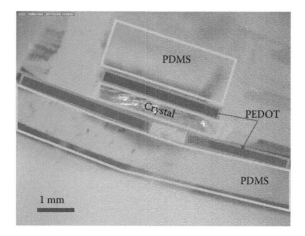

FIGURE 11.9 (**See color insert**) Optical microscopy image of a cross-sectional view of an optically transparent and bendable all-organic device with a 4HC crystal as the active sensing material. The PDMS substrate, the 4HCB crystal, and the PEDOT:PSS electrodes have been highlighted with color frames.

rather high-dose irradiation, the devices' electrical response to x-rays, ΔI, did not vary significantly; moreover, the same devices still provided a reliable response even after aging for 1 month (Figure 11.10). It is noteworthy that all the results reported here have been obtained in air and at room temperature, without any kind of encapsulation. Repeated x-ray-beam on/off cycles did not cause hysteresis or appreciable current drift at any of the bias voltages tested (Figure 11.10). The response time, shorter than 70 ms, is remarkably fast for organic electronic devices.

The response of 4HCB-based detectors is remarkably linear over quite a large range of dose rates (Figure 11.11a): the curves refer to the same single crystal probed along the vertical axis with two Ag electrodes and different applied bias. The fraction of e−h pairs collected at the electrodes following x-ray irradiation was evaluated

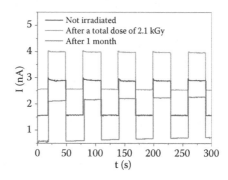

FIGURE 11.10 (**See color insert**) Electrical response of a 4HCB device to an on-off-switching x-ray beam at a bias voltage V = 100 V. The response is reported for an as-prepared device (black line) after it has been exposed for 3 h to a 170 mGy s^{-1} dose rate (total dose of 2.1 kGy) (red line) and after 1 month of storage in the ambient atmosphere (blue line).

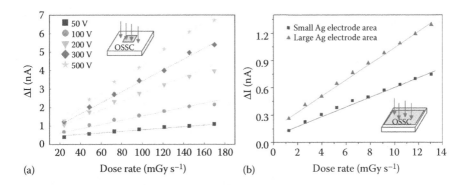

FIGURE 11.11 **(See color insert)** The x-ray-induced current variation $\Delta I = (I_{ON} - I_{OFF})$ is reported for increasing dose rates and different biases for the vertical axis of a 4HCB crystal on a quartz substrate with small-area (a) and large-area (b) Ag electrodes. (The lines are to guide the eye only.)

using the measured photocurrent $\Delta I = (I_{ON} - I_{OFF})$. We estimated that about 5% of the impinging photons are absorbed within a typical crystal thickness of 600 μm by calculating the absorption coefficients to the Mo K lines (17.9 and 19.5 keV) with the XCOM code [71], and we estimated a minimum effective efficiency f ≈ 2%, which takes into account both the pair-production efficiency and the collection efficiency of the electrodes. These last two factors will have to be further improved to enhance the organic radiation-detector performance. To this end, future work in this field will focus on dedicated molecular-crystal engineering and device-geometry design and architectures. As a major example, the sensitivity, defined as $S = \Delta I/D$, where D is the dose rate, can be significantly improved by simply improving the electrode geometry, and can reach up to 100 nC Gy^{-1} at 50 V (Figure 11.11b).

11.5 CONCLUSIONS AND OUTLOOK

The organic single-crystal radiation detectors presented here allow x-rays to be detected directly; that is, the incoming x-ray photon is converted into an electrical signal. They provide a linear response to the x-ray dose rate, even for driving voltages as low as a few tens of volts, and they respond to an on/off-switching x-ray beam with no hysteresis effects. They are based on low-cost solution-grown organic molecular crystals (e.g., 4HCB and NTI) and can operate at room temperature, in air, and under ambient light with high reproducibility.

The observed behavior can be ascribed to the combination of different relevant properties of OSSCs: large bandgap, low dark current, good transport properties, ease of growth (by solution), dimension tenability, robust structure, and relatively low environment sensitivity.

4HCB-based detectors have been described in detail as an example, but other OSSCs present similar properties and provide a reliable response even after exposure to significant doses of x-ray irradiation and after being aged for up to 1 month. All-organic (i.e., not containing any high-Z compound), optically transparent x-ray detectors have been fabricated and demonstrated to be as effective, radiation hard, and linearly responsive as metal-containing ones. The overall performances of the devices presented here indicate that OSSCs are promising candidates for a new generation of direct electrical sensors for ionizing radiation, and the anisotropic electrical response of the molecular crystals can be exploited to develop unprecedented device architectures, possibly coupled to the conformable properties of all-organic devices.

As a future outlook, the possibility of directly growing single crystals from solution on a patterned substrate via ink-jet printing processes (see Section 11.2), coupled with the reported possibility of fabricating all-organic transparent radiation detectors (see Section 11.4), opens the way for the development of novel flexible sensor systems. Moreover, the low voltage required to operate OSSC detectors suggests that it might be possible to couple the radiation sensor to an organic electronic readout circuit, all printable on flexible substrates and operated at low bias voltages (typically <40 V if pentacene CMOS logic is employed). Breakthrough applications can be envisaged for this revolutionary sensing system, relevant in health diagnostics, in industrial process control, in citizen security and border control monitoring, and in wearable electronics.

ACKNOWLEDGMENTS

The authors acknowledge financial support from the European Community under the FP7-ICT Project i-FLEXIS (2013–2016).

REFERENCES

1. M. Berggren, A. Richter-Dalfors, Organic bioelectronics, *Advanced Materials* 2007, 19, 3201.
2. K. Müllen, U. Scherf, *Organic Light-Emitting Devices: Synthesis, Properties and Applications*, Wiley-VCH, Weinheim, Germany, 2006.
3. T. Sekitani, U. Zschieschang, H. Klauk, T. Someya, Flexible organic transistors and circuits with extreme bending stability, *Nature Materials* 2009, 9, 1015.
4. C. Reese, Z. Bao, Organic single-crystal field-effect transistors, *Materials Today* 2007, 10, 20.
5. S. Gunes, H. Neugebauer, N. Sariciftci, Conjugated polymer-based organic solar cells, *Chemical Reviews* 2007, 107, 1324.
6. R. Capelli et al., Organic light-emitting transistors with an efficiency that outperforms the equivalent light-emitting diodes, *Nature Materials* 2010, 9, 496.
7. E. Roeling, et al. Organic electronic ratchets doing work, *Nature Materials* 2011, 10, 51.
8. L. Christophorou, J. Carter, Improved organic scintillators in 2-ethyl naphthalene, *Nature* 1966, 212, 816.
9. G. Hull, N. Zaitseva, N. Cherepy, J. Newby, W. Stoeffl, S. Payne, New organic crystals for pulse shape discrimination, *IEEE Transactions on Nuclear Science* 2009, 56, 899.
10. P. Keivanidis, N. Greenham, H. Sirringhaus, R. Friend, J. Blakesley, R. Speller, M. Campoy-Quiles, T. Agostinelli, D. Bradley, J. Nelson, x-ray stability and response of polymeric photodiodes for imaging applications, *Applied Physics Letters* 2008, 92, 23304.
11. T. Agostinelli, M. Campoy-Quiles, J. Blakesley, R. Speller, D. Bradley, J. Nelson, A polymer/fullerene based photodetector with extremely low dark current for x-ray medical imaging applications, *Applied Physics Letters* 2008, 93, 203305.

12. S. Tedde, J. Kern, T. Sterzl, J. Fürst, P. Lugli, O. Hayden, Fully spray coated organic photodiodes, *Nano Letters* 2009, 9, 980.
13. R. Newman, H. Sirringhaus, J. Blakesley, R. Speller, Stability of polymeric thin film transistors for x-ray imaging applications, *Applied Physics Letters* 2007, 91, 142105.
14. M. Atreya et al., Stability studies of poly(2-methoxy-5-(2'-ethyl hexyloxy)-*p*-(phenylene vinylene) [MEH-PPV], *Polymer Degradation and Stability* 1999, 65, 287.
15. E. Silva, J. Borin, P. Nicolucci, C. Graeff, T. Ghilardi Netto, R. Bianchi, Low dose ionizing radiation detection using conjugated polymers, *Applied Physics Letters* 2005, 86, 131902.
16. L. Zuppiroli, S. Bouffard, J. Jacob, Ionizing radiation dosimetry in the absorbed dose range 0.01–50 MGy based on resistance and ESR linewidth measurements of organic conducting crystals, *International Journal of Applied Radiation and Isotopes* 1985, 36, 843.
17. S. Graham, R. Friend, S. Fung, S. Moratti, High sensitivity radiation sensing by photo induced doping in PPV derivatives, *Synthetic Metals* 1987, 84, 903.
18. P. Beckerle, H. Ströbele, Charged particle detection in organic semiconductors, *Nuclear Instruments and Methods in Physics Research Section A* 2000, 449, 302.
19. F. Boroumand, M. Zhu, A. Dalton, J. Keddie, P. Sellin, J. Gutierrez, Direct x-ray detection with conjugated polymer devices, *Applied Physics Letters* 2007, 91, 33509.
20. A. Intaniwet, J. Keddie, M. Shkunov, P. Sellin, High charge-carrier mobilities in blends of poly(triarylamine) and TIPS-pentacene leading to better performing x-ray sensors, *Organic Electronics* 2011, 12, 1903.
21. A. Intaniwet, C. Mills, M. Shkunov, P.J. Sellin, J.L. Keddie, High-Z nanoparticles for enhanced sensitivity in semiconducting polymer x-ray detectors, *Nanotechnology* 2012, 23, 235502.
22. B. Fraboni, A. Ciavatti, F. Merlo, L. Pasquini, A. Cavallini, A. Quaranta, A. Bonfiglio, A. Fraleoni-Morgera, Organic semiconducting single crystals as next generation of low-cost, room-temperature electrical x-ray detector, *Advanced Materials* 2012, 24, 2289.
23. A. Ohtomo, H. Hwang, A high-mobility electron gas at the LaAlO$_3$/SrTiO$_3$ heterointerface, *Nature* 2004, 427, 4236.
24. A. Tsukazaki, et al., Observation of the fractional quantum Hall effect in an oxide, *Nature Materials* 2010, 9, 889.
25. J. Takeya, et al., Very high-mobility organic single-crystal transistors with in-crystal conduction channels, *Applied Physics Letters* 2007, 90, 102120.
26. A. Briseno, et al., Patterning organic single-crystal transistor arrays, *Nature* 2006, 444, 913.
27. N. Karl, Charge carrier transport in organic semiconductors, *Synthetic Metals* 2002, 649, 133.
28. V. Sundar, et al., Elastomeric transistor stamps: Reversible probing of charge transport in organic crystals, *Science* 2004, 303, 1644.
29. de Boer, et al., Organic single-crystal field-effect transistors, *Physica Status Solidi A* 2004, 201, 1302.
30. H. Yan, Z.H. Chen, Y. Zheng, C. Newman, J.R. Quinn, F. Dotz, M. Kastler, A. Facchetti, A high-mobility electron-transporting polymer for printed transistors, *Nature* 2009, 457, 679.
31. V. Podzorov, E. Menard, A. Borissov, V. Kiryukhin, J.A. Rogers, M.E. Gershenson, Intrinsic charge transport on the surface of organic semiconductors, *Physics Review Letters* 2004, 93, 086602.
32. H.Y. Li, B.C.K. Tee, J.J. Cha, Y. Cui, J.W. Chung, S.Y. Lee, Z.N. Bao, High-mobility field-effect transistors from large-area solution-grown aligned C60 single crystals, *Journal of the American Chemical Society* 2012, 134, 2760.
33. A.L. Briseno, et al., Patterning organic single-crystal transistor arrays, *Advanced Materials* 2006, 18, 2320–2324.

34. N. Karl, Growth and electrical properties of high purity organic molecular crystals, *Journal of Crystal Growth* 1990, 99, 1009–1016.
35. R. Zeis, C. Besnard, T. Siegrist, C. Schlockermann, X. Chi, C. Kloc, Field effect studies on rubrene and impurities of rubrene, *Chemistry of Materials* 2006, 18, 24.
36. O.D. Jurchescu, J. Baas, T.T.M. Palstra, Effect of impurities on the mobility of single crystal pentacene, *Applied Physics Letters* 2004, 84, 3061.
37. G. Müller, J.-J. Métois, P. Rudolph (eds), *Crystal Growth—From Fundamentals to Technology*, 2004, Elsevier, Amsterdam.
38. M. Arivanandhan, K. Sankaranarayanan, P. Ramasamy, Melt growth of novel organic nonlinear optical material and its characterization, *Materials Letters* 2007, 61, 4836.
39. K. Sankaranarayanan, P. Ramasamy, Unidirectional crystallization of large diameter benzophenone single crystal from solution at ambient temperature, *Journal of Crystal Growth* 2006, 292, 445.
40. N. Vijayan, R. Ramesh Babu, M. Gunasekaran, R. Gopalakrishnan, R. Kumaresan, P. Ramasamy, C.W. Lan, Studies on the growth and characterization of p-hydroxyaceto-phenone single crystals, *Journal of Crystal Growth* 2003, 249, 309–315.
41. S. Janarthanan, R. Sugaraj Samuel, S. Selvakumar, Y.C. Rajan, D. Jayaraman, S. Pandi, Growth and characterization of organic NLO crystal: β-Naphthol, *Journal of Materials Science and Technology* 2011, 27, 271.
42. R. Li, W. Hu, Y. Liu, D. Zhu, Micro- and nanocrystals of organic semiconductors, *Accounts of Chemical Research* 2010, 43, 529.
43. S.C.B. Mannsfeld, A. Sharei, S. Liu, M.E. Roberts, I. McCulloch, M. Heeney, Z. Bao, Highly efficient patterning of organic single-crystal transistors from the solution phase, *Advanced Materials* 2008, 20, 4044.
44. N. P. Zaitseva et al., Neutron detection with single crystal organic scintillators, *Proceedings of SPIE* 2009, 744911, 744911-(1-10).
45. G. Hull et al., New organic crystals for pulse shape discrimination, *IEEE Transactions on Nuclear Science* 2009, 56, 899.
46. S.K. Park, T.N. Jackson, J.E. Anthony, D.A. Mourey, High mobility solution processed 6,13-bis(triisopropyl-silylethynyl) pentacene organic thin film transistors, *Applied Physics Letters* 2007, 91, 063514.
47. B. Fraboni, C. Femoni, I. Mencarelli, L. Setti, R. Di Pietro, A. Cavallini, A. Fraleoni-Morgera, Solution-grown, macroscopic organic single crystals exhibiting three-dimensional anisotropic charge-transport properties, *Advanced Materials* 2009, 21, 1835.
48. E. Capria, L. Benevoli, A. Perucchi, B. Fraboni, M. Tessarolo, S. Lupi, A. Fraleoni-Morgera, Infrared investigations of 4-hydroxycyanobenzene single crystals, *Journal of Physical Chemistry A* 2013, 117, 6781.
49. B. Fraboni, R. Di Pietro, A. Castaldini, A. Cavallini, A. Fraleoni-Morgera, L. Setti, I. Mencarelli, C. Femoni, Anisotropic charge transport in organic single crystals based on dipolar molecules, *Organic Electronics* 2008, 9, 974.
50. B. Fraboni, A. Fraleoni-Morgera, A. Cavallini, Three-dimensional anisotropic density of states distribution and intrinsic-like mobility in organic single crystals, *Organic Electronics* 2010, 11, 10.
51. A. Fraleoni-Morgera, L. Benevoli, B. Fraboni, Solution growth of single crystals of 4-hydroxycyanobenzene (4HCB) suitable for electronic applications, *Journal of Crystal Growth* 2010, 312, 3466.
52. A. Fraleoni-Morgera, M. Tessarolo, A. Perucchi, L. Baldassarre, S. Lupi, B. Fraboni, Polarized infrared studies on charge transport in 4-hydroxycyanobenzene single crystals, *Journal of Physical Chemistry C* 2012, 116, 2563.
53. H. Minemawari, T. Yamada, H. Matsui, J. Tsutsumi, S. Haas, R. Chiba, R. Kumai, T. Hasegawa, Inkjet printing of single-crystal films, *Nature*, 2011, 475, 364.

54. Y. Diao, B.C.-K. Tee, G. Giri, J. Xu, D.H. Kim, H.A. Becerril, R.M. Stoltenberg, T.H. Lee, G. Xue, S.C.B. Mannsfeld, Z. Bao, Solution coating of large-area organic semiconductor thin films with aligned single-crystalline domains, *Nature Materials* 2013, 12, 665–671.

55. A. Dodabalapour, Organic and polymer transistors for electronics, *Materials Today* 2006, 9, 24.

56. C. Brabec, Organic photovoltaics: Technology and market, *Solar Energy Materials and Solar Cells* 2004, 83, 273.

57. B. Geffro, P. le Roy, C. Prat, Organic light-emitting diode (OLED) technology: Materials, devices and display technologies, *Polymer International* 2006, 55, 572.

58. A. Misra, P. Kumar, M. Kamalasanan, C. Chandra, White organic LEDs and their recent advancements, *Semiconductor Science and Technology* 2006, 21, R35.

59. T. Kawanishi et al., High-mobility organic single crystal transistors with submicrometer channels, *Applied Physics Letters* 2008, 93, 23303.

60. H. Alves, E. Molinari, H. Xie, A. Morpurgo, Metallic conduction at organic charge-transfer interfaces, *Nature Materials* 2008, 7, 574.

61. H. Li, G. Giri, J.B.-H. Tok, Z. Bao, Toward high-mobility organic field-effect transistors: Control of molecular packing and large-area fabrication of single-crystal-based devices, *MRS Bulletin* 2013, 38, 42.

62. M.L. Tang, J.H. Oh, A.D. Reichardt, Z.N. Bao, Chlorination: A general route toward electron transport in organic semiconductors, *Journal of the American Chemical Society* 2009, 131, 3733.

63. J.E. Anthony, A. Facchetti, M. Heeney, S.R. Marder, X. Zhan, n-type organic semiconductors in organic electronics, *Advanced Materials* 2010, 22, 3876.

64. B.J. Jung, N.J. Tremblay, M.L. Yeh, H.E. Katz, Molecular design and synthetic approaches to electron-transporting organic transistor semiconductors, *Chemistry of Materials* 2011, 23, 568

65. J.H. Oh, S.L. Suraru, W.Y. Lee, M. Konemann, H.W. Hoffken, C. Roger, R. Schmidt, Y. Chung, W.C. Chen, F. Wurthner, Z.N. Bao, High-performance air-stable n-type organic transistors based on core-chlorinated naphthalene tetracarboxylic diimides, *Advanced Functional Materials* 2010, 20, 2148.

66. K.P. Goetz, Z. Li, J.W. Ward, C. Bougher, J. Rivnay, J. Smith, B.R. Conrad, S.R. Parkin, T.D. Anthopoulos, A. Salleo, J.E. Anthony, O.D. Jurchescu, Effect of acene length on electronic properties in 5-, 6-, and 7-ringed heteroacenes, *Advanced Materials* 2011, 23, 3698.

67. J.K. Takimiya, S. Shinamura, I. Osaka, E. Miyazaki, Thienoacene-based organic semiconductors, *Advanced Materials* 2011, 23, 4347.

68. J. Lee, S. Roth, Y. Park, Anisotropic field effect mobility in single crystal pentacene, *Applied Physics Letters* 2006, 88, 25216.

69. Q. Tang, et al., Micrometer- and nanometer-sized organic single-crystalline transistors, *Advanced Materials* 2008, 20, 2947.

70. D. Braga, N. Battaglini, A. Yassar, G. Horowitz, M. Campione, A. Sassella, A. Borghesi, Bulk electrical properties of rubrene single crystals: Measurements and analysis, *Physical Review B* 2008, 77, 115205.

71. M.J. Berger, et al., XCOM: Photon cross sections database. NBSIR 87-3597. National Institute of Standards and Technology. http://www.nist.gov/pml/data/xcom/index.cfm.

12 Lanthanum Halide and Cerium Bromide Scintillators

Oliver J. Roberts

CONTENTS

12.1 MOTIVATION: THE NEED FOR NEW SCINTILLATORS

In developing a new scintillator crystal, a variety of important characteristics need to be addressed, such as the following:

- A high stopping power, related to the scintillator's density (ρ) and atomic number (Z).
- A high light output; that is, it should be able to readily convert the energy from incident radiation into light.
- Minimal afterglow; that is, the scintillator should be transparent to its own emitting wavelength, ensuring good light collection.
- Good linearity; that is, the conversion of quanta to light should be proportional and uniform throughout the crystal over a wide range of energies.
- Fast response; that is, induced luminescence should decay quickly.
- The material should be robust, low in cost, and available in a range of sizes that make it practical to use for various applications.

Currently, these requirements cannot be met by a single type of scintillator, and therefore interest still exists in developing a scintillator that will match all of the aforementioned criteria. In the 65 years since the discovery of NaI(Tl) for use as a scintillator material by Hofstadter [1], research and development of scintillator materials has led to the discovery of $Bi_4Ge_3O_{12}$ (BGO) [2], CsF [3], and BaF_2 [4]. High-efficiency bismuth-germanate or BGO crystals have been employed in fields such as high-energy physics [5] and are still commonly used as anti-Compton shields [6] around high-purity germanium detectors (HPGe) in nuclear spectroscopy. CsF and BaF_2 crystals allowed the first timing studies to be realized [7–9], which had previously been done only with plastic detectors.

The last two decades have seen significant progress in the development of scintillator crystals, driven largely by technological advances and a greater interest in applications. In industry and research, candidates such as lanthanum halide and $CeBr_3$ crystals have emerged as new inorganic crystals with attractive scintillator properties.

12.2   CHARACTERISTICS OF LANTHANUM HALIDE SCINTILLATORS

Inorganic scintillators are the most popular scintillators used in the field of nuclear spectroscopy due to their high light outputs and stopping powers. The scintillation mechanism in inorganic scintillators is due to the electron band structure in the crystals [10]. These types of scintillator crystals are principally alkali-halide crystals containing a small "activator" impurity, which plays an important role in the scintillation mechanism. The activator creates special sites in the crystal lattice, creating energy states within the forbidden gap through which an excited electron can de-excite back to the valence band. The photon generated from this de-excitation is responsible for the visible scintillation light. The de-excitation sites or luminescence centers determine the emission spectrum of the scintillator.

A nuclear or charged particle traversing through the detector medium will form a large number of electron–hole pairs as the electrons move from the valence to the conduction band. Positive holes are ionized by the activator site, which acts as an impurity trap. The electron will continue to pass through the medium until it encounters an ionized activator site, where it will create a neutral configuration with a set of excited energy states. All of these excited energy states are formed at once due to the short travel time of the electron. The time characteristics of the scintillation light emitted by the crystal are associated with the subsequent decay from these excited states [11].

In lanthanum halide crystals, the luminescence is due to the Ce^{3+} $5d{\rightarrow}4f$ transition, which is shown in Figure 12.1a. "Lattice relaxation" in the crystals causes a minimization in energy, resulting in the excited states populating the band-gap region of the crystal. The difference between the energy absorbed initially (prerelaxation) and emitted (postrelaxation) is called the *Stokes shift*, which is on the order of 0.55 eV in $LaBr_3$(Ce) [12,13].

Other competing processes exist, such as when the transition from an excited state (created by an electron and impurity trap) to the ground state is forbidden. In this case, an additional amount of energy in the form of thermal excitations allows a

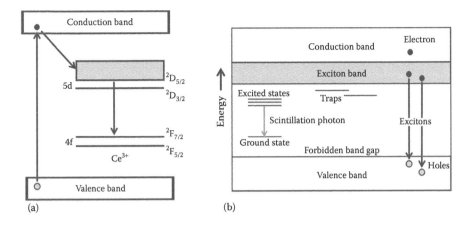

(a)   (b)

FIGURE 12.1 (a) A schematic representation of the energy levels of the 4f and 5d configurations of Ce^{3+}, responsible for the luminescence in lanthanum halide scintillators. (b) A diagram showing the typical electronic band structure found in scintillators. Variations in the size of the forbidden band-gap region result in different levels of conductivity and are ultimately responsible for each material's characteristic properties.

transition to a higher-lying state, from which de-excitation is allowed, to the ground state. This results in a slow component to the resulting scintillation light called phosphorescence, a problem in scintillators as it can be a notable source of "afterglow." Another process involves transitions between some of the excited states in the gap region, which emit no radiation as a result of an electron becoming captured by an impurity trap.

An electron–hole pair that travels after the excitation of an electron from the valence to the conduction band is known as an exciton. This delocalized configuration of a hole and electron moves freely within the material from the exciton band until it comes into contact with an activator site, which acts as a generator of electron levels in the forbidden energy-gap region (a schematic diagram of this process is shown in Figure 12.1b). The de-excitation from these states to the ground configuration ultimately results in the creation of scintillation light.

The inorganic halide salts $LaCl_3$ and $LaBr_3$ were first synthesized in 2001 by the Universities of Delft and Bern [14–16] and are now made commercially by Saint-Gobain under the product names BrilLanCe® 350 and BrilLanCe® 380, respectively [17]. The crystals are hexagonal in nature (similar in shape to the uranium chloride compound UCl_3), with a $P63/m$ space group [18,19]. $LaBr_3$ has a density of 5.07 g/cm^3 and a melting point of 783°C, and $LaCl_3$ has a density of 3.64 g/cm^3 and melts at 859°C [18]. The low melting points of both compounds allow the crystals to be grown via the Bridgman and Czochralski methods [20], using ultra-pure dry forms of the compounds' white powder.

The Bridgman crystal-growth technique requires these crystals to be sealed in quartz ampoules. Both $LaBr_3$ and $LaCl_3$ are rare-earth compounds and thus are easily contaminated by moisture and oxygen in the atmosphere. In order to prevent this, the crystal powders are loaded into an ampoule in a nitrogen-purged glove box, where the oxygen and moisture are purged [18]. The ampoule is then lowered into a vertical Bridgman furnace, in which independently controlled heater zones maintain a steady temperature gradient along the entire length of the furnace to facilitate large crystal growth. Failure to maintain an optimal, uniform temperature gradient along the furnace results in the build-up of internal stresses inside the crystal due to non-uniform thermal expansion, which is likely to cause fracturing. This is why lanthanum halide crystals are difficult to grow in large ingots [21]; the largest crystals currently available at the time of writing are Ø3.5"×8" (89×203 mm^2) [22]. The build-up of stress during the crystal-making process also makes the initially cylindrical ingots hard to cut, as they are likely to crack and split when cut against their natural cleaving planes. Despite this, different shapes and sizes such as cubes or tapered geometries are readily available. After the crystals are grown, they are subsequently stored under very dry environmental conditions in order to prevent contamination. More information about how these crystals are grown can be found in [18,23,24].

The scintillation properties of both LaBr$_3$ and LaCl$_3$ crystals change with different concentrations of Ce^{3+}, as shown in Table 12.1. Using x-ray-excited luminescence, the variation of the emitting wavelength at various temperatures for different concentrations of Ce^{3+} doping in LaBr$_3$(Ce) crystals can be studied. Pure LaBr$_3$ crystals were found to have two maxima in their emission light of approximately 340 and 430 nm at 100 K due to self-trapping exciton luminescence [25]. This type of luminescence is responsible for similar bands in other bromides [26–29] and pure LaCl$_3$ crystals [28,30]. As the temperature is increased to room temperature, however, the maximum near 340 nm disappears and the other maximum at 430 nm shifts to shorter wavelengths, resulting in an x-ray-excited optical luminescence spectrum being dominated by Ce^{3+} luminescence between 325 and 425 nm. Additional luminescence in LaBr$_3$(Ce:10%) crystals has been reported near 275 nm and is likely to be due to a defect or impurity related to Ce^{3+} [25].

12.2.1 Intrinsic Activity

Lanthanum halide crystals exhibit large amounts of intrinsic activity that limit the use of the crystals in low-counting experiments. These spurious lines arise due to the naturally abundant (0.09%) radioisotope ^{138}La and the α-decay from ^{227}Ac into its daughter nuclides. The former manifests itself in two ways: 66.4% of the time, ^{138}La undergoes electron capture resulting in the population of the 2$^+$ level in ^{138}Ba, which decays by emitting a 1436 keV γ-ray, summed with the 32 and 5 keV coincident x-rays; 33.6% of the time, ^{138}La β$^-$ decays into ^{138}Ce, resulting in a β-continuum with an end point of 255 keV and a peak at 789 keV due to the population and subsequent decay of the 2$^+$ level in ^{138}Ce. This 789 keV γ-ray line in the resulting energy spectrum is smeared to higher energies as it is in coincidence with the β electrons. All of these features are seen clearly below ~1.7 MeV in Figure 12.2, which was acquired over a period of 12 hours.

Above ~1.7 MeV, a large number of broad peaks are present up to ~3 MeV due to the α-decay of ^{227}Ac, which is shown in Figure 12.2. These α energies from the decay

TABLE 12.1

Scintillation Properties of Lanthanum Halide Scintillators When Doped with Different Amounts of Ce^{3+} Compared with Other Readily Available Scintillators

Crystal	Light Output (photons/MeV)	λ_{max} (nm)	E_{Res} (% FWHM at 662 keV)	Decay Time (ns)	References
$LaBr_3$	17,000	350–450	14.0	–	[25]
$LaBr_3$(Ce:0.2%)	60,700	360;380	–	23 (93.4%), 66 (6.6%)	[31]
$LaBr_3$(Ce:0.5%)	60,000	356;387	2.9	26 (93%), 66 (7%)	[25,31]
$LaBr_3$(Ce:1.3%)	47,000	356;387	–	16.5 (97%), 66 (3%)	[31]
$LaBr_3$(Ce:2.0%)	48,000	356;387	3.8	–	[25]
$LaBr_3$(Ce:4.0%)	48,000	356;387	3.5	–	[25]
$LaBr_3$(Ce:5.0%)	55,300	356;387	–	15 (97%), 66 (3%)	[31]
$LaBr_3$(Ce:10.0%)	45,000	356;387	3.9	15 (>99%)	[25, 32]
$LaBr_3$(Ce:20.0%)	64,000	375	–	17	[33]
$LaBr_3$(Ce:30.0%)	69,500	375	–	18.6	[33]
$LaBr_3$(Ce:5.0%,Sr:0.5%)	88,000	360–380	3.2	18.2	[34]
$LaBr_3$(Ce:5.0%,Ba:0.17%)	89,000	360–380	3.3	19.1	[34]
$LaCl_3$	–	405 (80%)	–	–	[24]
$LaCl_3$(Ce:0.1%)	50,500	350;430	–	20 (15%), 213 (85%)	[35]
$LaCl_3$(Ce:1.0%)	50,500	350;430	–	20 (33%), 213 (67%)	[35]
$LaCl_3$(Ce:5.0%)	48,000	350	–	–	[36]
$LaCl_3$(Ce:10.0%)	50,000	350;430	3.1	20 (70%), 213 (30%)	[35]
$LaCl_3$(Ce:20.0%)	38,000	350;430	–	25 (76%), 63 (13%), 213 (11%)	[35] [35]
$CeBr_3$	68,000	371	–	17	[37]
BaF_2	10,000 (slow)	310	9	630	[38–40]
	1,800 (fast)	220;195		0.6–0.8	[38]
SrI_2(Eu:0.5%)	68,000	435	5.3	1,100	[41]
SrI_2(Eu:5.0%)	120,000	435	3.0	1,200	[41]
SrI_2(Eu:6.0%)	115,000	435	2.8	1,200	[42]
SrI_2(Eu:8.0%)	80,000	435	6.7	1,100	[41]
$Cs_2LiLaBr_6$(Ce:10%) (CLLB)	50,000	390;420	2.9	55;>270	[43]

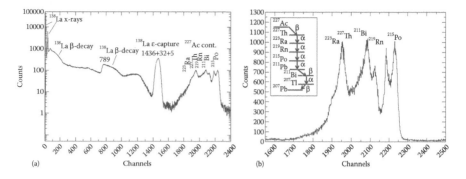

FIGURE 12.2 (a) A γ-ray spectrum showing the intrinsic activity of a single Ø1.5"×2" $LaBr_3(Ce)$ detector. (b) An energy spectrum showing the intrinsic activity of a single Ø1.5"×2" $LaBr_3(Ce)$ detector due to ^{227}Ac contamination. A schematic diagram of the decay path from ^{227}Ac is also shown. The data in both panels was acquired using "singles-mode" triggering over a period of 12 hours, in a lead castle.

of ^{227}Ac actually range from 4 to 8 MeV, meaning that their light is quenched by ~35% [44,45]. A list of the α energies, along with their intensities and calibrated energies as detected in a $LaBr_3(Ce)$ scintillator, is shown in Table 12.2. Discrimination between γ and α radiation in both $LaBr_3(Ce)$ and $LaCl_3(Ce)$ scintillators using pulse-shape analysis techniques has been investigated and has been found to be on the order of a maximum of 10% [46].

The counting rates are approximately 1–2 and 0.1 counts/s/cm³ for contamination from ^{138}La and ^{227}Ac, respectively [108]. In larger, more recently manufactured ingots, the contamination from ^{227}Ac has been reduced significantly in comparison with older crystals due to the purification of raw materials used in the crystal growing process [45,47].

12.2.2 NONPROPORTIONALITY AND ENERGY RESOLUTION

One of the main purposes of a radiation detector is to measure its response to a monoenergetic source of radiation. The resulting distribution of the detected radiation is known as the response function of the detector at that incident energy. A detector with perfect resolution has a δ-like response function. A realistic resolution has a larger variation of incoming pulses for the same incident energy, resulting in a peak with a greater width. The resolution of a detector for a particular energy can be described as being the width of the distribution at a level that is half the maximum (FWHM) of the peak divided by the centroid of the peak (the energy of the detected radiation). One of the dominant sources for fluctuations in the response function arises from statistical noise, which occurs due to a discrete number of charge carriers generated within the detector by a quantum of radiation [48]. The number of discrete carriers fluctuates from event to event, regardless of whether the incident energy is similar, and it can be estimated by describing the formation of each carrier as a Poisson process. Assuming this, the standard deviation of the number of charge carriers, N, is $\sqrt{N}$.

TABLE 12.2

Decays Responsible for Intrinsic α Radiation in LaBr$_3$(Ce) Scintillators

Decay Path	α Energy (keV)	I_γ of Decay (%)	Calibrated γ-Ray Energy (keV)
^{227}Th→^{223}Ra	5757	20	2015
	5978	24	2092
	6038	24	2113
^{223}Ra→^{219}Rn	5607	25	1962
	5716	52	2001
^{219}Rn→^{215}Po	6553	13	2294
	6819	79	2387
^{215}Po→^{211}Pb	7386	100	2585
^{211}Bi→^{207}Tl	6278	16	2197
	6623	84	2318

Source: Hartwell, J.K. and Gehrke, R.J., *Appl. Radiat. Isot.,* 63, 223, 2005; Quarati, F., et al., *Nucl. Instrum. Methods Phys. Res. A,* 574, 115, 2007.

Notes: The energies of the α particles in this table have an $I_\alpha > 10\%$. The final column shows the detected energies of these α particles, which are quenched due to the scintillation mechanism by ~35%.

The detector response function can usually be described by the Gaussian function:

$$G(x) = \frac{A}{\sigma\sqrt{2\pi}} \exp\left(-\frac{(x-x_0)^2}{2\sigma^2}\right), \tag{12.1}$$

where:

σ = width parameter
x_0 = peak centroid
A = area of the function

The width parameter σ defines the FWHM of the Gaussian function through the relationship FWHM = 2.35σ, which can be used to determine the Poisson limited resolution as

$$R_{\text{Poisson limit}} = \frac{2.35}{\sqrt{N}}. \tag{12.2}$$

This limiting resolution is what gives each detector type its characteristic energy response. If the number of charge carriers (N) increases, then the resolution improves due to a decrease in the limiting resolution. For a $R_{\text{Poisson limit}}$ of 1%, the average number of charge carriers ~55,000. This means that, in order to have the best energy resolution available (and thus the smallest limiting resolution possible), one needs to have as many charge carriers as possible. Semiconductor detectors such as CdZnTe and HPGe detectors are popular as these detectors generate a large number of charge carriers. However, discrepancies in the measured values of limiting resolutions have revealed that the total number of charge carriers cannot be completely described by Poisson statistics alone [48]. In an attempt to describe this discrepancy between the statistical fluctuation in the number of charge carriers and the number of charge carriers derived from Poisson statistics, the Fano Factor is introduced. Incorporating this factor of the observed and predicted variance in the number of charge carriers into Equation 12.2, the complete equation describing the nature of the limiting resolution of a detector is now

$$R_{\text{Poisson Limit}} = 2.35\sqrt{\frac{F}{N}}. \tag{12.3}$$

The energy resolution of a $\varnothing 1" \times 1"$ LaBr$_3$(Ce:0.5%) detector is 2.9% FWHM for a γ-ray photon energy of 662 keV, although this number varies depending on the concentration of the Ce^{3+} dopand, as shown in Table 12.1.

Many characteristics of a scintillator can influence the resulting energy resolution, such as the variation of the light output of the scintillator with respect to the incident energy (the photopeak position is proportional to the light output). This nonproportionality or non-linear dependence of the light output due to the variation in the absorbed amount of ionization energy, and thus the emitted number of photons per megaelectronvolts at different energies, generally occurs at energies in the x-ray region rather than at energies above 100 keV. The effect nonproportionality has on the resulting energy resolution [49–52] can be described such that

$$\left(\frac{\Delta E}{E}\right)^2 = R^2 = R_{nPr}^2 + R_{inh}^2 + R_p^2 + R_M^2. \tag{12.4}$$

where:

R_{nPr} = effect the nonproportionality of the scintillator has on the energy resolution
R_{inh} = inhomogeneous defects in the scintillator (which can cause fluctuations in the light output)
R_p = transfer resolution
R_M = contribution to the energy resolution from the photosensor (i.e., PMT) and Poisson statistics of detected photoelectrons [53–55]

The latter effect on the resolution can be derived from the equation

$$R_M = 2.35\sqrt{(1 + v(M))/N_{phe}^{PMT}}, \tag{12.5}$$

where:
$\nu(M)$ = variance from the electron multiplication process in the PMT (typically $1+\nu(M)=1.25$)
N_{phe}^{PMT} = photoelectron yield from the PMT [53]

For LaBr$_3$(Ce:5%), this photoelectron yield is ~21 keV. Inhomogeneities are responsible for an observed deterioration in the recorded energy resolution with increasing crystal size. In a pure, bare homogeneous crystal we assume the terms R_{inh} and R_p are negligible (<1% contribution to the overall resolution), leaving only the nonproportionality and photosensor terms R_{nPr} and R_M, respectively. Therefore, in order to obtain the best energy resolution possible, the nonproportionality of the crystal must be as low as possible, such that the energy response is mainly reliant on the gain and other characteristics of the PMT.

The processes behind nonproportionality can be explained in the following way. At the beginning of the scintillation process, the efficiency of transporting the charge to the luminescence centers is dependent on the ionization density, which is generated in the scintillator crystal by the interaction of incident radiation. This ionization density increases with the low-energy electrons subsequently generated after an interaction with an x-ray or γ-ray, resulting in an increase of competition with the scintillation process, which lowers the charge transport efficiency to the luminescence centers. This scintillation yield is consequently no longer proportional to the amount of ionizing x-ray or γ-ray radiation, thereby degrading the energy resolution [48,56].

The response of the nonproportional scintillation light in both LaBr$_3$(Ce) and LaCl$_3$(Ce) scintillators has been measured in various studies [51,57–60] down to a photon energy of 5 keV and an electron energy of 3 keV. Figure 12.3 shows the

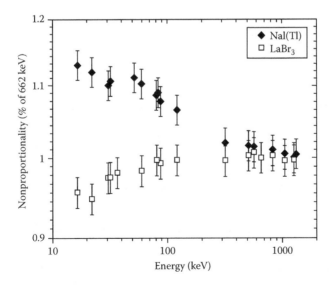

FIGURE 12.3 A comparison of the nonproportionality curves measured for both LaBr$_3$(Ce) and NaI(Tl) scintillators, each with crystal dimensions of Ø25×31 mm². (From Syntfeld, A., et al., *IEEE Trans. Nucl. Sci.*, 53, 3938. © (2006) IEEE. With permission.)

light yield as a function of γ- and x-ray energies relative to the yield at 662 keV. The nonproportionality of the scintillator response of a $\varnothing 25 \times 31$ mm^2 LaBr$_3$(Ce) crystal is compared with that of a NaI(Tl) crystal with similar dimensions. For energies above 1 MeV, the determination of the light yield (particularly for LaBr$_3$(Ce)), might depend on the non-linearity of the PMT. For energies above 100 keV, the LaBr$_3$(Ce) detector seems to be fairly proportional, with a similar degree of proportionality to the NaI(Tl) detector above ~300 keV [61]. At lower energies, the light production in LaBr$_3$(Ce) is much lower with decreasing energy, but is still seen to have a higher degree of light proportionality than NaI(Tl).

Nonproportionality has also been examined for various Ce^{3+} concentrations in LaBr$_3$ scintillators over a temperature range of 80–450 K [62]. In this study, it was found that for Ce^{3+} concentrations of 5% and 30%, the best proportionality and energy resolution were recorded at 80 K. For LaBr$_3$(Ce:0.2%), the lowest energy resolution and nonproportionality were obtained at room temperature [62].

12.2.3 Timing Resolution

The time resolution (pulse-resolving time) of a detector is another important criterion that influences the final choice of scintillator crystal. An ideal timing signal is one whose rising edge is almost vertical, as this gives a more precise moment in time (time pickoff). The criterion for a detector with good timing resolution is how quickly a timing signal is generated and decays, as the duration of the total time signal ultimately dictates the pileup rate. The insensitivity of a detector caused by a long decay time contributes to the dead time and limits the counting rate at which it can be operated.

The timing resolution can be influenced by the size, material, and outer surface condition of the scintillator, as well as by the properties of the light guide, photosensor, and subsequent electronics. Larger crystals will generally have longer timing responses due to the distance the light has to travel through the scintillator. The light guide in the case of lanthanum halides is particularly important, as the crystals are hermetically sealed due to their hygroscopic nature. The effect of a fluctuating gain and the spread in transit time of the accompanying photosensor on the resulting time resolution will be discussed in further detail in Section 12.2.5. In the case of lanthanum halide scintillators, the timing resolution of the detector also depends on the Ce^{3+} doping concentration, as shown in Figure 12.4.

Typically, the brighter and faster the light pulse generated by the scintillation mechanism in a crystal, the better its ability to resolve the arrival time of two light signals at the end of the scintillator. The uncertainty in the arrival-time difference for these two photons is known as the coincidence resolving time (CRT). In medical applications such as positron emission tomography (PET), the detection of two coincident 511 keV photons caused by electron-positron annihilation forms the basis of this method by using direct coincidences. However, the CRTs of current scintillators used in medical applications, such as LaBr$_3$(Ce), are so good that the emission point of these annihilation photons can be constrained to a small segment along the line of coincidence using time-of-flight (ToF) methods. The CRTs of such detectors are directly responsible for mapping the emission point, where a quicker or smaller CRT

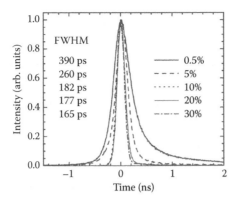

FIGURE 12.4 The effect of changing the dopant concentration on the timing resolution of a LaBr$_3$(Ce) scintillator. The timing signal shows the prompt detection of two 511 keV annihilation photons in coincidence. (From Glodo, J., et al., *IEEE Trans. Nucl. Sci.*, 52, 5. © (2005) IEEE. With permission.)

between two detectors results in a more accurate location of the initial emission point of the annihilation photons. Typically, the time resolutions in ToF measurements vary between 0.1 and 1 ns for high-quality scintillators. Small LaBr$_3$(Ce) crystals have been tested with silicon photomultipliers (SiPMs) and avalanche photodiodes (APDs) in such medical applications and have been found to give an exceptionally good CRT of ~100 ps, which corresponds to a ToF position resolution of 15 mm [64]. Other studies of CRTs involving small lanthanum halide detectors up to an inch in size can be found in [65,66]. The CRTs of larger LaBr$_3$(Ce) scintillators can be found in Table 12.3.

TABLE 12.3

CRTs of Detectors for Various Crystal Shapes and Sizes in Picoseconds

Dimensions	Geometry	T_{FWHM} at 511 keV	T_{FWHM} at 1332 keV
Ø2"×2"	Cylindrical	450 [17]	300 [67]
Ø1.5"×2"	Cylindrical	400 [68]	210 [68]
Ø1.5"×1.5"	Cylindrical	360 [17]	180 [67]
Ø1"×1"	Cylindrical	200 [69]	150 [67, 70, 71]
Ø1"×1.5"×Ø1.5"	Conical	–	160 [71]

Notes: Specific details on how the values were obtained and what equipment was used are given in the references cited. The CRT for conical detectors at an energy of 511 keV is not available in the literature at the time of writing.

Timing resolutions are also crucial in nuclear structure physics, as the electro-magnetic reduced matrix element $B(\lambda L)$ can be extracted by measuring the half-lives of excited nuclear states. The transition probability $T_{fi}(\lambda L)$ of a state decaying from a state with initial spin J_i to a final state with spin J_f is given by the formula [72]

$$T_{fi}(\lambda L) = \frac{8\pi(L+1)}{\hbar L((2L+1)!!)^2} \times \left(\frac{E_\gamma}{\hbar c}\right)^{2L+1} \times B(\lambda L : J_i \rightarrow J_f). \qquad (12.6)$$

The half-lives of excited states can vary considerably over many orders of mag-nitude and thus need to be determined using various techniques [73,74]. One of these techniques, the centroid shift method [75–78], compares the moments of the prompt- and delayed-response functions using the delayed-coincidence method [79]. Until recently, fast coincidences between discrete nuclear states were measured with BaF_2 detectors, which have a fast decay time component of 0.6–0.8 ns [38]. However despite having such a fast decay time, the detector suffers from weak light output (1800 photon/MeV [38]) and poor energy resolution of about 9% FWHM for a 662 keV γ-ray [39,40]. The use of superior, readily available scintillators such as $LaBr_3(Ce)$ has allowed for the construction of large arrays of fast-timing scintilla-tors [68,80]. It is envisaged that such arrays will measure the half-lives of excited states using the delayed-coincidence method down to tens of picoseconds [67,81–88]. In order to measure the half-lives in this sub-nanosecond regime effectively, it is necessary to have a good understanding of the prompt response function, the width of which depends on the sum of the CRTs between pairs of fast-timing $LaBr_3(Ce)$ detectors [81]. The CRTs of various $LaBr_3(Ce)$ detectors of different shapes and sizes are shown in Table 12.3.

12.2.4 NOVEL CODOPED LANTHANUM HALIDE SCINTILLATORS

In an effort to improve the currently acclaimed $LaBr_3(Ce)$ crystals, ionic co-doping of $LaBr_3(Ce)$ was investigated [34]. Small (Ø60×80 mm²) $LaBr_3(Ce: 5\%)$ crystals were co-doped with Sr^{2+} and Ba^{2+}, resulting in an improvement of the light output of ~25% and a better energy resolution over a dynamic range of 122–2615 keV, due to the increase in the light output combined with improved energy proportional-ity. Despite this, co-doping $LaBr_3(Ce:5\%)$ scintillators with 0.5% of Sr and 0.17% of Ba degrades the initial decay time of the $LaBr_3(Ce:5\%)$ scintillator by 3–4 ns. Deconvolving the time profiles using an impulse-response function, however, results in the extraction of "true" decay times of 18.2 and 19.1 ns for the Sr and Ba co-doped $LaBr_3(Ce)$ scintillators, respectively [34]. Similar studies using other co-dopands such as Li^+, Na^+ Mg^{2+}, and Ca^{2+} with $LaBr_3(Ce)$ scintillators can be found in [89,90].

12.2.5 PMT SELECTION AND OTHER READOUT DEVICES

Selecting a PMT is crucial in order to maintain the anticipated performance of the scintillator by adhering to its intrinsic properties. In order to do this, the PMT has to be compatible with the scintillator in various ways:

- It needs to be matched in light; that is, the peak emission wavelength of the crystal needs to match the highest quantum efficiency offered by the PMT at a similar wavelength.
- It needs to be properly gain-matched in order to maintain linearity in the response.
- The grease used to couple the scintillator and PMT window must ensure maximum light transmission.

It is important to match the peak emission wavelength (λ_{max}) of the PMT with the crystal to ensure that all or most of the light at the emitted wavelength is registered. As we know from Section 12.2, the peak emission wavelength of LaBr$_3$(Ce) and LaCl$_3$(Ce) scintillation light occurs principally between 350 and 420 nm, although changes in the concentration of Ce^{3+} changes the light output by producing variations in the scintillation process. In order to reduce the loss of the collected scintillation light, a PMT needs to be sensitive to blue ultraviolet (UV) light (with a range of 13–15 µA/lmF), and have a high quantum efficiency (such as a Bi-alkali PMT).

The PMT used with a very high-light-yielding crystal (like LaBr$_3$(Ce) or LaCl$_3$(Ce)), needs to be chosen carefully as space charge effects and saturation can occur, resulting in the degradation of the signal. This can be solved by reducing the gain on the PMT or the number of PMT stages used in the cascade. However, although reducing the bias voltage lowers the gain and current of the PMT by slowing down the electrons during the multiplication process, non-linearity in the generated energy spectrum is likely to occur. Taking the signal from one of the dynode stages (reducing the number of PMT stages used during the multiplication process) is seen to be the most favorable option in avoiding this; however, a greater amount of variance in the signal is produced due to the higher voltages between each dynode stage. Changing the impedance in the voltage divider on the stage used to take the signal is another alternative in trying to improve the energy resolution of a detector [68].

For timing applications, it is important that the selected PMT has a high number of photoelectrons generated by the photocathode and a low spread in gain due to the electron multiplier, and that it limits the transit time jitter associated with the cascading electrons travelling to the first dynode from the photocathode. In some PMTs, a screening grid is placed inside the last dynode, resulting in an improvement in the charge collection due to the reduced ToF of the electrons between the last dynode and the anode [70]. In doing so, a "parasitic" component is induced in the resulting anode signal, as the travelling electrons in the last stage of the PMT increase the triggering point in the fast discriminator due to the charge from this shifted component. This is found to improve the rise time of the anode pulse as the triggering point is much higher in comparison to the main component (generated by the collection of electrons travelling from the last dynode), where the properties of the scintillation detector require only a small fraction of the anode-pulse height in order to achieve the best time resolution.

Optical grease needs to be chosen to allow for maximum light transmission between the PMT window and the crystal (or light guide in the case of hygroscopic LaBr$_3$(Ce) and LaCl$_3$(Ce) scintillators). The coupling grease needs to have a similar refractive index to the crystal or light guide and the PMT window, and it needs to be transparent. Air bubbles need to be eliminated during the coupling process, as these will affect the performance of the detector.

In some applications, the use of other readout devices can have advantages over conventional photosensors such as PMTs. One such field is medical physics, where $LaBr_3(Ce)$ and $LaCl_3(Ce)$ scintillators are coupled to SiPMs, solid-state sensors that are insensitive to magnetic fields, unlike PMTs. Additionally, SiPMs are transparent to 511 keV γ-rays and are very compact detectors. Such a detector system enables novel designs of compact arrays that are high resolution and allow for depth-of-interaction correction [91]. Moreover, SiPMs are compatible with magnetic resonance imaging (MRI) devices, allowing the possibility of hybrid (PET and MRI) devices to be explored [92,93]. Other applications of SiPMs and $LaBr_3(Ce)$ scintillators, such as Compton cameras, can be found in [94].

12.2.6 PROTON AND NEUTRON ACTIVATION

In applications that involve high proton and neutron fluxes (such as space science), the durability of the detector is challenged. In such hostile environments, radiation damage can compromise the performance of the detector and thus needs to be tested.

Exposing $LaBr_3(Ce)$ and $LaCl_3(Ce)$ scintillators to proton fluxes of 10^{8-12} protons/cm^2 results in an activation spectrum due to the proton capture of the constituent elements of the crystal. Although the proton inelastic reaction cross section for the lighter elements (Br and Cl) is less than the $^{138,139}La+p$ cross section, Br and Cl are more abundant. Consequently, the resulting activation spectrum for a $LaBr_3(Ce)$ scintillator is dominated by γ-ray transitions from ^{77}Br and ^{79}Kr [95]. Similarly, the activation spectrum of a $LaCl_3(Ce)$ scintillator is dominated by ^{33}Cl and ^{32}Cl. Other nuclides that feature in the activation spectrum include ^{140}Cs and ^{139}Ce from the $^{138,139}La+p$ reaction channel. Measurements also report that the proton light yield is 44% less than the light yield of γ-rays, due to the higher ionization density that is generated, which quenches the light [45].

Neutron activation of lanthanum halide crystals also occurs due to the large (n, γ) cross sections of stable ^{139}La (9.0 b) and the two stable isotopes in bromine: ^{79}Br and ^{81}Br (11.0 and 2.4 b, respectively) [96,97]. Neutron inelastic scattering of thermal neutrons off these isotopes results in the population of excited levels in ^{140}La and $^{80,82}Br$, which subsequently undergo β$^-$-decay and electron capture. Similar results in the case of thermal neutron activation have been found [98].

Pulse shapes of thermal and fast neutron signals have also been compared with those of signals from γ-ray detections in $LaBr_3(Ce)$ scintillators using "software CFD," a method in which the values for the slow and fast components are analyzed in a similar manner to a nuclear instrumentation module (NIM) discriminator module. The fast and slow signals are integrated and the values for the fast and slow components deduced via the zero crossover method. Figure 12.5 shows the sum of the fast and slow components from γ-ray and neutron-anode pulses plotted against the slow component only, showing little or no quenching of the light from the signals of neutrons detected in $LaBr_3(Ce)$ scintillators. However, other investigations report that there is modest discrimination between neutrons and γ-rays, which are detected

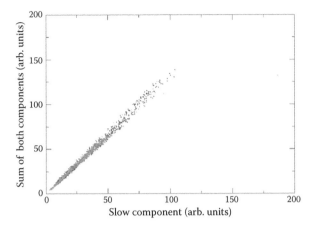

FIGURE 12.5 (**See color insert**) A spectrum showing a comparison between acquired neutron pulses (which also include coincident γ-rays) from an AmBe source (black) and pulses acquired from a ^{60}Co source (red). No discrimination between both sets of pulse data is possible.

in LaCl$_3$(Ce) scintillators [99]. Further coincidence data with a beamline would aid in verifying these results.

12.3 CHARACTERISTICS OF CERIUM TRI-BROMIDE (CeBr$_3$) SCINTILLATORS

Cerium tri-bromide (CeBr$_3$) crystals were initially developed by Radiation Monitoring Devices in 2004 [33] and offer a cheaper alternative to expensive LaBr$_3$(Ce) and LaCl$_3$(Ce) scintillators. They are currently marketed by Scionix Holland B.V. and can now be grown up to Ø3"×4" in size [56]. CeBr$_3$ (like lanthanum halide scintillators) is grown via the Bridgman and Czochralski techniques [20] as it melts at 722°C and has a UCl$_3$ lattice structure. Growing the crystals using these melt-based techniques is advantageous as it allows for the growth of larger ingots [100]. The crystal has an asymmetrical hexagonal structure and a tendency to crack due to the build-up of internal stresses from non-uniform thermal expansion inside the crystal, similar to lanthanum halide scintillators. However, compared to the lanthanum in LaBr$_3$(Ce) and LaCl$_3$(Ce) scintillators, the ionic radius of cerium is smaller (122 vs. 120 pm, respectively) [101] and is responsible for CeBr$_3$ having a larger "effective" atomic number. Consequently, the density of CeBr$_3$ is 0.11 g/cm^3 larger than LaBr$_3$(Ce:5%) at 5.18 g/cm^3 [56]. Due to the hygroscopic nature of the crystal, it has to be prepared using nonaqueous slurries of Al$_2$O$_3$ grit and mineral oil [37]. They are also encapsulated in aluminum containers, with a quartz window for transmitting the light from the scintillator to a photosensor device.

The crystal has a high photon-light yield of 68,000 photons/MeV [37,102,103] (although other findings suggest that the absolute light yield of various large encapsulated CeBr$_3$ crystals varies from 40,000 to 47,000 photons/MeV [56]), with a peak emission wavelength that is reported to be from ~370 nm [37,56] to 390 nm [102], making it compatible with bi-alkali, blue-sensitive PMTs. The luminescence in

$CeBr_3$ is similar to that observed in the lanthanum halide crystals, its doublet structure attributed to the transition from the lowest 5d level to the spin-orbit splitting of the $4f^2F_{5/2}$ and $^2F_{7/2}$ levels. This doublet structure is seen to be less noticeable in $CeBr_3$ than $LaBr_3(Ce:5\%)$ [102], which might be due to the smaller lattice configuration and site size, shifting the emission of Ce^{3+} by several nanometers [56].

The energy resolution of $CeBr_3$ has been measured to be between ~3.4% and ~4.4% (FWHM) [37,104,105]. These values for the energy resolution are similar to energy resolutions for smaller $CeBr_3$ crystal sizes, indicating little loss in the light yield when the size of the crystal is increased. The characterization of these detectors over a dynamic range of γ-ray energies from 81 to 2615 keV, which is determined using ^{133}Ba, ^{137}Cs, ^{60}Co, and ^{232}Th sources, is shown in Figure 12.6a.

The energy dependence of the data in Figure 12.6 can be fitted using the power-law function $E^{-1/2}$, similar to that used in [106] for $LaBr_3(Ce)$ crystals. Reference [104] also reports that $CeBr_3$ is linear over the majority of this energy range, with a linear correlation coefficient of $R^2 = 0.99975$, although an exponential term is needed in order to take into account non-linearity above ~1.7 MeV. Other studies report a similar $1/\sqrt{E}$ dependence for $CeBr_3$ detectors [56], but add that the nonproportionality component of the overall resolution (defined as R_{nPr} in Equation 12.4) strongly contributes to the limit of the resulting resolution, R. Therefore, the amount of variance in the electron multiplication process of the PMT cannot be attributed entirely to the degradation in the energy resolution [56,90].

Section 12.2.2 commented on how the amount of nonproportionality as a function of energy in the light yield can affect the intrinsic performance of a detector. The light yield of a $CeBr_3$ scintillator is influenced by self-absorption and re-emission processes, which result in the increased likelihood of photon loss within encapsulated

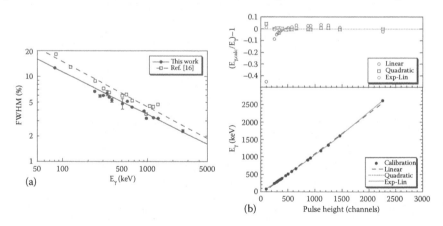

FIGURE 12.6 **(See color insert)** (a) The measured energy resolutions for γ-ray energies between 81 and 2615 keV. The solid line shows the fit of the power function $E^{-0.4993}$. The dashed line shows the anticipated $E^{1/2}$ behavior. (b) The residuals for the different fits to the calibration curves described. (c) The relationship between the pulse height and γ-ray energy using linear (dashed), quadratic (dotted), and "exponential-linear" (solid) fits. (From Billnert, R., Oberstedt, S., Andreotti, E., Hult, M., Marissens, G., and Oberstedt, A., *Nucl. Instrum. Methods Phys. Res. A*, 647, 94, 2011.)

crystals, unlike lanthanum halide crystals. Samples of $CeBr_3$ that are not encapsulated are found to have a higher light yield, but without a corresponding improvement in the energy resolution. The light yield from $CeBr_3$ scintillators is very proportional over an energy range of 122–1274 keV, with the nonproportionality of the light yield being ~4% [37]. The contribution to the energy resolution of ~4.0% (typically recorded for these scintillators for an incident energy of 662 keV) is dominated by the photoelectron yield (13 phe/keV for $CeBr_3$ compared to 21 phe/keV reported for $LaBr_3(Ce:5\%)$ [56]), and the nonproportionality with energy (R_{nPr}). The contribution from the nonproportionality term is greater for $CeBr_3$ than for $LaBr_3(Ce:5\%)$ [49,53,56]. At present, a model to assess the energy resolution due to the nonproportionality of the crystal is still being devised.

12.3.1 PROTON ACTIVATION AND INTRINSIC ACTIVITY OF $CeBr_3$ SCINTILLATORS

The robustness of $CeBr_3$ scintillators when subjected to high proton fluxes was assessed by replicating energies typically observed from solar events [107], generated by a superconducting cyclotron. The fluxes ranged from 10^9 to 10^{12} protons/cm^2, where subsequent activation of $CeBr_3$ was found to be similar to that observed in $LaBr_3(Ce)$. The resulting spectrum was dominated by ^{77}Kr, ^{79}Kr, and ^{140}Cs from activated bromine and cerium [56]. This study by Quarati et al. also found that during the irradiation of protons on $CeBr_3$, little degradation in the energy resolution or the light yield of the detector was observed [56]. This makes $CeBr_3$ detectors a very promising scintillator to use for future space missions.

Unlike the ^{138}La radioisotope in lanthanum halide detectors, cerium and bromine do not have any long-lived metastable states or radioactive isotopes, and therefore the intrinsic activity in $CeBr_3$ crystals can be ascribed to the decay of ^{227}Ac into its daughter nuclides, similar to that seen in lanthanum halide crystals. The presence of this contamination is likely to be due to the fact that the elements Ac, La, and Ce are chemically similar, or may be due to insufficient purification of the raw materials that go into the making of the crystal. However, the source of the contamination is most likely to originate from ^{227}Ac rather than from raw materials such as ^{235}U ore, due to the structure of the peaks in both lanthanum halide and $CeBr_3$ crystals appearing to be very similar in structure [56]. The levels of ^{227}Ac contamination are largely attributed to the choice of material that is used to grow the crystal, where levels vary from 0.001 cts/s/cm^3 to similar levels observed in $LaBr_3(Ce)$ scintillators (0.1 cts/s/cm^3). However, the level of ^{227}Ac can be reduced by growing the $CeBr_3$ crystals from selected batches, although the long-term availability of batches with low ^{227}Ac cannot currently be guaranteed [56]. Unlike lanthanum halide crystals (which show different levels of intrinsic activity for various crystal sizes due to the escape probabilities and attenuation lengths of the ^{138}La products), the intrinsic activity observed in $CeBr_3$ scintillators is not very dependent on the crystal size [108]. Light from the α radiation registered in the detector due to the intrinsic ^{227}Ac contamination is found to be quenched by 26% when compared to γ radiation. The ratio of this α/γ light yield observed with $CeBr_3$ is thus ~1.33 times lower than that typically observed for $LaBr_3(Ce)$ scintillators [56], reinforcing the fact that the charge-transport efficiency to the luminescence centers in $CeBr_3$ is strongly influenced by the higher ionization density.

12.3.2 TIMING PROPERTIES OF CEBR₃ SCINTILLATORS

CeBr₃ scintillators have been found to be almost as fast as cerium-doped lanthanum halide scintillators, the short lifetime of the emission of Ce^{3+} luminescence being responsible for the 1/e nature of the decay time constant (τ). For CeBr₃, a decay constant of $\tau = 17$ ns has been obtained by fitting an exponential component over the integrated light output and subtracting the background [37]. Similarly, the rise time of the timing signal was found to be 0.7 ns [37]. On average, the decay constant for CeBr₃ has been reported to be longer than LaBr₃(Ce) due to the self-absorption/ re-emission cycle.

Timing resolutions using CeBr₃ scintillators have been measured using various types of PMTs to help establish the best combination of scintillator and photosensor. In a study by Fraile et al. [109], a combination of a CeBr₃ scintillator with a Hamamatsu (R9779) PMT was found to give a timing resolution (FWHM) of 119 ± 2 and 164 ± 2 ps when used with a reference BaF₂ detector at ^{60}Co and ^{22}Na energies, respectively. These values are better than those achieved with a CeBr₃-Photonis (XP20D0) detector combination, which was found in the same study to give timing resolutions (FWHM) of 146 ± 2 and 210 ± 2 ps at ^{60}Co and ^{22}Na energies, respectively [109]. This result is interesting as, traditionally, Photonis tubes such as the XP20D0 model (the properties of which are given alongside the Hamamatsu R9779 model in Table 12.4) have been found to perform very well with LaBr₃(Ce) crystals in previous studies [70]. Despite the non-linearity of the energy spectrum when these PMTs are used with CeBr₃ crystals at an operating voltage above 800 V, the timing resolutions of these scintillators are still 20%–30% worse than results obtained when using a CeBr₃ scintillator coupled to a Hamamatsu R9779 PMT [109].

TABLE 12.4
Main Properties of PMTs Used with CeBr₃ Crystal

PMT	τ_{rise} (ns)	FWHM TTS (ns)	Blue Sensitivity (μA/lmF)
XP20D0	1.60	0.52 ± 0.03 [70]	12.00, 11.60
R9779	1.80	0.25 [110]	9.55

Source: Fraile, L.M., Mach, H., Vedia, V., Olaizola, B., Paziy, V., Picado, E., and Udías, J.M., *Nucl. Instrum. Methods Phys. Res. A*, 701, 235, 2013. The blue sensitivity for the R9779 is taken from Hamamatsu Photonics. Photomultiplier tube R9779 specifications, 2009.

Notes: The PMTs each have eight stages and borosilicate glass windows. The transit time spread (TTS) is shown in the third column and is given in nanoseconds.

12.4 SUMMARY

This chapter began by introducing the requirements that need to be met in order to have an ideal scintillator and briefly looked at the history of scintillator development up to the present day. Although not perfect, the lanthanum halide crystals $LaBr_3(Ce)$ and $LaCl_3(Ce)$ have been shown to be popular crystals in the field of radiation detection. The good energy (2.9% FWHM at 662 keV [25,31]) and excellent timing (~110 ps FWHM at ^{60}Co energies with a Ø1"×1" crystal [70]) resolutions of these scintillators allow for many opportunities to be realized in medical and space applications, as well as in scientific research, when matched with a good photosensor. However, these crystals suffer from intrinsic activity due to the decay of the radioisotope [138]La and α-decay from [227]Ac (with rates of approximately 0.7 and 0.1 counts/s/cm^3, respectively [45,47]).

More recently, an alternative to lanthanum halide scintillators, $CeBr_3$, has been shown to perform almost as well as its sister crystals. Although $CeBr_3$ does not match the excellent energy resolution offered by $LaBr_3(Ce)$ crystals (3.5%–4.5% FWHM at 662 keV [37,104,105]), the timing resolution is comparable (~120 ps FWHM at ^{60}Co energies for a Ø1"×1" crystal [109]). The most desirable quality of these new scintillators is the absence of a large intrinsic background, making them ideal for low rate/background counting applications.

As technologies and techniques in the field of radiation detection advance, research continues to seek the ideal scintillator detector, with garnet ceramics [42] and other inorganic scintillators (i.e., $SrI_2(Eu)$ [41,42], $Cs_2LiYCl_6(Ce)$ (CLYC) [111–115], $Cs_2LiLaCl_6(Ce)$ (CLLC) [116], and $Cs_2LiLaBr_6(Ce)$ (CLLB) [43,116]) presently leading the way.

REFERENCES

1. R. Hofstadter, Alkali halide scintillation counters, *Phys. Rev.*, 74:100, 1948.
2. O.H. Nestor and C.Y. Huang, Bismuth germanate: A high-Z gamma-ray and charged particle detector, *IEEE Trans. Nucl. Sci.*, NS-22:68, 1975.
3. M. Moszyński, et al., Properties of CsF, a fast inorganic scintillator in energy and time spectroscopy, *Nucl. Instrum. Methods Phys. Res.*, 179:271, 1981.
4. M. Laval and M. Moszyński, et al., Barium fluoride—Inorganic scintillator for subnanosecond timing, *Nucl. Instrum. Methods Phys. Res.*, 206:169, 1983.
5. S.A. Wender, Bismuth germanate as a high-energy gamma-ray detector, *IEEE Trans. Nucl. Sci.*, NS-30:1539, 1983.
6. P.J. Nolan and D.W. Gifford, The performance of a bismuth germanate escape suppressed spectrometer, *Nucl. Instrum. Methods Phys. Res. A*, 236:95, 1985.
7. M. Moszyński and H. Mach, A method for picosecond lifetime measurements for neutron-rich nuclei, *Nucl. Instrum. Methods Phys. Res. A*, 277:407, 1989.
8. H. Mach, R.L. Gill, and M. Moszyński, A method for picosecond lifetime measurements for neutron-rich nuclei, *Nucl. Instrum. Methods Phys. Res. A*, 280:49, 1989.
9. H. Mach, et al., A method for picosecond lifetime measurements for neutron-rich nuclei, *Nucl. Instrum. Methods Phys. Res. A*, 280:49, 1989.
10. P. Lecoq, A. Annenkov, A. Gektin, M. Korzhik, and C. Pedrini. *Inorganic Scintillators for Detector Systems: Particle Acceleration and Detection.* Springer-Verlag, Berlin, 2006.
11. G. Knoll. *Radiation Detection and Measurement.* Wiley, New York, 3rd edn, 232, 2000.

12. J. Andriessen, et al., Lattice relaxation study of the 4f-5d excitation of Ce^{3+}-doped $LaCl_3$, $LaBr_3$, and $NaLaF_4$: Stokes shift by pseudo Jahn-Teller effect, *Phys. Rev. B*, 76:7, 2007.

13. K.W. Krämer, et al., Development and characterization of highly efficient new cerium-doped rare earth halide scintillator materials, *J. Mater. Chem.*, 16:2773, 2006.

14. J. Andriessen, O.T. Antonyak, and P. Dorenbos, et al., Experimental and theoretical study of the spectroscopic properties of Ce^{3+} doped $LaCl_3$ single crystals, *Opt. Commun.*, 178:355, 2000.

15. E.V.D. van Loef, P. Dorenbos, C.W.E. van Eijk, K. Krämer, and H.U. Güdel, High-energy-resolution scintillator: Ce^{3+} activated $LaCl_3$, *Appl. Phys. Lett.*, 77:1467, 2000.

16. E.V.D. van Loef, P. Dorenbos, C.W.E. van Eijk, K. Krämer, and H.U. Güdel, High-energy-resolution scintillator: Ce^{3+} activated $LaBr_3$, *Appl. Phys. Lett.*, 79:1573, 2001.

17. Saint-Gobain. BrilLanCe™ scintillators performance summary. Saint-Gobain. 2009. http://www.crystals.saint-gobain.com/uploadedFiles/SG-Crystals/Documents/Technical/SGC%20BrilLanCe%20Scintillators%20Performance%20Summary.pdf.

18. W.M. Higgins, et al., Bridgman growth of $LaBr_3$:Ce and $LaCl_3$:Ce crystals for high-resolution gamma-ray spectrometers, *J. Cryst. Growth*, 287:239, 2006.

19. B. Morosin, Crystal structures of anhydrous rare-earth chlorides, *J. Chem. Phys.*, 49(7):3007, 1968.

20. J. Czochralski. Ein neues Verfahren zur Messung des Kristallisationsgeschwindigkeit der Metal, *Z. Phys. Chem.*, 92:219, 1918.

21. P.R. Menge et al., Performance of large lanthanum bromide scintillators, *Nucl. Instrum. Methods Phys. Res. A*, 579:6, 2007.

22. A. Giaz, et al. *2012 IEEE NSS/MIC Record*, 331, 2012.

23. H. Chen, C. Zhou, P. Yang, and J. Wang. Growth of $LaBr_3$:Ce^{3+} single crystal by vertical Bridgman process in nonvacuum atmosphere, *J. Mater. Sci. Technol.*, 25:753, 2009.

24. Y. Pei, X. Chen, R. Mao, and G. Ren, Growth and luminescence characteristics of undoped $LaCl_3$ crystal by modified Bridgman method, *J. Cryst. Growth*, 279:390, 2005.

25. E.V.D. van Loef, P. Dorenbos, C.W.E. van Eijk, K.W. Krämer, and H.U. Güdel, Scintillation properties of $LaBr_3$:Ce^{3+} crystals: Fast, efficient and high-energy-resolution scintillators, *Nucl. Instrum. Methods Phys. Res. A*, 486:254, 2002.

26. P. Dorenbos, et al., Scintillation properties of $RbGd_2Br_7$:Ce^{3+} crystals: fast, efficient, and high density scintillators, *Nucl. Instrum. Methods Phys. Res. B*, 132:728, 1997.

27. O. Guillot-Noël, et al. *Nucl. Instrum. Methods Phys. Res. B*, 132:728, 1997.

28. O. Guillot-Noël, et al., Optical and scintillation properties of cerium-doped $LaCl_3$, $LuBr_3$ and $LuCl_3$, *J. Lumin.*, 85:21, 1999.

29. E.V.D. van Loef, P. Dorenbos, C.W.E. van Eijk, K. Krämer, and H.U. Güdel, Optical and scintillation properties of pure and Ce^{3+} doped $GdBr_3$, *Opt. Commun.*, 189:297, 2001.

30. E.V.D. van Loef, P. Dorenbos, C.W.E. van Eijk, K. Krämer, and H.U. Güdel, Scintillation properties of $LaCl_3$:Ce^{3+} crystals: Fast, efficient, and high-energy resolution scintillators, *IEEE Trans. Nucl. Sci.*, NS-48:341, 2001.

31. K.S. Shah, et al., $LaBr_3$:Ce scintillators for gamma-ray spectroscopy, *IEEE Trans. Nucl. Sci.*, 50:2410, 2003.

32. K.S. Shah, et al., High-energy resolution scintillation spectrometers, *IEEE Trans. Nucl. Sci.*, NS-51(5):2395, 2004.

33. K.S. Shah, $Cebr_3$ scintillator. US Patent, 2008. US 7405404.

34. K. Yang, J.J. Buzniak P.R. Menge, and V. Ouspenski. Performance improvement of large Sr^{2+} and Ba^{2+} co-doped LaBr:Ce^{3+} scintillation crystals. *Nucl. Sci. Symp. Med. Imag. Conf.*, 308, 2012.

35. K.S. Shah, et al., $LaCl_3$:Ce scintillator for γ-ray detection, *Nucl. Instrum. Methods Phys. Res. A*, 505:76, 2003.

36. J.T.M. DeHaas and P. Dorenbos, Advances in yield calibration of scintillators, *IEEE Trans. Nucl. Sci.*, 55:1086, 2008.
37. K.S. Shah, J. Glodo, W. Higgins, E.V.D. van Loef, W.W. Moses, S.E. Derenzo, and M.J. Weber, $CeBr_3$ scintillators for gamma-ray spectroscopy, *IEEE Trans. Nucl. Sci.*, NS-52(6):3157, 2005.
38. Saint-Gobain. Baf_2 barium fluoride scintillation material, Saint-Gobain Ceramics and Plastics, Hiram, OH, 2014. http://www.crystals.saint-gobain.com/uploadedFiles/SG-Crystals/Documents/Barium%20Fluoride%20Data%20Sheet.pdf.
39. L.M. Fraile, et al., Fast-timing study of a $CeBr_3$ crystal: Time resolution below 120 ps at 60Co energies, *Nucl. Instrum. Methods Phys. Res. A*, 701:235, 2013.
40. H. Mach and L.M. Fraile. *Hyperfine Interactions*, forthcoming, 2012.
41. E.V.D. van Loef, et al., Crystal growth and scintillation properties of strontium iodide scintillators, *IEEE Trans. Nucl. Sci.*, 56:869, 2009.
42. N.J. Cherepy, et al., Scintillators with potential to supersede lanthanum bromide, *IEEE Trans. Nucl. Sci.*, 56:873, 2009.
43. U. Shirwadkar, et al., Scintillation properties of $Cs_2LiLaBr_6$ (CLLB) crystals with varying Ce^{3+} concentration, *Nucl. Instrum. Methods Phys. Res. A*, 652:268, 2011.
44. J.K. Hartwell and R.J. Gehrke, Observations on the background spectra of four $LaCl_3$ Ce scintillation detectors, *Appl. Radiat. Isot.*, 63:223, 2005.
45. F. Quarati, et al., X-ray and gamma-ray response of a $2'' \times 2''$ $LaBr_3$:Ce scintillation detector, *Nucl. Instrum. Methods Phys. Res. A*, 574:115, 2007.
46. F.C.L. Crespi, et al., Alpha–gamma discrimination by pulse shape in $LaBr_3$:Ce and $LaCl_3$:Ce, *Nucl. Instrum. Methods Phys. Res. A*, 602:520, 2009.
47. B.D. Milbrath, et al., Characterization of alpha contamination in lanthanum trichloride scintillators using coincidence measurements, *Nucl. Instrum. Methods Phys. Res. A*, 547:504, 2005.
48. G. Knoll, *Radiation Detection and Measurement*. Wiley, New York, 3rd edn, 330, 2000.
49. P. Dorenbos, J.T.M. DeHaas, and C.W.E. Van Eijk, Non-proportionality in the scintillation response and the energy resolution obtainable with scintillation crystals, *IEEE Trans. Nucl. Sci.*, 42:2190, 1995.
50. E.V.D. van Loef, W. Mengesha, J.D. Valentine, P. Dorenbos, and C.W.E. Van Eijk, Non-proportionality and energy resolution of a $LaCl_3$:10% Ce^{3+} scintillation crystal, *IEEE Trans. Nucl. Sci.*, 50:155, 2003.
51. M. Moszyński, L.S. Widerski, and T. Szczesniak, et al., Study of $LaBr_3$ crystals coupled to photomultipliers and avalanche photodiodes, *IEEE Trans. Nucl. Sci.*, 55:1774, 2008.
52. M. Moszyński, Inorganic scintillation detectors in γ-ray spectrometry, *Nucl. Instrum. Methods Phys. Res. A*, 505:101, 2003.
53. I. Khodyuk and P. Dorenbos, Nonproportional response of $LaBr_3$:Ce and $LaCl_3$:Ce scintillators to synchrotron x-ray irradiation, *J. Phys. Condens. Matter*, 22:485402, 2010.
54. P. Dorenbos, Light output and energy resolution of Ce^{3+}-doped scintillators, *Nucl. Instrum. Methods Phys. Res. A*, 486:208, 2002.
55. P. Dorenbos, Fundamental limitations in the performance of Ce^{3+}-, Pr^{3+}-, and Eu^{2+}-activated scintillators, *IEEE Trans. Nucl. Sci.*, 57:1162, 2010.
56. F.G.A. Quarati, et al., Scintillation and detection characteristics of high-sensitivity $CeBr_3$ gamma-ray spectrometers, *Nucl. Instrum. Methods Phys. Res. A*, 729:596, 2013.
57. A. Owens, A.J.J. Bos, S. Brandenburg, and P. Dorenbos, et al., The hard x-ray response of Ce-doped lanthanum halide scintillators, *Nucl. Instrum. Methods Phys. Res. A*, 574:158, 2007.
58. S. Kraft, E. Maddox, and A. Owens, et al., Development and characterization of large La-halide gamma-ray scintillators for future planetary missions, *IEEE Trans. Nucl. Sci.*, 54:873, 2007.

59. C. D'Ambrosio, F. de Notaristefani, G. Hull, V.O. Cencelli, and R. Pani, Study of LaCl$_3$:Ce light yield proportionality with a hybrid photomultiplier tube, *Nucl. Instrum. Methods Phys. Res. A*, 556:187, 2006.

60. S.A. Payne, N.J. Cherepy, G. Hull, J.D. Valentine, W.W. Moses, and W.-S. Choong, Nonproportionality of scintillator detectors: Theory and experiment, *IEEE Trans. Nucl. Sci.*, 56:2506, 2009.

61. A. Syntfeld, et al., Comparison of a LaBr$_3$ (Ce) scintillation detector with a large volume CdZnTe detector, *IEEE Trans. Nucl. Sci.*, 53:3938, 2006.

62. I.V. Khodyuk, F.G.A. Quarati, M.S. Alekin, and P. Dorenbos, Energy resolution and related charge carrier mobility in LaBr$_3$:Ce scintillators, *J. Appl. Phys*, 114:123510, 2013.

63. J. Glodo, et al., Effects of Ce concentration on scintillation properties of LaBr$_3$:Ce, *IEEE Trans. Nucl. Sci.*, 52:5, 2005.

64. D.R. Schaart, et al., LaBr$_3$:Ce and SiPMs for time-of-flight PET: Achieving 100 ps coincidence resolving time, *Phys. Med. Biol.*, 55:N179, 2010.

65. I. Deloncle, et al., Fast timing: Lifetime measurements with LaBr$_3$ scintillators, *J. Phys. Conf. Ser.*, 205:012044, 2010.

66. C.P. Allier, et al., Readout of a LaCl$_3$(Ce^{3+}) scintillation crystal with a large area avalanche photodiode, *Nucl. Instrum. Methods Phys. Res. A*, 485:547, 2002.

67. N. Mărginean, et al., In-beam measurements of sub-nanosecond nuclear lifetimes with a mixed array of HPGe and LaBr$_3$:Ce detectors, *Eur. Phys. J. A*, 46:329, 2010.

68. O.J. Roberts, et al. Development of a LaBr$_3$(Ce) fast-timing array for FAIR. *EPJ Web Conf.*, 63, 2013.

69. J.-M. Régis, et al., The mirror symmetric centroid difference method for picosecond lifetime measurements via γ–γ coincidences using very fast LaBr$_3$(Ce) scintillator detectors, *Nucl. Instrum. Methods Phys. Res. A*, 622:83, 2010.

70. M. Moszyński, et al., New Photonis XP20D0 photomultiplier for fast timing in nuclear medicine, *Nucl. Instrum. Methods Phys. Res. A*, 567:1, 2006.

71. L.M. Fraile, et al. Fast Timing Collaboration. Fast Timing Studies at ISOLDE: Highlights and Perspectives. Universidad Complutense, Madrid, 2009. https://indico.cern.ch/event/67060/session/8/contribution/36/material/slides/0.pdf.

72. K.S. Krane. *Introductory Nuclear Physics*. Wiley, New York, 1988.

73. P.J. Nolan and J.F. Sharpey-Shafer, The measurement of the lifetimes of excited nuclear states, *Phys. Rep.*, 42:1, 1979.

74. A.Z. Schwarzschild and E.K. Warburton, The measurement of short nuclear lifetimes, *Ann. Rev. Nucl. Sci.*, 18:265, 1968.

75. Z. Bay, Calculation of decay times from coincidence experiments, *Phys. Rev.*, 77:419, 1949.

76. W. Andrejtscheff, et al., The generalized centroid shift method for lifetime measurements in heavy ion reactions, *Nucl. Instrum. Methods Phys. Res. A*, 204:123, 1982.

77. P. Petkov, et al., Complex time distributions from isomers in cascade: A case in [176]Lu, *Nucl. Instrum. Methods Phys. Res. A*, 321:259, 1992.

78. H. Mach, et al., Retardation of B(E2; $0_1^+ \to 2_1^+$) rates in [90–96]Sr and strong subshell closure effects in the $A \sim 100$ region, *Nucl. Phys. A*, 523:191, 1991.

79. P.H. Regan, et al., Precision lifetime measurements using LaBr$_3$ detectors with stable and radioactive beams, *EPJ Web Conf.*, 63, 2013.

80. O.J. Roberts, et al., A LaBr$_3$:Ce fast-timing array for DESPEC at FAIR, *Nucl. Instrum. Methods Phys. Res. A*, 748:91, 2014.

81. J.-M. Régis, et al., The generalized centroid difference method for picosecond sensitive determination of lifetimes of nuclear excited states using large fast-timing arrays, *Nucl. Instrum. Methods Phys. Res. A*, 726:191, 2013.

82. T. Alharbi, et al., Electromagnetic transition rates in the $N=80$ nucleus $^{138}_{58}$Ce, *Phys. Rev. C*, 87:014323, 2013.

83. T. Alharbi, et al., Gamma-ray fast-timing coincidence measurements from the 18O+18O fusion-evaporation reaction using a mixed LaBr$_3$-HPGe array, *Appl. Radiat. Isot.*, 70:1337, 2012.

84. P.J.R. Mason, et al., Half-life of the I-pi=4(-) intruder state in P-34: M2 transition strengths approaching the island of inversion, *Phys. Rev. C*, 85:064303, 2012.

85. P.J.R. Mason, et al., Half-life of the yrast 2$^+$ state in ^{188}W: Evolution of deformation and collectivity in neutron-rich tungsten isotopes, *Phys. Rev. C*, 88:044301, 2013.

86. S. Kisyov, et al., Fast-timing measurements in ^{95,96}Mo, *J. Phys. Conf. Ser.*, 2012.

87. S. Kisyov, et al., In-beam fast-timing measurements in ^{103,105,107}Cd, *Phys. Rev. C*, 84, 2011.

88. O.J. Roberts, et al., Half-life measurements of excited states in ^{132}Te, ^{134}Xe, *Acta Phys. Pol.*, B44, 2013.

89. M.S. Alekhin, et al., Improvement of γ-ray energy resolution of LaBr$_3$:Ce^{3+} scintillation detectors by Sr^{2+} and Ca^{2+} co-doping, *App. Phys. Lett.*, 102:161915, 2013.

90. M.S. Alekhin, D.A. Biner, K.W. Krämer, and P. Dorenbos, Improvement of LaBr$_3$:5%Ce scintillation properties by Li$^+$, Na$^+$, Mg^{2+}, Ca^{2+}, Sr^{2+}, and Ba^{2+} co-doping, *J. Appl. Phys.*, 113:224904, 2013.

91. S. Espana, et al., Performance evaluation of SiPM photodetectors for PET imaging in the presence of magnetic fields, *Nucl. Instrum. Methods Phys. Res. A*, 613:308, 2010.

92. C. Catana, et al., Simultaneous acquisition of multislice PET and MR images: Initial results with a MR-compatible PET scanner, *J. Nucl. Med.*, 47:1968, 2006.

93. M.S. Judenhofer, et al., Simultaneous PET-MRI: A new approach for functional and morphological imaging, *Nat. Med.*, 14:459, 2008.

94. G. Llosa, et al., First Compton telescope prototype based on continuous LaBr$_3$-SiPM detectors, *Nucl. Instrum. Methods Phys. Res. A*, 718:130, 2013.

95. E.-J. Buis and F. Quarati, et al., Proton induced activation of LaBr$_3$:Ce and LaCl$_3$:Ce, *Nucl. Instrum. Methods Phys. Res. A*, 580:902, 2007.

96. O.J. Roberts, et al. Neutron response of 1.5" LaBr$_3$:Ce crystal scintillators for PARIS. 2008. http://paris.ifj.edu.pl/documents/detectors/Ce.pdf.

98. A.A. Naqvi, et al., Detection efficiency of low levels of boron and cadmium with a LaBr$_3$:Ce scintillation detector, *Nucl. Instrum. Methods Phys. Res. A*, 665:74, 2011.

99. C. Hoel, et al., Pulse-shape discrimination of La halide scintillators, *Nucl. Instrum. Methods Phys. Res. A*, 540, 2005.

100. J.C. Brice. *Crystal Growth Processes*. Blackie Halsted, London, 1986.

101. R.D. Shannon, Revised effective ionic radii and systematic studies of interatomic distances in halides and chalcogenides, *Acta Crystallogr.* A32:751, 1976.

102. W. Drozdowski, P. Dorenbos, A. Bos, G. Bizarri, A. Owens, and F. Quarati, CeBr$_3$ scintillator development for possible use in space missions, *IEEE Trans. Nucl. Sci.*, NS-55(3):1391, 2008.

103. P. Guss, M. Reed, D. Yuan, A. Reed, and S. Mukhopadhyay, CeBr$_3$ as a room-temperature, high-resolution gamma-ray detector, *Nucl. Instrum. Methods Phys. Res. A*, 608(2):297, 2009.

104. R. Billnert, S. Oberstedt, E. Andreotti, M. Hult, G. Marissens, and A. Oberstedt, New information on the characteristics of 1in.×1in. cerium bromide scintillation detectors, *Nucl. Instrum. Methods Phys. Res. A*, 647:94, 2011.

105. F. Quarati. LaBr$_3$ gamma-ray spectrometers for space applications. PhD thesis, Technische Universiteit Delft.

106. M. Ciemala, et al., Measurements of high-energy-rays with detectors, *Nucl. Instrum. Methods Phys. Res. A*, 608:76, 2009.

107. A. Owens, et al., Assessment of the radiation tolerance of LaBr$_3$:Ce scintillators to solar proton events, *Nucl. Instrum. Methods Phys. Res. A*, 572:785, 2007.

108. F.G.A. Quarati, et al., Study of ^{138}La radioactive decays using LaBr$_3$ scintillators, *Nucl. Instrum. Methods Phys. Res. A*, 683:46, 2012.

109. L.M. Fraile, H. Mach, V. Vedia, B. Olaizola, V. Paziy, E. Picado, and J.M. Udías, Fast timing study of a CeBr$_3$ crystal: Time resolution below 120ps at ^{60}Co energies, *Nucl. Instrum. Methods Phys. Res. A*, 701:235, 2013.

110. Hamamatsu. Photomultiplier tube R9779 specifications. Hamamatsu Photonics, Iwata City, Japan, 2012. http://www.hamamatsu.com/resources/pdf/etd/R9779_TPMH1297E06.pdf.

111. C.M. Combes, et al., Optical and scintillation properties of pure and Ce^{3+}-doped Cs$_2$LiYCl$_6$ and Li$_3$YCl$_6$:Ce^{3+} crystals, *J. Lumin.*, 82:299, 1999.

112. E.V.D. van Loef, et al., Optical and scintillation properties of Cs$_2$LiYCl$_6$:Ce^{3+} and Cs$_2$LiYCl$_6$:Pr^{3+} crystals, *IEEE Trans. Nucl. Sci.*, 52:1819, 2005.

113. A. Bessiere, et al., Luminescence and scintillation properties of CS$_2$LiYCl$_6$:Ce^{3+} for γ and neutron detection, *Nucl. Instrum. Methods Phys. Res. A*, 537:242, 2005.

114. A. Bessiere, et al., New thermal neutron scintillators: Cs$_2$LiYCl$_6$:Ce^{3+} and Cs$_2$LiYBr$_6$:Ce^{3+}, *IEEE Trans. Nucl. Sci.*, 51:2970, 2004.

115. J. Glodo, et al., Scintillation properties of 1 inch Cs$_2$LiYCl$_6$ crystals, *IEEE Trans. Nucl. Sci.*, 55:1206, 2008.

116. J. Glodo, et al., Selected properties of Cs$_2$ LiYCl$_6$, Cs$_2$ LiLaCl$_6$, and Cs$_2$ LiLaBr$_6$ scintillators, *IEEE Trans. Nucl. Sci.*, 58:333, 2011.

13 Novel X- and Gamma-Ray Detectors Based on Metamaterials

Paul Lecoq

CONTENTS

13.1 INTRODUCTION

The detection of x-rays, gamma-rays (γ-rays), and ionizing particles plays an important role in a wide range of applications, such as high- and medium-energy physics and astrophysics detectors, spectroscopy, medical imaging, industrial nondestructive control systems, and homeland security. Present x and γ detectors are mostly based on inorganic scintillating materials or direct-conversion semiconductor materials. They generally aim to measure the coordinate of the conversion point and the total energy released in the detector volume. If information on the direction of the incoming particle is desired, it is obtained by packing several detector planes.

These ionizing radiation detectors convert the energy of an incoming particle into light in the case of a scintillator or into electronic carriers in the case of a solid-state detector. In both cases, these materials must be dense enough to allow an efficient x- or γ-ray conversion in a small detector volume, have a fast response time to sustain high acquisition rates, and have a linear response as a function of the energy deposit. A good energy and position resolution is generally required in the majority of the applications. However, it is difficult to have both at the same time, and most of the detector designs have to compromise between these two parameters. Moreover, present systems provide global information on the total energy deposit in the detector but have no particle-identification capability and give no details on the cascade mechanism of this energy deposition and on the physical signal-generation process.

New scintillator production processes and recent advances in nanotechnologies, in particular in the domain of photonics crystals and nanocrystals, open interesting perspectives for the development of new detector concepts capable of delivering much richer information about x- or γ-ray energy deposition. They are based on metamaterials to simultaneously record with high precision the maximum amount of information on the cascade conversion process, such as its direction, the spatial distribution of the energy deposition, and its composition in terms of electromagnetic, charged, and neutral hadron contents in the case of high-energy incoming particles.

13.2   MECHANISMS OF ENERGY CONVERSION

Charged and neutral particles interact with absorbing materials following the well-known mechanisms of radiation interactions in matter described by many authors [1,2]. A scintillator is an absorbing material, and has the additional property of being able to convert into light a fraction of the energy deposited by ionizing radiation. Charged particles continuously interact with the electrons of the medium through Coulomb interactions, resulting in atomic excitation or ionization. Neutral particles will first have to undergo a direct interaction with either an electron or a nucleus, producing secondary electrons, recoil protons, or fission fragments, which will then transfer their energy to the medium in the same way as primary charged particles.

The rate of energy loss ($-dE/dx$) for charged particles is strongly energy dependent. It is well described by the Bethe–Bloch formula (Equation 13.1) for incoming particles in the megaelectronvolt to gigaelectronvolt range, with atomic shell corrections at lower energy and radiative loss corrections at higher energy:

$$-\frac{dE}{dx} = \frac{4\pi e^4 z^2}{m_0 v^2} NZ \left[\ln \frac{2m_0 v^2}{I} - \ln\left(1 - \frac{v^2}{c^2}\right) - \frac{v^2}{c^2} \right] \qquad (13.1)$$

where:
  v and ze = velocity and the electric charge of the primary particle
  m_0 and e = rest mass and the electric charge of the electron
  N and Z = atom-number density and atomic number of the absorbing material
  I = mean excitation energy of the absorber

For heavy materials currently used as scintillators with a density of 6–8 g/cm^3, the energy loss is typically on the order of 10 MeV/cm for a minimum ionizing particle, but it can be a factor of up to 100 more at very low or very high energy.

In the case of x- or γ-rays, the three fundamental mechanisms of electromagnetic interaction are [3]

- Photoabsorption
- Compton scattering
- Electron–positron pair production

The dominant process at low energy (up to a few hundred kiloelectronvolts for heavy materials) is the photoelectric absorption. The interacting photon is completely absorbed and transfers its energy to an electron from one of the electron shells of the absorber atom (usually from a deep shell). The resulting photoelectron is ejected with a kinetic energy corresponding to the incident photon energy minus the binding energy of the electron on its shell. This is followed by a rapid reorganization of the electron cloud in order to fill the electron vacancy, which results in the emission of characteristic x-rays or Auger electrons. The photoelectric process has the highest probability when the incident photon has an energy comparable to the kinetic energy of the electron on its shell. This is the origin of the typical peaks observed in the cross-section curve corresponding to resonances for the different electron shells (Figure 13.1). The general trend of this cross section is a rapid decrease with energy and a strong dependence on the atomic number (Z) of the absorber, which explains the preponderance of high-Z materials for x- or γ-rays shielding and detecting materials:

$$\sigma_{ph} \propto \frac{Z^5}{E_\gamma^{7/2}} \tag{13.2}$$

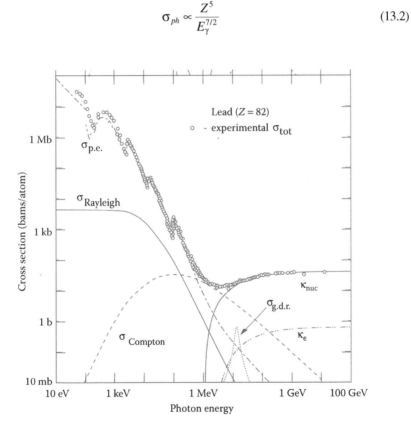

FIGURE 13.1 Energy dependence of photon total cross sections in lead. (Extracted from Particle Data Group. Particle Physics Booklet. Particle Data Group, Berkeley, CA, 2014. http://pdg.lbl.gov/2014/download/rpp-2014-booklet.pdf [4].)

At energies above a few hundred kiloelectronvolts, Compton scattering becomes predominant. In this case, the incident photon transfers only part of its initial energy E_γ to an electron of the atomic shells and is scattered at an angle θ with respect to its original direction. The recoil electron is then rapidly absorbed by the scintillator and releases an energy according to Equation 13.3:

$$E_e = E_\gamma - E'_\gamma - E_{ebinding} \tag{13.3}$$

where E'_γ is the energy of the scattered photon given by (with m_0 the rest mass of the electron)

$$E'_\gamma = \frac{E_\gamma}{1 + \dfrac{E_\gamma}{m_0 c^2}(1 - \cos\theta)} \tag{13.4}$$

The energy released in the scintillator by the recoil electron is distributed on a continuum between zero and a maximum of up to $E_\gamma - m_0 c^2 / 2 = E_\gamma - 256$ keV (for gamma energy that is large compared to the rest mass of the electron).

The probability of Compton scattering is related to the electron density in the medium and increases linearly with the atomic number of the absorber. As this Z dependence is much smaller than for the photoelectric effect, high-Z materials will be therefore favored when a high fraction of photoelectric conversion is required, as is the case for positron emission tomography (PET) scanners, for instance.

Above a threshold of 1.02 MeV (twice the rest mass of the electron), the mechanism of e^+e^- pair production can take place, predominantly in the electric field of the nuclei and to a lesser extent in the electric field of the electron cloud (κ_{nuc} and κ_e, respectively, in Figure 13.1). Similarly to photoabsorption and Compton scattering, this process has a higher probability for high-Z materials as the cross section is approximately given by Equation 13.5 [5]:

$$\sigma_{pair} \propto Z^2 \ln(2E_\gamma) \tag{13.5}$$

As long as the energy of particles is high enough for multiple scattering and electron–positron pair creation, their energy is progressively distributed to a number of secondary particles of lower energy, which form an electromagnetic shower. Below the threshold of electron–positron pair creation, electrons will continue to lose energy, mainly through Coulomb scattering.

In the case of an ordered material like a crystal, another mechanism takes place at this stage. In the process of energy degradation within a shower, the electrons in the kiloelectronvolt range start to couple with the electrons and atoms of the lattice and to excite the electrons from the occupied valence or core bands to different levels in the conduction band. Each of these interactions results in an electron–hole pair formation. If the energy of the electron is high enough to reach the ionization threshold, free carriers are produced, which will move randomly in the crystal until they are trapped by a defect, collected on electrodes (semiconductors), or recombine on a luminescent center (scintillators). In the event that the ionization threshold is not reached, the electron and hole release part of their energy by coupling to the lattice vibration modes until they reach the top of the valence band for the hole and the bottom of the conduction band for the electron. They can also be bound and form an exciton whose energy is in general slightly smaller than the band-gap between the valence and the conduction bands. At this stage, the probability is maximum for their relaxation on luminescent centers through an energy- or a charge-transfer mechanism.

For a material to be a scintillator, it must contain luminescent centers. They are either extrinsic, generally doping ions; or intrinsic, that is, molecular systems of the lattice or of defects of the lattice, which possess a radiative transition between an excited and a lower energy state. Moreover, the energy levels involved in the radiative transition must be smaller than the forbidden energy band-gap, in order to avoid reabsorption of the emitted light or photoionization of the center.

In a way, a scintillator can be defined as a wavelength shifter. It converts the energy (or wavelength) of an incident particle or energetic photon (ultraviolet [UV]-, x-, or γ-ray) into a number of photons of much lower energy (or longer wavelength) in the visible or near-visible range, which can be easily detected by current photomultipliers, photodiodes, or avalanche photodiodes.

Due to the complexity of the conversion mechanism, there is generally a trade-off between the spatial and the energy resolution, as a good spatial resolution requires a high segmentation, whereas a good energy resolution is obtained in a large enough detector volume to contain all the cascade interactions generated by the incoming particle. The increasing demand for better spatial resolution in all three dimensions for the majority of applications may lead to a huge increase in the number of readout channels, with all the associated problems of connectivity, detector integration, and heat dissipation.

Moreover, present detectors provide very little or no information about the way the energy is released in the detector, that is, the spatial distribution of the cascade events and the physics mechanisms of the energy loss (ionization, multiple scattering, nuclear interaction, and Cherenkov). This is particularly important for high-energy physics experiments wherein the high-precision measurement of hadrons and jets resulting from the decay of heavy short-lived particles produced in high-energy collisions is one of the detector challenges in future high-energy colliders. Figure 13.2 illustrates the complexity of a 3 TeV electron collision producing pairs of W and Z bosons. The simulated event shows the distribution of the multiple tracks in jets in the central tracking section of the detector and the showering of these tracks in the calorimeter section.

One of the difficulties is related to the lateral extension of the showers associated with the individual tracks in a jet. These showers strongly overlap and produce a signal, which is dependent on the relative amount of electromagnetic and nuclear interaction events in the energy-conversion process. Indeed, these two mechanisms are characterized by different mechanisms of nonradiative energy losses in the majority of known scintillators and have therefore a different scintillation efficiency. As a consequence of large event-to-event fluctuations in the relative contribution of these two mechanisms, the resulting energy resolution becomes deteriorated. One possible

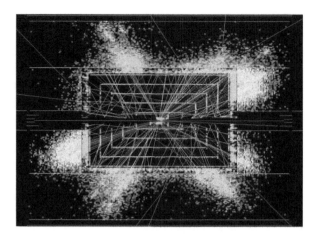

FIGURE 13.2 **(See color insert)** Simulated WW/ZZ event resulting from a 3 TeV electron collision at the future linear collider CLIC. (Courtesy of C. Grefe.)

solution is to measure independently the scintillation light produced in an active medium, which is proportional to the total energy deposited by the shower particles, and the Cherenkov light, which is only produced by the charged, relativistic shower particles. Since the latter are almost exclusively found in the electromagnetic (em) shower component (dominated by π^0s produced in hadronic showers), a comparison of the two signals makes it possible to measure the energy fraction carried by this component, f_{em}, event by event. As a result, the effects of fluctuations in this component, which are responsible for significant loss of performance in the majority of calorimeters (nonlinearity, poor energy resolution, and non-Gaussian response function), can be eliminated or at least reduced in order to improve substantially the energy resolution of hadronic showers in general and of jets in particular. This leads to an important improvement in the performance of the hadronic calorimeter. This approach, also known as dual readout, has been developed by R. Wigmans [5].

Although apparently less complex, the conversion events at low energy are also far from punctual, even in the case of a photoelectric conversion producing a recoil electron and a cascade of x-rays and Auger electrons as a result of the reorganization of the atomic electron cloud after the initial event. Indeed, one of the limitations of currently available PET and single photon emission tomography (SPECT) cameras is a poor spatial resolution (5–6 mm in whole body clinical cameras), which is partly related to the need to integrate on a sufficiently large detector pixel the primary Compton-scattering interactions (70% of the cases in the currently used conversion materials such as lutetium orthosilicate [LSO]) and the final photoelectric interaction in order to reconstruct the full energy of the event with sufficient precision. In spite of this, a large number of Compton events still escape the pixel where the primary interaction took place. These events are generally rejected by the image reconstruction algorithm, thereby reducing by a large amount the overall sensitivity of the camera. The possibility of making use of this large fraction of rejected events, for which at least one Compton interaction took place, would considerably increase the sensitivity of the camera. The same image quality could therefore be obtained in

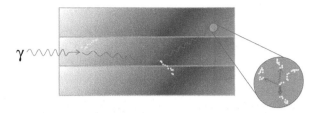

FIGURE 13.3 (**See color insert**) Example of a 511 keV gamma conversion in three crystal pixels of a PET camera, with two successive Compton interactions followed by a photoelectric event. Note that a large fraction of energy escapes the central crystal, where the first interaction took place.

a shorter time or with a lower dose injected into the patient. Similar arguments apply for the control of industrial systems or for homeland security: a detailed recording of the whole Compton-photoelectric interaction chain would allow the separation of the spatial resolution from the energy-resolution requirement and would have a strong impact on both the reconstructed image quality and on the sensitivity of the imaging device. An example of a 511 keV conversion event in a PET detector made up of several pixel crystals (typically $2 \times 2 \times 10$ mm^3) is shown in Figure 13.3. Even for such a simple event, it appears that a large fraction of the energy escapes the central pixel, where the first Compton interaction took place.

13.3 HEAVY CRYSTAL FIBER TECHNOLOGY

Shaped crystalline fibers grown from the melt have a number of advantages compared to more traditional technologies involving bulk crystal growth followed by cutting and polishing. First of all, fibers can be grown with cross sections approaching the final geometry, thereby minimizing the work and cost of mechanical processing. Furthermore, the structural quality of the surface is not damaged by the use of abrasives. The growth rate is generally high, which not only has an impact on the production costs but also on the homogeneity of the distribution of the doping ions. Such methods generally induce less thermal stresses with a resulting smaller probability of cracks.

The analysis of the potential of different melt growth techniques shows that micro-pulling-down technology from a shape-controlled capillary die allows the production of elongated crystals with dimensions that are not accessible using traditional cutting and polishing of bulk crystals grown using the more standard Czochralski or Bridgman methods. The size of the melting zone in the pulling-down technique is up to one order of magnitude smaller than that observed in the Czochralski method. Therefore, it is believed that the pulling-down process can be considered a good way to achieve stationary pulling conditions and can facilitate the growth process, allowing, for instance, much faster growth and higher concentration of doping ions, even for those with high segregation coefficient.

Since 1993, several groups around the world have contributed to impressive progress in the development of the micro-pulling-down scintillating fiber technology (see [6,7] and references therein), mainly driven by laser-rod production. New applications in the field of ionizing radiation detection are now being pursued in the frame of the Crystal Clear collaboration [8], particularly for particle physics experiments as well as for medical imaging instrumentation. High-quality fibers of lutetium aluminum garnet (LuAG) have been grown in collaboration with the company Fibercryst (Lyon) and the Laboratoire de Physico-Chimie des Matériaux Luminescents (LPCML) at Université Claude Bernard (Lyon). Other well-known heavy scintillating crystals such as bismuth germanate (BGO), LSO and lutetium-yttrium orthosilicates (LYSO) [9], and yttrium and lutetium aluminum perovskites (YAP and LuAP) have also been grown in fibers of different sizes, and lead tungstate (PWO) is under study. These high-quality scintillating fibers can open attractive possibilities for the design of future detectors for high-energy physics, medical imaging, or other applications. Table 13.1 summarizes the most relevant parameters of these crystals.

In the present state of the art, the micro-pulling-down technique allows the growth of fiber (rod)-shaped crystals with a controlled diameter of between 0.3 and 3.0 mm and up to 2 m in length. By modifying the shape of the capillary die, it is also possible to produce elongated crystalline materials with more complex noncylindrical cross sections (square, rectangular, or hexagonal) for easier integration of the crystal in complex detectors. The procedure based on micro-pulling-down technology was recently improved at the LPCML laboratory and the Fibercryst Company in Lyon in order to grow both single-crystal fibers and shaped-bulk crystals. The crystals are produced from the melt obtained at the capillary die positioned at the center bottom of a cylindrical iridium crucible as illustrated in Figure 13.4. Once the melt drop is formed, the growth process is initiated after the seed is connected with the drop at the bottom of the crucible (capillary die). Then the seed is pulled down continuously with a pulling rate ranging from 0.1 to 0.5 mm/min (about 10 times faster than Czochralski and 50 times faster than Bridgman). Ongoing developments involve a multiple-capillary die crucible, which allows several fibers to be grown in parallel. A study into the industrial optimization and cost-effectiveness of this process still needs to be made, but this should lead to a comparable or even smaller cost-per-unit volume of grown crystal than with standard crystal-growth approaches.

TABLE 13.1
Some Parameters of Crystalline Fibers Grown by Fibercryst

Crystal	Light Yield (ph/MeV)	Decay Time (ns)	Peak Emission Wavelength (nm)	Density (g/cm³)
BGO	8,000	300	480	7.13
GSO:Ce	14,000	60	460	6.71
YSO:Ce	14,000	37 and 82	420	4.45
LYSO:Ce	35,000	40	420	7.40
YAG:Ce	20,000	70	550	4.57
LuAG:Ce	20,000	70	535	6.73
LuAG:Pr	20,000	20	290–350	6.73

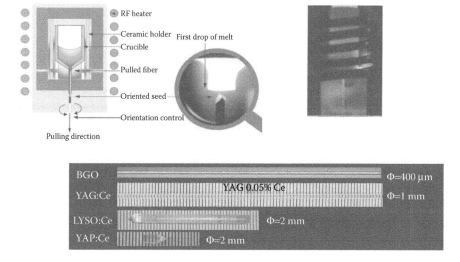

FIGURE 13.4 (See color insert) The micro-pulling-down crystal-growth technology. (Courtesy of Fibercryst.)

13.4 PHOTONIC CRYSTALS AND QUANTUM DOTS

Similarly impressive progress in the field of micro- and nanotechnologies today offers new perspectives for the development of novel detector designs based on metamaterials. The microstructuration of bulk materials and of their surfaces allows the consideration of macroscopic structures, in which light production and propagation is governed by quantum effects, giving them unusual and very attractive properties (e.g., negative refractive index, control of the direction of propagation, possibility of gating, and perfect antireflection at all angles). This is the domain of nanocrystals, quantum dots, and photonics crystals. Coupled with the parallel developments in micro- and nanoelectronics, these new approaches will undoubtedly generate revolutionary changes in paradigms in detector concepts.

13.4.1 QUANTUM DOTS-BASED MATERIALS

Quantum dots are nanoparticles that have attracted widespread interest, particularly in biology and medicine, because of their unique optical and electronic properties. Quantum dots are made out of semiconductor materials. The electrons in quantum dots have a range of energies, and the concepts of energy levels, band-gap, conduction band, and valence band still apply as in any semiconductor material. However, there is a major difference. Excitons (bound electron–hole pairs) have an average physical separation between the electron and hole, referred to as the exciton Bohr radius, given by Equation 13.6:

$$r_B = \frac{\varepsilon h^2}{\pi m_e e^2}$$

(13.6)

where:

ε   = permittivity of the medium

h = Planck constant

m_e and e = mass and electric charge of the electron, respectively

In bulk, the dimensions of the semiconductor crystal are much larger than the exciton Bohr radius, allowing the exciton to extend to its natural limit. However, if the size of a semiconductor crystal becomes small enough that it approaches the size of the material's exciton Bohr radius (typically a few nanometers for most of the known semiconductors), the electron energy levels become discrete through the effect of quantum confinement. This has major repercussions for the absorptive and emissive behavior of the semiconductor material, which can be easily tuned by the radius of the quantum dots [10,11]. Indeed, an interesting property of quantum dots is their very fast emission (in the nanosecond range) at a wavelength directly coupled to the size of the nanosphere. Moreover, the quantum efficiency of this emission can reach as much as 70%–80% of absorbed energy as compared to about 15% for the majority of known scintillators.

New technologies for preparing transparent ceramics from nanopowders can be exploited to determine the conditions for which these optical properties can be maintained in a partially synthesized material. Another approach is based on the choice of a host material, whether inorganic or organic, as a substrate for quantum dots. They are suspended in an organic solvent so that they are castable from a solution—for example, by spin-coating techniques—onto such a substrate without necessarily requiring lattice matching. An interesting avenue to explore is to dope heavy host materials (such as fluoride glasses [12]) with quantum dots of different diameters with a known spatial distribution within the detection block. An appropriate distribution of quantum dots of different sizes (and therefore different emission wavelength) in the host volume can help encoding the x- or γ-ray conversion-point coordinates with the emission wavelength of the quantum dots. The recording of the emission spectrum from this detector, when hit by ionizing radiation, would then give an exact representation of the three-dimensional (3-D) distribution of the energy deposit within this block, opening the way for an imaging calorimeter concept with a minimum number of readout channels. This approach is challenging but has probably the highest potential in terms of the density of information such metamaterials can achieve.

13.4.2 PHOTONIC CRYSTALS FOR LIGHT TRANSPORT AND LIGHT EXTRACTION

Photonic crystals are periodically structured electromagnetic media, generally possessing photonic band-gaps: ranges of frequencies in which light cannot propagate through the structure. This periodicity, whose length scale is proportional to the wavelength of light in the band-gap, is the electromagnetic analogue of a crystalline atomic lattice, where the latter acts on the electron wave function to produce the familiar band-gaps, well known to solid-state physicists. Photonics crystals are usually made of a block of transparent dielectric material that contains a number of tiny air holes arranged in a lattice pattern. To be able to create photonic crystals for optical devices, state-of-the-art semiconductor microfabrication techniques are used. The patterned dielectric material will block light with wavelengths in the photonic band-gap, while allowing other wavelengths to pass freely.

One interesting application of photonic crystals concerns the coupling of optical propagation modes to extraction modes at the interface between two media with different indices of refraction. This approach is actively pursued in our group [13] to improve the light-extraction efficiency from heavy and therefore highly refractive bulk materials, such as the majority of commonly used scintillators, for which the light collection efficiency rarely exceeds 30% with standard techniques. The basic idea is to exploit three important characteristics of 2-D periodic photonic crystal slab geometries. The first is the possibility of designing a photonic band-gap within a range of frequencies for the in-plane guided modes of the system. The second is the possibility of coupling to resonant modes above the band-gap. The third is an intrinsic upper cutoff frequency for waveguiding in the plane. In all cases, these frequency ranges can be utilized to forbid radiation from propagating into the dielectric slab, thus forcing the light into the light cone of the air region.

By combining the properties of quantum dots and of photonics crystals in properly engineered microstructured dielectric materials, it is indeed possible to tune the light-emission parameters and the light-propagation properties in a bulk material at a small scale and to guide the light through complex structures made of assembled scintillating fibers or nanocrystals. In other words, it means that a high segmentation of a bulk system and spatial encoding of the light signals would be possible, allowing a high level of granularity without having to multiply the number of readout electronics channels.

In order to extract these encoded signals from complex metamaterial structures, the concept of photonic crystal fibers is of particular interest. Also called hollow-core band-gap fibers, they consist of a regular lattice of air cores running along the fiber, which can then transmit a wide range of wavelengths without suffering from dispersion (Figure 13.5). They are made by packing a series of hollow glass capillary tubes in a material with a different dielectric constant. This structure is then heated and stretched to create a long fiber that is only a few microns in diameter. The fiber has the unusual property of being able to transmit a single mode of light, even if the

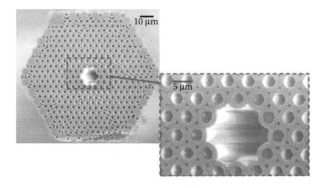

FIGURE 13.5 **(See color insert)** Photonic crystal fiber.

diameter of the core is very large. Unlike standard fibers, in which light is guided in a core of higher refraction index than for the cladding, photonic crystal fibers transmit light through the hollow cores. The bulk of the fiber is only a support for the hollow-core lattice and does not participate in the transfer of the electromagnetic energy. Such fibers are therefore ideal in a high-radiation environment as the light transmission does not become deteriorated by radiation damage.

The development of photonics crystal fibers is a very active field in modern photonics as the market perspectives on this segment are large for telecommunications and chip-to-chip interconnect for highly intergrated electronics. Several manufacturers offer such fibers at a moderate price, which is likely to significantly decrease if the market opens up, as expected.

13.5 DETECTOR CONCEPT

The design of the majority of x- and γ-ray detectors today results from a trade-off between spatial- and energy-resolution requirements. It does not provide any detail about the complexity of the energy-conversion mechanism, which carries important information on the spatial distribution as well as on the energy-loss processes of the shower resulting from the interaction of the incoming particle with the detection medium. The basic idea for improving this situation is to structure the standard detector block or pixel in such a way as to extract more information than the total energy deposit in the block. In electromagnetic calorimeters, the dimensions of the unit-detection block are typically 1 Moliere radius (1–3 cm for commonly used scintillators) and 25 radiation lengths (20–30 cm) in lateral and longitudinal directions, respectively. For medical imaging devices, the pixel size is generally one order of magnitude smaller, a few millimeters in section, and 1–3 cm in length.

The proposed approach is based on scintillating fibers packed together to form trunks of "cables." The variety of fiber section shapes that can be produced allows the detector design to be tuned as a function of the requirements: hexagonal fibers if ultimate homogeneity is required, rectangular-section fibers for different granularity requirements in two directions, and cylindrical fibers if free channels are needed for photonic crystal fiber–based light-collection systems or for other services. Notice also that the relative mechanical flexibility of these scintillating fibers (depending on the material) allows them to be twisted in the cable in a similar way as in a rope, thereby minimizing the impact of interfiber gaps for incoming particles. Moreover, various scintillators can be selected to build these cables, having different emission wavelength and different scintillation yield and decay time, so that a direct encoding of the light can be made as a function of the point of emission. A single photodetector can then decode this complex signal and provide much more complete information about the x- and γ-ray conversion. For high-energy calorimeters, materials with different ultraviolet (UV) transmission cutoff can be selected so that the fraction of Cherenkov emission in the detected light can be determined. The direct extraction of the Cherenkov signal in a scintillator from the pulse-shape or wavelength analysis may, however, be difficult if the scintillating signal is much higher than the Cherenkov one, which is usually the case. Moreover, Cherenkov and scintillation signals are not independent, as part of the UV Cherenkov emission is absorbed in the crystal to excite the scintillation activator ions. An alternative approach is to select scintillating materials activated by a doping ion instead of self-activated scintillators such as BGO or PWO. For example, cerium-doped LSO or LYSO, which is both very fast (40 ns decay time) and dense (7.4 g/cc); LuAG (70 ns decay time and 6.73 g/cc density); and LuAP (17 ns decay time and 8.34 g/cc density) are excellent candidates for mixing cerium-doped fibers. These would then behave as scintillators with undoped fibers of the same material, which would only produce Cherenkov light. One could then obtain a very homogeneous, dense, and compact calorimeter with a uniform radiation length, Moliere radius, and interaction length in the whole volume of the detector. This detector would have the additional feature of being able to sample the shower with a number of Cherenkov fibers conveniently distributed in the cable in order to directly measure the electromagnetic-to-hadronic ratio of a shower on an event-to-event basis. Moreover, the assembly of the fibers can be organized in a flexible way, allowing a multitude of detector geometries.

If the detection of neutrons proves to substantially improve the overall jet-energy resolution, neutron-sensitive scintillation fibers can also be inserted in the cables. No attempts have been made yet to grow such fibers, but scintillators like lithium tetraborate LBO ($Li_2B_4O_6$), lithium fluoride–based materials like LiCAF ($LiCaAlF_6$), or more generally the cerium-doped elpasolite family ($Cs_{2-x}Rb_xLiMX_6$, where $X = Sc$, Y, La, Lu and $X = Cl$, Br, I) are very attractive candidates because of the presence of high-neutron-capture cross-section lithium and boron and because of their low density (2.42 for LBO), which makes them rather insensitive to gamma conversions.

Figure 13.6 shows a possible concept for such a detector block presently being developed for future linear collider calorimeters [14,15]. It is made of three types of fibers (scintillating, Cherenkov, and neutron sensitive) arranged in a cable, possibly in a twisted configuration to minimize cracks seen by the incoming particles. The fiber's section is hexagonal so as to allow a compact packing with minimum dead

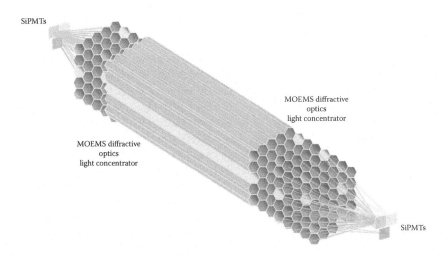

FIGURE 13.6 **(See color insert)** Concept of a metacable for calorimetry in future linear colliders.

space between the fibers. Such a configuration is quasihomogeneous, allowing the best possible energy resolution for electromagnetic calorimetry. Moreover, it provides useful information about the composition of hadronic showers (charged light and heavy particles and neutrons), allowing a significant improvement in the resolution of jets. The light produced by each type of fiber is collimated by diffractive optics or microlens plates on three small size solid-state photodetectors (avalanche photodiode [APD] or Geiger-mode silicon photomultiplier [SiPMT]) at both sides of the detector block to further provide depth-of-interaction information, with a precision of typically one tenth of the fiber length. The optical transfer system from the fibers to the photodetectors requires special attention, as it must allow high specificity in multiplexing and measuring the different components with the minimum of cross talk and must also be as compact as possible. It must be noticed that similar optical systems are under development for adaptive optics devices to be used in future large telescopes or telescope arrays. Fortunately, we can take advantage here of the ongoing large-scale academic and industrial effort to transfer microelectronics technologies to the development and production of micro-opto-electro-mechanical systems (MOEMS). Microlens arrays are a vital part of today's optical systems in a large range of domains, from telecommunications to machine vision. Today's technology allows the production of either refractive or diffractive lenses as small as a few tens of microns in diameter arranged in large-size arrays with a position accuracy of just a few microns. Standard integrated-circuit production methods, such as optical lithography, are used to mass-produce diffractive optics components with a high level of functionality and integration from photomasks imprinted by holographic techniques. A master component is created, from which a negative mold is fabricated. The mold is then used for the embossing or injection molding of plastic plates, similarly to the way holographic Christmas wrapping papers and CD-Roms are duplicated. Such microlens arrays are therefore thin and mechanically flexible, allowing easy coupling to the fiber bundle, and can be mass-produced at low cost. Furthermore, there is a large variety of mechanical supports available, which is important for high-energy physics applications, where reasonably radiation-hard materials must be selected.

An alternative approach is being developed in our group, more oriented to low x- and γ-ray energies below the pair-production threshold. The objective is to design a detector head for a PET scanner on the basis of photonic crystals and quantum dots. PET scanners measure in coincidence two 511 keV γ-rays resulting from the decay of a positron emitted by a radioactive isotope (generally ^{18}F) labeling a molecule involved in some metabolic process. Once injected into the patient, this molecule will concentrate in the organs where this metabolic function is active (e.g., cancer cells concentrate sugar more than healthy cells). The recording of the two 511 keV γ-rays associated with each ^{18}F positron emission allows the 3-D reconstruction of the functionally active parts of the organs under examination.

The detector block is made of a matrix of heavy scintillating fibers. The scintillating material is selected as a function of its intrinsic light yield, which has to be high enough to allow the splitting of the light through different readout channels while minimizing the variance on the number of detected photons and its negative impact on the energy resolution. Moreover, a good linearity of the response is requested to achieve good energy resolution. We are presently exploring different materials, among which LuAG is a good candidate with a light yield of 20,000 ph/MeV (about one half of sodium iodide); a decay time of 70 ns or 20 ns, whether it is cerium or praesodymium doped; and reasonable response linearity. The technology for growing such fibers using micro-pulling-down technology is also well mastered, and fibers of several meters in length and a diameter ranging between 300 μm and 3 mm can be grown with consistent quality. The diameter of the fibers is defined as a function of the desired spatial resolution and of the detector's ability to identify the first interaction point in the conversion process. The angular distribution of photons diffused by Compton scattering of 511 keV incident γ-rays being strongly peaked forward, a precise determination of the depth of each interaction in the conversion chain associated to one 511 keV γ conversion should allow determining with high precision where the first interaction took place.

If the scintillating fibers have a cylindrical shape, the tiny gaps between them can be filled with photonic crystal fibers of a few hundred microns in diameter. As previously explained, photonic crystal fibers transmit light through a regular lattice of holes arranged in a material of a higher dielectric constant than air. This supporting material can be loaded with quantum dots with a different diameter range from fiber to fiber and/or distributed with a gradient of their diameter along each individual fiber. The spatial encoding is provided by the emission wavelength of these quantum dots. The principle is to organize the readout through two different channels, one dedicated to energy-deposit measurement with the best possible resolution and the second optimized for 3-D spatial resolution of the different components of the conversion cascade. A conversion event in a scintillating fiber, whether photoelectric or Compton, will produce an isotropic emission of scintillating photons. The photons emitted in the forward or backward direction will propagate along the scintillating fiber and eventually be collected at both ends and summed on a sufficiently large number of fibers to measure the total energy deposited in the cascade of primary and secondary events produced by the conversion of the incident x- or γ-ray. On the other hand, some of the photons emitted laterally will escape the scintillating fiber and excite the quantum dots of the photonic crystal fiber, allowing the fiber hit to be identified and the depth of interaction to be determined with high precision by measuring the spectrum of the light collected on miniaturized spectrophotometers. Diffractive optics components will be used in the same way as for the previous detector to collect the light from the different types of fibers and to collimate it on specific photodetectors. A schematic of this low-energy detector block is shown in Figure 13.7.

13.6   CONCLUSION

The underlying idea of the work presented in this chapter is to develop metamaterials for the purpose of extracting the maximum information from the conversion of x-rays or γ-rays in a material. Two proof-of-concept systems are described, aimed at different application domains. The first proposes the concept of a dual-readout (scintillation + Cherenkov) or even triple-readout (with the possibility of measuring neutrons) imaging calorimeter aiming at an excellent

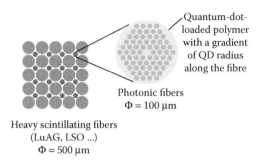

FIGURE 13.7   **(See color insert)** Concept of a metacable for low-energy x- and γ-rays.

jet-energy resolution for high-energy-physics experiments. The second is oriented toward low-energy x-ray or γ-ray detection and is intended to be used particularly for a finely 3-D granulated PET detector head. But the two approaches are complementary, and the related generic work on heavy scintillating fibers, photonic crystals, quantum dots, and refractive optics is expected to open useful perspectives for the development of novel metamaterial-based x-ray or γ-ray detectors in a wide range of energy.

ACKNOWLEDGMENT

This contribution would not have been possible without many exchanges of views and discussions with all my colleagues, both physicists and medical doctors, from the Crystal Clear collaboration.

REFERENCES

1. B. Rossi, *High Energy Particles*, Prentice-Hall, Englewood Cliffs, NJ (1952).
2. U. Fano, Penetration of protons, alpha particles, and mesons, *Ann. Rev. Nucl. Sci.*, 13, 1 (1963).
3. J.D. Jackson, *Classical Electrodynamics*, 2nd ed.Wiley, New York (1975), chap. 13.
4. Particle Data Group. Particle Physics Booklet. Particle Data Group, Berkeley, CA (2014). http://pdg.lbl.gov/2014/download/rpp-2014-booklet.pdf.
5. N. Akchurin, et al., Hadron and jet detection with dual-readout calorimeter, *Nucl. Instrum. Methods A*, 537, 537–561 (2005).
6. J. Ricard, U.S. Patent 4 565 600, 21 January, 1986.
7. K. Lebbou, G. Boulon, Fiber crystals growth from the melt, *Advances in Materials Research Series*, eds T. Fukuda, P. Rudolp and S. Uda, Springer Verlag, Berlin (2003), pp. 219–254.
8. R&D proposal for the study of new fast and radiation hard scintillators for calorimetry at LHC, Crystal Clear Collaboration, CERN/DRDC/P27/91–15, project RD 18.
9. P. Anfré, C. Dujardin, J.M. Fourmigué, B. Hautefeuille, K. Lebbou, C. Pedrini, D. Perrodin, and O. Tillement, Evaluation of fiber-shaped LYSO for double readout gamma photon detection, *IEEE Trans. Nucl. Sci*, 54(2), 391–397 (2007).
10. Y. Shen, C.S. Friend, Y. Jiang, D. Jakubczyk, J. Swiatkiewicz, and P. N. Prasad, Nanophotonics: Interactions, materials and applications, *J. Phys. Chem. B*, 104, 7577–7587 (2000).
11. Y. Kayakuma, Quantum size effects of interacting electrons and holes in semiconductor microcrystals with spherical shape, *Phys. Rev. B*, 38, 9797 (1988).

12. E. Auffray, et al., Cerium doped heavy metal fluoride glasses, a possible alternative for electromagnetic calorimetry, *Nucl. Instrum. Methods. A*, 380, 524–536 (1996).

13. M. Kronberger, E. Auffray, P. Lecoq, X. Letartre, C. Seassal, and J.L. Leclercq, Probing the concepts of photonic crystals on scintillating materials, *IEEE Trans. Nucl. Sci*, 55(3), 1102–1106 (2008).

14. P. Lecoq, New crystal technologies for novel calorimeter concepts, in *Proceedings of CALOR2008, International Conference on Calorimetry in High Energy Physics*, Pavia, May 2008.

15. E. Auffray and P. Lecoq, Dual readout calorimetry with heavy crystal fibers, *IEEE NSS/MIC Conference Records*, Dresden, 2009.

14 Gamma-Imaging Devices Based on Position-Sensitive Photomultipliers

Michael Seimetz

CONTENTS

14.1 INTRODUCTION

Miniaturized imaging devices for γ-radiation have become increasingly important in medical and preclinical applications throughout the last decade. They commonly consist of two main parts: the radiation sensor and an (external) collimator. The sensor allows for a precise measurement of the position of interaction of incident γ photons in an optically dense medium (typically inorganic scintillation crystals). Its main parts are schematically depicted in Figure 14.1a. The crystal is coupled to a photomultiplier in which the scintillation light is converted into amplified electronic pulses. These, in turn, give rise to a (smaller or larger) number of digital output channels by means of onboard electronics. The generation of a sharp image requires, in addition, knowledge of the incidence angle of γ-rays. Although the choice of suitable collimation systems is an integral part of the imaging equipment, we will restrict the scope of this overview to a detailed description of the radiation sensor.

The technical feasibility of compact gamma-imaging devices is strongly entangled with the development of position-sensitive photomultiplier tubes (PSPMTs). Early models, such as the Hamamatsu R2486 and 3292 series (3 in./5 in. diameter, respectively), provided two-dimensional (2-D) readout through crossed wire anodes while maintaining the traditional, circular PMT shape. They have been applied in gamma camera prototypes with continuous [1] or pixelated crystals [2,3].

(a)   (b)

FIGURE 14.1 (a) Schematic view of the small gamma camera Sentinella 102, with a flat crystal coupled to a PSPMT. The readout electronics and power supply are mounted on four circuit boards. (b) PSPMT Hamamatsu H8500.

Second-generation PSPMTs such as the Hamamatsu R7600-C12 were of multianode type; however, they suffered from significant dead space at the borders of the photocathode [3]. The Hamamatsu H8500 multianode series has become by far the most popular model (Figure 14.1b). Its square shape with 52 mm side length makes it suitable for individual use, but also for larger arrays with small dead space. The 49×49 mm^2 active area is divided into 8×8 channels with 12 stages. Examples for the use of this third-generation PSPMT (called "flat panel" for its 33 mm height) in numerous devices will be presented in the following sections. Finally, the Hamamatsu H9500 is of the same size, but provides enhanced intrinsic spatial resolution by 16×16 individual channels.

14.2   SCINTILLATION CRYSTALS

The choice of a suitable scintillator material and geometry for a gamma-imaging application comprises various aspects. Each of the commonly available inorganic crystals (Table 14.1) has its own advantages and drawbacks with respect to density, light output, decay time, and hygroscopicity. In general, medical applications of planar gamma or single-photon emission computerized tomography (SPECT)-type

TABLE 14.1

Physical Properties of Frequently Used Inorganic Crystals

Material	Density (g/cm^3)	Decay Time (ns)	Light Output (% of NaI(Tl))	Hygroscopicity
NaI(Tl)	3.7	200	100	Yes
CsI(Tl)	4.5	1000	45	Slightly
CsI(Na)	4.5	630	85	Yes
LaBr$_3$(Ce)	5.3	25	160	Very hyg.
GSO	6.7	56	35	No
LSO	6.5	40	75	No
LYSO	7.1	41	75	No

are based on radioisotopes with energies in the 100–200 keV range. For these γ-rays, NaI(Tl) or CsI crystals of a few millimeters in thickness provide sufficiently high stopping powers. For example, at 140 keV the sensitivity of 4 mm thick CsI(Na) is 67%. At higher energies, the use of denser materials is mandatory, although for continuous crystals the image quality degrades with increasing thickness. $LaBr_3(Ce)$ combines a comparatively high density with a very short decay time and very high light output, which make it potentially attractive for many applications. Single- and multipad prototype cameras based on monolithic $LaBr_3(Ce)$ have been demonstrated [8,9,16]. However, the strong hygroscopicity of $LaBr_3$ imposes serious constraints on its handling and lifetime [9]. For the detection of 511 keV annihilation photons in positron emission tomography (PET) modules, denser materials such as LSO or LYSO with more than 10 mm thickness are necessary. These provide high light output and the very short decay time that is required for high count rates, but suffer from background events due to the natural radioactivity of lutetium. Single-pad modules with 10 mm thick crystals can be arranged in ring structures with the FOV adapted to special applications like small-animal [10] or breast imaging [11].

Many gamma detectors are based on pixelated crystals [4]. These may simply have the same size as the PSPMT pads (6 mm width for the H8500) with each anode corresponding to a single pixel [5]. However, smaller pixels allow for much higher spatial resolution. This is possible if the scintillation light of each pixel is detected by several neighboring PSPMT channels. If most of the light is collected by only one PSPMT pad position, nonlinearities are observed in the reconstructed image. This undersampling can be corrected by a larger number of pads, as is demonstrated by direct comparison between the H8500 and H9500 models [6]. Some light spread is always provided due to the thickness of the PSPMT entrance window. In addition, the crystal array may be sealed with a glass window (Figure 14.2), which can improve the localization of pixels, but introduces a blurring of the image. Systematic trials with different pixel sizes [3] show that CsI(Tl) pixels as small as 1 mm^2 can be resolved by the H8500.

An alternative method relies on continuous scintillation crystals. Here, light can freely spread throughout the entire detector surface (Figure 14.2b). Figure 14.3 shows typical images obtained during the calibration procedure [7] of a commercialized version of the single-pad gamma camera of Figure 14.1. They represent the raw event distribution (without software correction) of a ^{99m}Tc source placed 40 cm in front of the camera, one of them with a tungsten mask with 11×11 holes defining a 40 mm

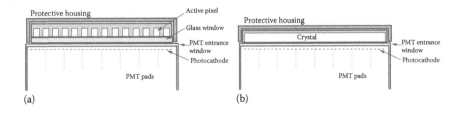

FIGURE 14.2 Light detection in pixelated (a) and continuous (b) crystals.

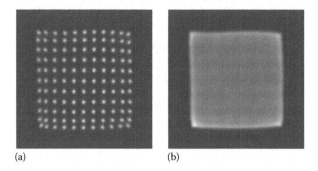

FIGURE 14.3 Raw images from a gamma camera with continuous CsI(Na) crystal, obtained with a 11×11 hole mask (a) and a uniform flood field (b).

wide field of view (FOV). The 4 mm thick CsI(Tl) crystal has 49.5 mm side length (matching the PSPMT active area). The outer margin is eliminated from image reconstruction because the spatial resolution decreases significantly due to internal light reflection, a common drawback of monolithic crystals with respect to pixelated ones. However, the image quality over the active area is very good, with 1.6 mm intrinsic spatial resolution. The mask hole positions are used as reference points for the reconstruction of the entire image plane, with a polynomial interpolation algorithm described in [7].

The uniform event distribution (Figure 14.3b) shows remainders of the PSPMT pad structure. These nonuniformities at the level of a few percent are software corrected in real time by the multiplication of pixel coefficients, which are calibrated for each camera.

The limitations of continuous crystals due to internal light reflection, especially at the borders, can partly be overcome by painting the uncoupled surfaces. While pixelated crystals usually are embedded in white reflectors to maximize light output, for monolithic ones it is advantageous to cover the lateral faces [1] or, for PET applications, all external faces with black absorber [10,11].

14.3   READOUT ELECTRONICS AND IMAGE RECONSTRUCTION

For the H8500 PSPMT with 8×8 output channels, basically three different readout schemes have been applied. Either all 64 anode signals can be digitized and used for data processing, or they can be summarized into a much smaller number of output channels, typically 16 (8 rows and 8 columns), or even just 4. A direct comparison has been presented in [5] for an 8×8 square crystal array coupled to the H8500 in the way that each individual crystal corresponds to one anode.

When the event rate is small compared to the sampling rate, multiple hits are very rare, and the 2-D signal distribution can be approximated by two 1-D projections. To this purpose, the 64 anode currents are equally split and then summarized along pad rows and columns, resulting in 8 x and 8 y outputs. These may be digitized to reconstruct the hit coordinates via

$$x = \frac{\sum_i q_x^i x_i}{\sum_i q_x^i}, y = \frac{\sum_i q_y^i y_i}{\sum_i q_y^i}$$

(14.1)

and the event energy

$$E = \sum_i \left(q_x^i + q_y^i\right)$$

(14.2)

An additional resistor may be coupled to each anode individually to compensate for inherent gain differences [12]. For the H8500C/D series, these amount typically to a factor of 2–3. A map of gain distributions for each PSPMT is provided by the manufacturer. The effort of adapting the circuitry may be worthwhile for the setup of prototypes. For mass-production, however, flexible software solutions are clearly preferable.

The 64 anode currents may also be injected into a common, 2-D resistor matrix network with only four output channels (X_A, X_B, X_C, X_D). These correspond geometrically to the four corners of the image. Here, the energy and hit position are reconstructed via

$$E = X_A + X_B + X_C + X_D$$

(14.3)

and

$$x = [(X_A + X_B) - (X_C + X_D)] / E, \, y = [(X_B + X_C) - (X_A + X_D)] / E$$

(14.4)

The so-called center-of-gravity (COG) method (Equations 14.1 and 14.2) and the resistive chain algorithm (Equations 14.3 and 14.4) are equivalent. If gain differences are not too big, they allow for a reasonable reconstruction of the event position, provided the scintillation light is distributed symmetrically around the maximum. At the borders of the PSPMT and the crystal, this is usually not the case, leading to image compression, as can be seen in Figure 14.3. This effect can be corrected for by posterior software processing as described in the following paragraphs. However, the corresponding loss of spatial resolution cannot easily be cured. Some improvement can be achieved by the truncated center-of-gravity (TCOG) method. Here, a constant fraction of the integrated charge in rows and columns, Σq_x^i and Σq_y^i, is subtracted from each of the corresponding x and y outputs before applying Equations 14.3 and 14.4. Thereby, the low-energy tail far from the hit position is cut off, and only the symmetric part of the light distribution is utilized. This correction can be implemented by software or directly in the readout electronics [13].

The raw event distribution and energy spectrum, obtained with a four-channel output logic, are used as a starting point for the software correction process. An example for the necessary steps has been published for a gamma camera with a continuous CsI(Na) crystal [7]. First, the true hit position, (X, Y), is calculated as a function of the raw coordinates from Equation 14.4

$$X = P_X(x, y), Y = P_Y(x, y) \tag{14.5}$$

P_X and P_Y are 2-D polynomials that have to be calibrated for each gamma detector, starting from a point distribution like the one in Figure 14.3. Second, in a similar manner, the event energy, E, from Equation 14.3 is corrected as

$$E_c = E / P_E(X, Y) \tag{14.6}$$

compensating for relative differences of photopeak positions throughout the sensitive area of the PSPMT. The absolute energy scale, relating analog-to-digital converter (ADC) channels to γ energies (in kiloelectronvolts), is obtained from several reference sources within the interesting interval. The resulting energy resolution over the entire sensor area is around 13%.

After these steps, the hit position throughout the entire crystal can be reconstructed and background can be eliminated through a cut around the photopeak position ("energy window"). Figure 14.4 shows the results for the same raw data as displayed in Figure 14.3. The 40×40 mm^2 sensitive area is represented in 300×300 pixels. The hole positions of the tungsten mask are very accurately reproduced. Close to the borders, the points appear somewhat elliptical, an artifact of the compression in the initial image. The uniform flood field shows remainders of the PSPMT pad structure and, possibly, further deviations due to local differences of the crystal sensitivity, reflectivity, and optical coupling efficiency.

The image of Figure 14.4b is used as input for the final calibration step. After application of a gaussian filter (for smoothing), at each pixel (i, j) an inverse factor

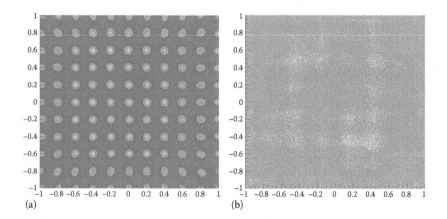

FIGURE 14.4 **(See color insert)** 40 mm wide images from a gamma camera with continuous CsI(Na) crystal, obtained with a 11×11 hole mask (a) and a uniform flood field (b), after the first reconstruction steps.

$$U(i, j) \sim 1 / N(i, j) \tag{14.7}$$

is calculated, which is utilized as weight for all subsequent acquisitions. In this way the overall nonuniformity [14] is reduced to 6%–8% typically. Note that the recommended limit for the nonuniformity of medical gamma cameras is 10%.

The four-channel readout followed by the reconstruction process described here gives good results for scintillation crystals of a few millimeters in depth. For thicker ones, image deformations at the borders become more and more pronounced. LYSO crystals of 10–12 mm thickness, used in PET applications [10,11], have to be entirely painted black to reduce internal reflections to a minimum. The performance may be improved by readout of all 64 PSPMT channels. With the light distribution known over the entire PSPMT area, the hit position can be reconstructed with off-line algorithms; for example, by adjusting gaussian functions [2] or similar models [16]. However, this implies higher hardware and computational costs.

A remarkable extension of the readout logic is the possibility of measuring the hit position not only on the photodetector plane but also in the normal direction. This so-called depth of interaction (DOI) is a key parameter for improving the resolution of PET applications where thick crystals are mandatory and the parallax error for γ photons under oblique incidence is important. Its measurement is based on the width of the light cone hitting the PSPMT surface; for a γ-ray detected close to the photocathode, it will be much narrower than for an event close to the crystal entrance face. Interestingly, it can be implemented not only in the full processing of 64 anode signals, but also in the four-channel version by electronic determination of the second moment of light distribution and readout via a fifth channel [15].

14.4   LARGE-AREA DETECTORS MADE OF PSPMT ARRAYS

The 49×49 mm^2 active area of the H8500 has been very successfully exploited for medical and preclinical applications (see Section 14.5). Its effective FOV at the object under study may be increased by using magnifying collimators; for example, of the pinhole type. In addition, flat-panel PSPMTs offer the very interesting opportunity to merge several modules together on a common, larger crystal and thereby to build sensors with inherently larger active area. A systematic study with pixelated as well as continuous crystals of various materials coupled to a 2×2 array of H8500 PSPMTs has been published in [17]. The extension of the readout electronics to multiple PSPMTs is not a major problem. For example, if the 64 channels of a single PSPMT are merged into 8 rows and 8 columns, the 2×2 array may provide 16×16 channels treated with the same logic. It is also possible to join all 256 anodes together into only four readout signals. This has been demonstrated in [7], where a wide-field camera has been developed in complete analogy to the single-pad version described in the same publication. Some care has to be taken with the power supply because of the inherent gain differences of the PSPMTs. They either need individual high voltages (which may be provided by a common source and a specially adapted resistor chain), or they have to be preselected with similar gains as specified by the manufacturer. A compromise of both has been realized for a commercialized version of the model described in [7], where two pairs of photomultipliers with equal, overall amplification share common, built-in high-voltage supplies. This gamma camera, called Sentinella™ 108, will be discussed in some detail in the following paragraphs. Its raw-image quality has been significantly improved with respect to the original prototype due to optimized optical coupling between the crystal and the PSPMTs.

A common feature of multiple-pad sensors is the lack of sensitivity in the space between adjacent PSPMTs (Figure 14.5). This region often lies right at the center of the active area and imposes a challenge to image uniformity and spatial resolution. An example is illustrated in Figure 14.6. Both images have been taken with the same camera with 2×2 H8500 photomultipliers glued onto a common, continuous CsI(Na) crystal with 102.4 mm width. Hole masks made of tungsten have been placed in front to define a regular array of reference points throughout a 91×91 mm² FOV (the crystal borders are excluded from image reconstruction). The most obvious feature is the asymmetric, butterfly-like shape originating from the design of the readout resistor network. Similar to Figure 14.3, the separation of mask points is very good in the central regions of individual PSPMTs, while at the outer borders, internal reflections lead to image compression. However, all the maxima can clearly be identified, even with only 3.5 mm spacing. In the central row and column, additional phenomena can be observed. Here, all points are slightly shifted toward the image center, an effect that is especially pronounced at the borders. Further, the lateral separation between the neighboring rows and columns appears larger than those inside the individual PSPMTs. These effects are combined with an overall loss of event energy (Figure 14.7): smooth variations of photopeaks measured in distinct positions can be observed over the entire active area, but in the central region an additional, sharp difference with respect to the nearest neighbors is obvious.

These observations can be understood from loss of scintillation light in the insensitive areas between the PSPMTs. Its consequences for the image reconstruction are discussed in detail in [18]. The global polynomial method of Equations 14.5 and 14.6 alone is not flexible enough to properly manage abrupt differences such as those of Figure 14.6. However, it may be extended by an additional, intermediate step in

FIGURE 14.5   2×2 array of H8500 photomultipliers mounted on a common analog board.

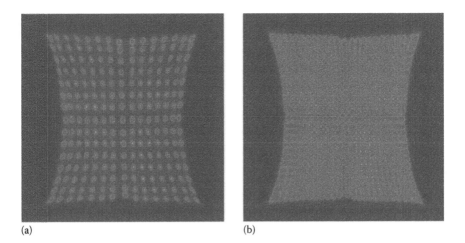

(a)   (b)

FIGURE 14.6   (See color insert) (a,b) Images of calibration masks with 15×15 and 27×27 holes, respectively, taken with a 2×2 PSPMT array coupled to a monolithic CsI(Na) crystal.

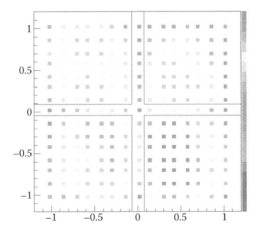

FIGURE 14.7   (See color insert) Color map of photopeak positions in 15×15 reference positions. Points in the central region are systematically shifted toward lower energy values with respect to their direct neighbors.

which the light loss for each of the four input channels (X_A, X_B, X_C, X_D) is modeled as a function of the raw hit position (x, y) from Equation 14.4. If scintillation light is detected in a circular intersection between the light cone and the photocathode, it may be assumed that a constant fraction of the light hitting the central region is lost. It may be "recovered" for each event by purely geometric considerations based on the width of the central area and the light cone. The necessary computational effort is sufficiently small to guarantee the real-time operation of the camera. The overall image quality (uniformity, energy resolution) of Sentinella 108 is comparable to that of the one-pad version, Sentinella 102. This proves that, despite the gap between PSPMTs, multiple-pad configurations are feasible, even with a common, monolithic crystal.

Similar effects to those described here have been reported from PSPMT arrays with pixelated crystals [17,19]. In this case, if the size of the pixels is similar to or smaller than the width of the gap between PSPMTs, hardly any light at all may be detected at the image center. In order to achieve a more uniform light spread, a glass window may be mounted between the crystal and the PSPMT. An unwanted side effect is, however, a general loss of spatial resolution over the entire FOV. Another option for recollecting light from the central region is to place thin strips of reflecting material inside the insensitive areas [20].

14.5   EXAMPLES

As mentioned in the previous sections, several research groups have developed γ-ray imagers based on position-sensitive photomultiplier tubes, and a few of these devices have given rise to commercial products for medical or preclinical purposes. Sentinella™ 102 (Oncovision, Valencia, Spain), the marketed version of the small-area prototype described in [7], has been extensively used in radio-guided surgery, especially for the localization of sentinel lymph nodes [25–27]. Further examples for intraoperative, handheld gamma cameras are a prototype reported in [31], Imaging Probe™ [4,28] (Li-Tech, Naples, Italy), and IHGC [29,30]. Thanks to their small size and flexibility, these devices allow for real-time acquisition of 2-D gamma images during a surgical intervention, even in cases where classical gamma probes fail to identify spots of enhanced tracer activity. Sentinella 102 has also been implemented in the dual-head SPECT module of the Albira™ small-animal imaging system (Oncovision) [10]. More recently, the successful implementation of Sentinella 108 (Oncovision) in Albira has been demonstrated, with significantly enhanced FOV, higher spatial resolution, and higher sensitivity. Despite the difficulties due to dead space between PSPMTs, this multipad device provides excellent image quality. Preclinical SPECT applications with a single-PSPMT sensor have also been reported in [23]. A 2×1 array, adapted to the size of a mouse, has been presented by [21,32]. The sensor with the largest size (3×3 PSPMTs) has been marketed in the Siemens Inveon™ system [24].

Table 14.2 summarizes some basic characteristics of published single- or multi-PSPMT sensors. We restrict the selection to crystals with less than 10 mm thickness, which are commonly used for planar gamma or SPECT applications. They are often used individually and their properties are relatively similar, while thicker crystals, such as those for PET (511 keV), may show significantly different intrinsic characteristics. Not all authors provide data for the intrinsic spatial resolution, Δx. Some present a single, mean value, while others give a range corresponding to the variation between central and border areas of the FOV. Note that the intrinsic spatial resolution may differ significantly from the actual resolution of the entire device, which is normally dominated by the properties of external collimators.

14.6   CONCLUSIONS AND OUTLOOK

Throughout the last decade, a considerable number of research groups have reported on attempts to develop compact gamma-imaging sensors based on position-sensitive PSPMTs and inorganic crystals. The Hamamatsu H8500 series PSPMTs have been

TABLE 14.2
Intrinsic Characteristics of Single- and Multi-PSPMT Sensors

Author	PSPMTs	Crystal, Pitch (mm)	Active Area (mm²)	$\Delta E/E$ (%)	Δx (mm)
Sekiya [5]	1 H8500	CsI(Tl), 6.1	49 × 49	18[b]	
Trotta [4]	1 H8500	CsI(Tl), 2.1	49 × 49	19.1[a]	2.0
Trotta [6]	1 H9500	CsI(Tl), 1.1	43.2 × 43.2	24.2[a]	1.2
Lage [23]	1 H8500	NaI(Tl), 1.6	48 × 48	9.4–10.4[a]	
Fabbri [16]	1 H8500-Mod8	NaI(Tl), 2.0	49 × 49	9.5[b]	
Fabbri [16]	1 H8500-Mod8	CsI(Tl), 1.65	24 × 24	20[b]	
Fabbri [16]	1 H8500-Mod8	LaBr₃(Ce), cont.	51 × 51	7.4–8.5[b]	
Yamamoto [9]	1 H8500	LaBr₃(Ce), cont.	50.8 × 50.8	8.9[b]	0.6
Salvador [31]	1 H8500	GSO(Ce), cont.	50 × 50		
Sánchez [7]	1 H8500	CsI(Na), cont.	40 × 40	11[a]	1.3–1.6
Loudos [21]	2 × 1 H8500	NaI(Tl), 1.2	96 × 48	15.6[a]	1.6
Xi [32]	2 × 1 H8500	NaI(Tl), 2.2	92 × 42	10.8[a]	2.2
Tamda [19]	2 × 2 H8520	NaI(Tl), 2.2	51.4 × 51.4	10–21[a]	
Qi [22]	2 × 2 H8500	NaI(Tl), 1.4	102 × 102	20[a]	
Fabbri [16]	2 × 2 H8500-Mod8	LaBr₃(Ce), cont.	100 × 100	9.0[a]	
Sánchez [7]	2 × 2 H8500	CsI(Na), cont.	91 × 91	13[a]	2.0–2.3
Stolin [20]	2 × 2 H8500	NaI(Tl), 1.6	100 × 100	15[a]	1.6
Stolin [20]	2 × 2 H8500	NaI(Tl), 1.5	100 × 100	20[a]	1.5
Stolin [20]	3 × 4 H8500	NaI(Tl), 1.2	150 × 200	23[a]	1.4
Austen [24]	3 × 3 H8500	NaI(Tl), 2.2	150 × 150	12.5[a]	

[a] The energy resolution, $\Delta E/E$, is measured at 140 keV.
[b] The energy resolution, $\Delta E/E$, is measured at 122 keV.

applied in the large majority of published examples. With readout electronics of similar structure, the sensors are scalable from single photomultipliers to arrays with FOV up to 150 mm in width. Systems with continuous or pixelated crystals provide similar intrinsic spatial resolutions, typically around 1.5 mm full width at half maximum (FWHM). Some systems have been successfully commercialized in medical gamma cameras or preclinical SPECT devices. For PET applications, modular systems have been realized with modules made from a single PSPMT and thicker crystals of denser materials than those for lower gamma energies.

Despite the maturity and success of this technology, the current trend points toward a somewhat different direction. Silicon photomultipliers (SiPMs) have been on the rise during recent years and may gain importance with respect to "classical"

PSPMTs in the near future. These can now be mounted on printed circuit boards in arrays of 50×50 mm^2 size (or other convenient formats) to be used in a similar way to the PSPMTs. SiPMs are of smaller size than PSPMT pads; that is, when arranged in arrays, they offer higher intrinsic spatial resolution than the pads of the H8500. Further, they offer the advantages of lower costs and low-voltage operation. They are almost unaffected by magnetic fields and thus can be applied in combined PET nuclear magnetic resonance (NMR) systems. However, the temperature dependence of the SiPM gain has to be kept under control, and cooling significantly below room temperature is often necessary. A convenient readout scheme for an array of SiPMs is based on application-specific integrated circuits (ASICs), on which the large number of input channels is reduced in a flexible, user-defined manner. The first studies for PET applications with continuous crystals have been published [33,34] and are ongoing.

REFERENCES

1. F. Sánchez et al., Design and tests of a portable mini gamma camera, *Med. Phys.* **31**, 1384–1397 (2004).
2. D. Thanasas et al., An analytical position correction algorithm for γ-camera planar images from resistive chain readouts, *2009 IEEE Nuclear Science Symposium Conference Record (NSS/MIC 2009)*, 2766–2769 (2009).
3. R. Pani et al., A novel compact gamma camera based on at panel PMT, *Nucl. Instrum. Meth.* A **513**, 36–41 (2003).
4. C. Trotta et al., New high spatial resolution portable camera in medical imaging, *Nucl. Instrum. Meth.* A **577**, 604–610 (2007).
5. H. Sekiya et al., Studies of the performance of different front-end systems for flat-panel multi-anode PMTs with CsI(Tl) scintillator arrays, *Nucl. Instrum. Meth.* A **563**, 49–53 (2006).
6. C. Trotta et al., High-Resolution Imaging System (HiRIS) based on H9500 PSPMT, *Nucl. Instrum. Meth.* A **593**, 454–458 (2008).
7. F. Sánchez et al., Performance tests of two portable mini gamma cameras for medical applications, *Med. Phys.* **33**, 4210–4220 (2006).
8. V. Orsolini Cencelli et al., A gamma camera with the useful field of view coincident with the crystal area, *2009 IEEE Nuclear Science Symposium Conference Record (NSS/MIC 2009)*, 1886–1890 (2009).
9. S. Yamamoto et al., Development of a compact and high spatial resolution gamma camera system using LaBr$_3$(Ce), *Nucl. Instrum. Meth.* A **622**, 261–269 (2010).
10. F. Sánchez et al., ALBIRA: A small animal PET/SPECT/CT imaging system, *Med. Phys.* **40**, 051906 (2013).
11. L. Moliner *et al.*, Design and evaluation of the MAMMI dedicated breast PET, *Med. Phys.* **39**, 5393–5404 (2012).
12. V. Popov et al., A novel readout concept for multianode photomultiplier tubes with pad matrix anode layout, *Nucl. Instrum. Meth.* A **567**, 319–322 (2006).
13. Q. Dai et al., Investigation of high resolution compact gamma camera module based on a continuous scintillation crystal using a novel charge division readout method, *Chinese Phys.* C **34**, 1148–1152 (2010).
14. NEMA Standards Publication NU 1-2007, *Performance Measurements of Gamma Cameras*, Rosslyn, VA: National Electrical Manufacturers Association (2007).
15. Ch.W. Lerche et al., Depth of interaction detection for γ-ray imaging, *Nucl. Instrum. Meth.* A **600**, 624–634 (2009).

16. A. Fabbri et al., A new iterative algorithm for pixelated and continuous scintillating crystal, *Nucl. Instrum. Meth.* A **648**, S79–S684 (2011).

17. S. Majewski et al., Optimization of a mini-gamma camera based on a 2×2 array of Hamamatsu H8500 PSPMTs, *Nucl. Instrum. Meth.* A **569**, 215–219 (2006).

18. M. Seimetz et al., Correction algorithms for signal reduction in insensitive areas of a small gamma camera, *J. Instrum.* **9**, C05042 (2014).

19. N. Tamda et al., Feasibility study of a γ-ray detector based on square PSPMT array for breast cancer imaging, *Nucl. Instrum. Meth.* A **557**, 537–543 (2006).

20. A.V. Stolin et al., Characterization of imaging gamma detectors for use in small animal SPECT, *2003 IEEE Nuclear Science Symposium Conference Record (NSS 2003)*, 2085–2089 (2004).

21. G. Loudos et al., Performance evaluation of a mouse-sized camera for dynamic studies in small animals, *Nucl. Instrum. Meth.* A **571**, 48–51 (2007).

22. Y. Qi et al., Development of a simplied readout for a compact gamma camera based on 2×2 H8500 multi-anode PSPMT array, *2010 IEEE Nuclear Science Symposium Conference Record (NSS/MIC 2010)*, 2210–2212 (2010).

23. E. Lage et al., rSPECT: A compact gamma camera based SPECT system for small-animal imaging, *2009 IEEE Nuclear Science Symposium Conference Record (NSS/MIC 2009)*, 3126–3131 (2009).

24. D.W. Austin et al., Design and performance of a new SPECT detector for multimodality small animal imaging platforms, *2006 IEEE Nuclear Science Symposium Conference Record (NSS/MIC 2006)*, 3008–3011 (2006).

25. S. Vidal-Sicart et al., Added value of intraoperative real-time imaging in searches for difficult-to-locate sentinel nodes, *J. Nucl. Med.* **51**, 1219–1225 (2010).

26. S. Vidal-Sicart et al., The use of a portable gamma camera for preoperative lymphatic mapping: A comparison with a conventional gamma camera, *Eur. J. Nucl. Med. Mol. Imaging* **38**, 636–641 (2011).

27. L. Vermeeren et al., A portable γ-camera for intraoperative detection of sentinel nodes in the head and neck region, *J. Nucl. Med.* **51**, 700–703 (2010).

28. A. Soluri et al., Radioisotope guided surgery with imaging probe, a hand-held high-resolution gamma camera, *Nucl. Instrum. Meth.* A **583**, 366–371 (2007).

29. P.D. Olcott et al., Performance characterization of a miniature, high sensitivity gamma ray camera, *IEEE Trans. Nuclear Sci.* **54**, 1492–1497 (2007).

30. P.D. Olcott et al., Clinical evaluation of a novel intraoperative handheld gamma camera for sentinel lymph node biopsy, *Physica Medica* **30** (2013), 340–345.

31. S. Salvador et al., An operative gamma camera for sentinel lymph node procedure in case of breast cancer, *J. Instrum.* **2**, P07003 (2007).

32. W. Xi et al., MONICA: A compact, portable dual gamma camera system for mouse whole-body imaging, *Nucl. Med. Biol.* **37**, 245–253 (2010).

33. A.J. González, et al., Innovative PET detector concept based on SiPMs and continuous crystals, *Nucl. Instrum. Meth.* A **695**, 213–217 (2012).

34. P. Conde, et al., Results of a combined monolithic crystal and an array of ASICs controlled SiPMs, *Nucl. Instrum. Meth.* A **734**, 132–136 (2014).

15 Silicon Photomultipliers for High-Performance Scintillation Crystal Read-Out Applications

Carl Jackson, Kevin O'Neill, Liam Wall, and Brian McGarvey

CONTENTS

15.1   INTRODUCTION

Silicon photomultipliers (SiPMs) have emerged as the solid-state alternative to the vacuum tube photomultiplier tube (PMT) in a number of scintillation detection applications ranging from medical imaging to radiation detection and isotope identification. The primary technical and commercial driver for the development of the SiPM has been medical imaging, specifically positron emission tomography (PET). There is a simultaneous requirement in PET for higher-quality images and a lower raw component sensor cost compared with the PMT. The radiation detection market requires higher-performance sensors for better isotope identification, increased physical robustness for handheld and field operations, low operating voltage compared with the high voltage of the PMT, and reduced sensor cost for high-volume homeland security applications. A typical generic application example of an SiPM detecting the light emission from a scintillating crystal is shown in Figure 15.1. This figure shows the scintillation crystal coupled to an SiPM sensor that is directly connected to a printed circuit board (PCB). The signal from the SiPM is typically processed via an amplifier or shaper or both, followed by data acquisition. The scintillation crystal is capable of converting the incident gamma rays to visible photons and the SiPM sensor has the sensitivity to convert

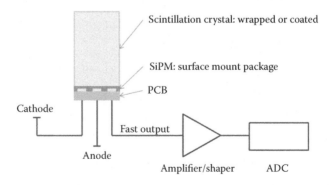

FIGURE 15.1   Simplified diagram of a detector and read-out block diagram for scintillation crystal readout.

the visible photons to electrical charge, which can be amplified, shaped, and read out by the electronics. The SiPM is typically mounted on a PCB and due to its compact nature it is possible to place the read-out electronics on the same PCB.

The SiPM can be used to replace the PIN photodiode, the avalanche photodiode (APD), or the PMT currently used in these systems. For all applications the main benefits of SiPM technology include: high optical sensitivity, low operating voltage, unsurpassed uniformity across the sensor, uniformity from sensor to sensor, small form factor, and low cost. Specifically, the sensors discussed in this chapter will be complementary metal–oxide–semiconductor (CMOS)-manufactured SiPM sensors, which have been developed to meet the performance and cost requirements of medical imaging PET and radiation detection markets.

15.2   A SHORT HISTORY OF THE SiPM

The SiPM sensor, now in mass-production, has been developed through knowledge of the Geiger-mode operation of photodiodes, which can be traced back to the original work of Haitz [1–4] and Oldham et al. [5]. The Geiger-mode photodiode was used as a tool to investigate diode breakdown properties and microplasmas in diodes and it helped to develop the diode impact ionization and avalanche breakdown models. The work of McIntyre [6,7] developed equations allowing for a full understanding of the device physics in a practical and workable model of breakdown and above breakdown Geiger-mode operation. Further published work on Geiger-mode technology during the 1980s and 1990s centered on using the Geiger-mode photodiode for single-photon counting applications [8]. This work was largely focused on individual photodiode sensors with external quenching circuitry [9,10]. In the late 1990s and early 2000s, the SiPM concept emerged from Russia through work published by Saveliev [11], Bisello [12], and Golovin [13]. However, the development of the SiPM concept can be traced back to papers published in the 1970s by Krachenko et al. [14,15] in which arrays of Geiger-mode avalanche sensors, passively quenched, were used for low-light sensing. At the time of publication in 1978, the authors noted a limitation of PN photodiodes as an avalanche-based sensor in the presence of microplasmas in the photodiode depletion region, which limits the useful size of the photodiode's active area. This limitation was removed in the late 1990s and 2000s, as Geiger-mode photodiodes were developed utilizing higher-quality start material and processing conditions and architectures were developed that reduced the presence of defects in the depletion region of the photodiode. This advance allowed large Geiger-mode photodiode arrays to be fabricated. These developments allowed the SiPM to be manufactured in CMOS foundries, creating a cost-competitive alternative to the PMT.

15.3   SILICON PHOTOMULTIPLIERS

In this section, the SiPM will be described in more detail, from the basic theory of operation to the major parameters that govern sensor performance. The basic theory of the SiPM will be developed from the theory of Geiger-mode operation of single microcells, to arrays of microcells that form the model for the SiPM.

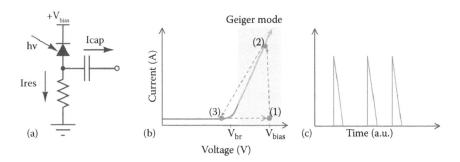

FIGURE 15.2 (a–c) Basic schematic and operation of a Geiger-mode avalanche photodiode.

15.3.1 GEIGER-MODE AVALANCHE PHOTODIODE

The basic building block of an SiPM is a photodiode that is sensitive to single photons, possesses high internal gain, and self regulates its output current. A circuit schematic of the basic structure that accomplishes this is shown in Figure 15.2a. The photodiode in Figure 15.2a is designed to operate in Geiger mode so that it can detect single photons. The resistor in series with the photodiode provides the means of limiting the current flow and resetting the photodiode after each photon is detected. Geiger-mode operation can be explained by reviewing the current versus voltage characteristic shown in Figure 15.2b. This figure shows the ideal reverse bias characteristics for a photodiode and overlays the mode of operation, known as Geiger mode, which allows for single-photon detection. A photodiode has a breakdown voltage (V_{br}) at which point the current flow in the diode significantly increases. For operation in Geiger mode, a voltage in excess of V_{br} is applied (V_{bias}), as shown in Figure 15.2b. The difference in voltage between V_{bias} and V_{br} is termed *the overvoltage* and typical values are between 1 and 5 V. Initially, on application of V_{bias}, no current flows in the photosensitive region of the photodiode. There is always a small parasitic current, shown in Figure 15.2b, flowing in the photodiode, which to the first order can be neglected. On arrival of a photon and conversion of the photon into an electron–hole pair, the charge carriers undergo impact ionization, leading to avalanche multiplication. During the avalanche process, the photodiode current changes from a no current state, labeled as (1) in Figure 15.2b, to a high current state, labeled as (2). The current generated in the photodiode follows two paths in Figure 15.2a, shown as Ires and Icap. Once Ires begins to flow through the photodiode, the voltage is dropped across the series resistor. The voltage drop causes the photodiode voltage and the electric field to decrease and the operating point shown in Figure 15.2b moves from (2) to (3). As a current is no longer flowing across the resistor, the voltage bias across the photodiode increases to the original V_{bias} and the photodiode returns to a state that can detect the next photon. The output from this type of sensor is shown in Figure 15.2c and can be used to explain both Ires and Icap.* The output current is characterized by a sharp and fast rise time at

---

* Icap is available in products from SensL to allow for a low capacitance and fast output. Other SiPMs will not contain Icap.

the onset of the incident photon, relating to the transition from (1) to (2) and a slower recovery time as the photodiode transitions from (3) back to (1). As will be shown in subsequent sections, the addition of a capacitor to the photodiode and resistor provides a simple means of accessing the fast switching behavior inside the SiPM.

The output signal from the Geiger-mode APD consists of nearly identical current pulses in time. Each current pulse relates directly to the detection of a photon or thermal noise event in the sensor. The thermal noise, known as the dark count rate, has the same properties as a single photon and will be shown to be a key parameter for the understanding of SiPM operation.

15.3.2 SILICON PHOTOMULTIPLIER

The Geiger-mode APD provides single-photon sensitivity and it outputs a current pulse that signals the arrival time of the incident photon. However, the Geiger-mode APD is a binary sensor and does not have the ability to sense multiple-photon events since the output can be considered identical regardless of the photon number. Additionally, the size of the photodiode is limited by two effects: firstly, the characteristic dark count noise increases with the active area; and secondly, the recovery time (during which the sensor is insensitive to further incident photons) increases with the increasing capacitance of the diode, which in turn increases with the active area. Therefore, only relatively small Geiger-mode APDs are used, typically tens to hundreds of micrometers in diameter. To allow for both a large-area sensor and to detect multiple photons, the SiPM was created. The structure of the SiPM is shown in Figure 15.3a. A key feature of the SiPM architecture is the large array of parallel-connected Geiger-mode APDs with quenching elements and output capacitors* connected together. Each photodiode, resistor, and capacitor element is collectively called the microcell. The schematic in Figure 15.3a shows a simplified SiPM demonstrated as a 4×3 array of microcells. A commercial SiPM would have from hundreds to thousands of

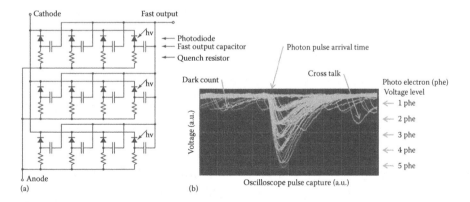

FIGURE 15.3   (a) Silicon photomultiplier schematic. (b) Characteristic output response of a SiPM as viewed on an oscilloscope with the persistence setting engaged.

---

* Fast output capacitors are only included in SiPMs provided by SensL.

microcells per millimeter squared. The SiPM is biased with a positive voltage on the cathode with respect to the anode. The SiPM output can be measured as the current flow from the cathode to the anode, or from the fast output. A representation of the output of an SiPM in response to a low-level light pulse is shown in Figure 15.3b. This image was captured on an oscilloscope with the persistence setting set to allow for displaying multiple detected events. To create the voltage signal, the output of the SiPM was connected to an external resistive load.

Figure 15.3b shows several key features of the SiPM output, such as (1) the ability to detect multiple-photon events; (2) the dark count signal, which is at the same level as the single photoelectron signal; and (3) cross talk, which is an unwanted feature caused by cotriggering inside the SiPM. The detection of multiple photons is shown as the clearly discernible voltage levels labeled 1–5 phe. These represent the detection of between 1 and 5 single photons. The photon number detection is a desired characteristic while the dark count and cross talk are negative features of the SiPM operation. Reviewing the architecture in Figure 15.3a, it is clear that in an SiPM there is an optically active photodiode as well as optically inactive structures such as the resistor and capacitor. The necessity to provide a physical separation for the individual microcell photodiode elements required for independent Geiger-mode operation means that there is a dead space between photodiodes, which limits the total active area. The percentage of active area in an SiPM is termed *the fill factor* and must be considered with the number of microcells that are in each SiPM. A higher fill factor increases photon detection efficiency (PDE) at the expense of the dynamic range. The different aspects of this issue will be covered in more detail in subsequent sections.

15.3.3 SiPM Output

The SiPM has three outputs, as shown in Figure 15.3a. The connection for biasing the photodiode is made through the cathode and anode terminals. In this example, the anode is directly connected to the quenching resistor. However, this is still termed *the anode* since it assists the user in the application of the correct bias voltage for reverse bias operation. The additional fast output is shown as a capacitively coupled output, which is directly connected to the microcell at the intersection between the anode and the quench resistor. The output signal is shown in Figure 15.4 for a fast-pulsed laser with a pulse width of less than 100 ps. The anode–cathode output is characterized by a fast onset time followed by the recovery of the output signal, which is governed by the internal impedance of the SiPM and the external circuitry directly connected to the sensor. The recovery time increases with the photodiode area and the quench resistor and parasitic impedance. The fastest recovery times are achieved with the smallest microcell size, but at the expense of the fill factor. The fast output signal is characterized as a short fast pulse, which is the derivative of the switching behavior of the photodiode during Geiger-mode operation detailed in Figure 15.2b (1 and 2). The decay time, which is a feature of the anode–cathode signal, is not present; instead, this is replaced by a small undershoot to the baseline voltage as the microcell recovers. The fast output accurately represents the signal and since the recovery time is removed from the output signal it is easier to identify additional incident photons within the recovery period. It will be shown that

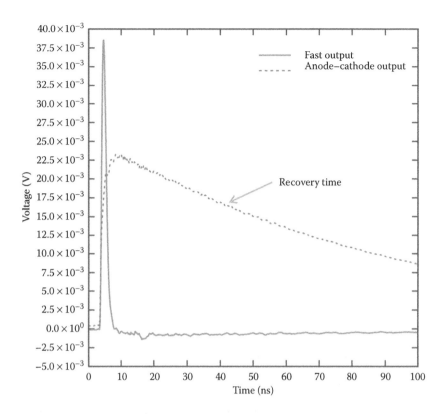

FIGURE 15.4 Pulse output a from 3×3 mm² SiPM with 35 µm microcells. Fast output and anode–cathode output are shown superimposed on the same graph.

scintillation crystal readout can be achieved with excellent energy and coincidence timing resolution using only the signal from the fast output.

15.3.4 SiPM Operation in Detail

Sections 3.1 through 3.3 described the basic operation of the SiPM. In this section, the theory of operation will be reviewed in more detail. As has been shown, the SiPM provides the ability to detect individual photons and provides an output current that is proportional to the number of incident photons. This is a key factor for the read out of the light emission from a scintillation crystal, as discussed in Section 15.8.1. For the SiPM to be able to detect the photons emitted from scintillation crystals and output current pulses as shown in Figure 15.3b and Figure 15.4, the following processes must occur:

1. Photon capture: The photon must be absorbed and converted into an electron–hole pair.
2. Amplification: The electron–hole pair must be converted to a measurable current through the internal gain mechanism of the SiPM.

15.3.4.1 Photon Capture

Photons entering the SiPM are absorbed according to the Beer–Lambert equation for photon absorption given as

$$I(z, \lambda) = I_0 e^{-\alpha(\lambda)z}, \tag{15.1}$$

where:

I_0 = incident photon flux (typically photons per second)
$\alpha(\lambda)$ = attenuation coefficient of silicon at the wavelength of interest
z = distance into the SiPM

This equation determines the ability of an SiPM to detect photons of a given wavelength. Using the scintillator peak emission values given in Figure 15.5 and combining these with the attenuation coefficients for crystalline silicon [16], it is possible to calculate the percentage of photons absorbed as a function of the distance into the SiPM. This calculation is shown in Figure 15.6 for various peak intensities of scintillators relevant to this work. LaBr$_3$ [17] emits photons with a peak wavelength at 380 nm and Figure 15.6 shows that almost 100% of these photons are absorbed at a distance of 0.2 μm into the SiPM. Conversely, at 0.2 μm

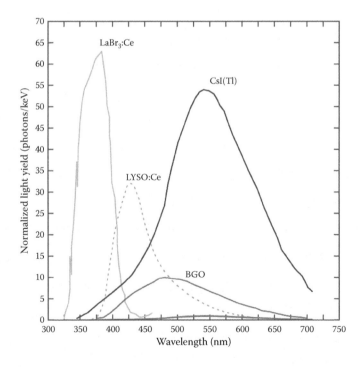

FIGURE 15.5 Crystal emission light yield from common scintillation crystals used in radiation detection.

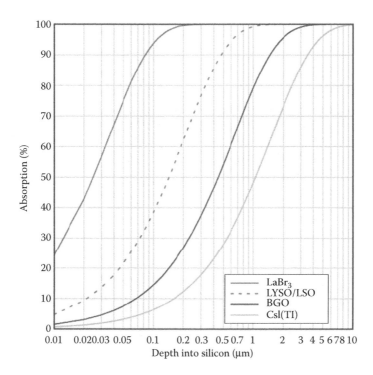

FIGURE 15.6 Photon absorption in SiPM sensors for various crystal peak emission wavelengths.

only 10% of the photons from CsI(Tl), which has an emission peak at 550 nm, are absorbed.

A longer emission wavelength requires an increased depletion region width to allow the efficient capture of the incident photons. However, all the scintillators described here emit photons in a range that can be sensed by SiPM devices, which possess shallow junctions on the order of 0.2–1 μm, and are the subject of this work. These SiPM rely on the planar, shallow structures that CMOS processing is ideally suited to fabricate.

15.3.4.2 Amplification

Once a photon has been absorbed by the SiPM, the next step is the amplification of the generated electrons and holes. This amplification is governed by the impact ionization of electrons and holes in the depletion region of the photodiode. The depletion region of a P-on-N photodiode, shown in Figure 15.7a, is the region of high electric field established where the N-type and P-type doped silicon layers come together. This dopant structure is fully compliant with standard CMOS processing, which is used for the production of high-quality SiPMs in volume. The anode of the photodiode for a P-on-N structure is located close to the surface beneath an antireflection coating (ARC), which has been designed to maximize photon transmission into the silicon. The cathode of the photodiode is below the

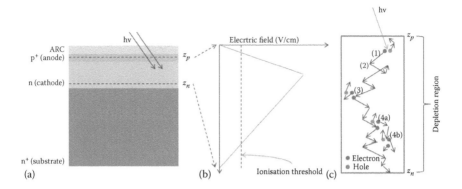

FIGURE 15.7 **(See color insert)** (a) Simplified cross section of a P-on-N Geiger-mode APD structure. (b) Electric field profile across the photodiode depletion region. (c) A pictorial representation of an avalanche breakdown in a photodiode across the depletion region.

anode and is electrically connected to a low-resistance substrate of the same dopant type. To put the physical sizes in perspective, a typical silicon wafer that is used to manufacture an SiPM is on the order of 750 μm thick while the depletion region with a breakdown voltage of 24.5 V is on the order of 1 μm thick. Most of the active volume of the SiPM that is designed for radiation detection is located close to the surface and the substrate typically provides a low ohmic contact to the cathode and necessary structural support.

The SiPM avalanche process is governed by the electric field across the depletion region of the photodiode as shown in Figure 15.7b. With knowledge of the cathode and anode dopant levels, Poisson's equation can be used for the calculation of the electric field across the junction. For the discussion here, it is sufficient to know that the depletion region electric field is increased through the application of a reverse bias voltage. When a reverse bias is applied to the SiPM in excess of the breakdown voltage, the electric field increases beyond the ionization threshold. At this point, charge carriers gain sufficient energy to create further electron–hole pairs through collisions, a process called impact ionization. If the electric field is sufficiently high then avalanche breakdown will occur. Avalanche breakdown is pictorially represented in Figure 15.7c as a succession of charge-carrier movements labeled (1) through (4). As the charge carrier passes through the depletion region, it is accelerated by the electric field. The electric field accelerates electrons toward the cathode and holes toward the anode. The acceleration increases the carrier kinetic energy but carriers cannot pass through the depletion region without interacting with the underlying silicon structure. Two main types of interactions can occur: phonon generation or impact ionization. Phonon generation is shown as (1) and (2) in Figure 15.7c and results in a scattering event that reduces the kinetic energy of the charge carrier. Ionization events are shown as (3), (4a), and (4b) in Figure 15.7c. The ionization causes the generation of additional electron–hole pairs, which can themselves generate ionization events, as shown in (4b). With a sufficiently high electric field, the impact ionization process can, in theory, continue indefinitely. However, with the inclusion of the quench resistor in the microcell, as shown in Figure 15.2a, the electric charge at each breakdown event, and therefore the current, is limited to only the capacitance of the microcell's photodiode. The avalanche process provides the output current and the characteristic breakdown voltage of the SiPM.

15.4   SiPM SENSOR PARAMETERS

It is important to discuss the major sensor parameters that impact the performance characteristics of the SiPM. The following are considered the basic SiPM sensor parameters:

- Breakdown voltage: the voltage at which the SiPM current increases rapidly with increasing reverse bias
- Photon detection efficiency (PDE): the absolute PDE of the SiPM including all geometrical effects
- Dark count: the number of false single pulse height signals measured per second
- Gain: the internal gain of the SiPM in response to a single photon
- Cross talk: the output signal caused by one microcell randomly firing one or more additional microcells in the array
- Afterpulsing: the secondary avalanche events that are triggered by stored charge during a previous event
- Dynamic range: the ability of an SiPM to detect incoming photons without saturation

To aid in the understanding of the sensor parameters discussed, Table 15.1 is used. This details the various SiPM types available, including the active area, the microcell size, the number of microcells, and the fill factor. It is important to note that the microcell size used in this table is the size of the active area of the microcell and not the pitch of the microcells in the SiPM. These parameters are from the current commercial SiPM produced by SensL [18].

Each of these parameters will impact the radiation detection performance when the SiPM is coupled to a scintillating crystal. The individual parameters are discussed next, including information on their accurate measurement.

TABLE 15.1
Physical Parameters for the SiPM Discussed in This Chapter

Active Area Dimensions (mm²)	SiPM Type	Microcell Size (µm)	No. of Microcells	Fill Factor (%)	Recovery Time (1/e) (ns)
1×1	10035	35	576	64	82
3×3	30020	20	10,998	48	41
	30035	35	4,774	64	82
	30050	50	2,668	72	159
6×6	60035	35	18,980	64	96

15.4.1 Photon Detection Efficiency

The PDE is a key parameter for SiPM operation. This parameter determines the percentage of photons that will be detected by an SiPM. The PDE of an SiPM can be defined by the following equation:

$$\text{PDE} = QE(\lambda) * AIP(V) * FF, \tag{15.2}$$

where:

$QE(\lambda)$ = internal quantum efficiency of the SiPM
AIP = avalanche initiation probability
FF = fill factor of the SiPM

The QE is wavelength dependent and includes any losses at the optical interface of the sensor. The AIP is set by the doping levels and the physics of the depletion region of the photodiode and increases with voltage. The FF is a geometrical parameter that is governed by the ratio of the active area to the nonactive area in the SiPM microcell.

To measure the PDE, an accurate measurement must be made that utilizes photon statistics. In photon statistics that are inherent in SiPM sensors there is a specific relationship between the ratio of events that trigger zero photons, N_0, and the total number of events that include any total number of photons, N_{tot}. This relationship is determined through the chance of observing a zero photon event in a Poisson probability distribution with a mean number of photons, n_{ph}, and can be written as $N_0 / N_{tot} = e^{-n_{ph}}$. To correctly take account of background pulses, a measurement in the dark must be obtained. The parameters N_0^{dark} and N_{tot}^{dark} must be measured, and then the mean number of photons, n_{ph}, arriving at the SiPM can be calculated using

$$n_{ph} = -In\left(\frac{N_0}{N_{tot}}\right) + In\left(\frac{N_0^{\text{dark}}}{N_{tot}^{\text{dark}}}\right). \tag{15.3}$$

The PDE can then be calculated by dividing the mean rate of detected photons by the recorded rate of arriving photons, given by the optical power recorded by a calibrated photodiode P and the period T of the light pulses:

$$\text{PDE} = \frac{(n_{ph}R)/T}{P/(\lambda/hc)}. \tag{15.4}$$

Here, R is the geometrical optical power ratio between the light falling on the SPM and the calibrated photodiode. An example of this measurement technique that is extensively referenced is the work of Eckert [19]. This technique produces the true SiPM PDE that does not contain the effects of cross talk or afterpulsing.

Figure 15.8 shows the PDE for a 30035 type SiPM across the wavelength range of 250–1100 nm. The peak emission is shown at 420 nm with a maximum peak detection efficiency of 41% achieved at 5.0 V overvoltage.

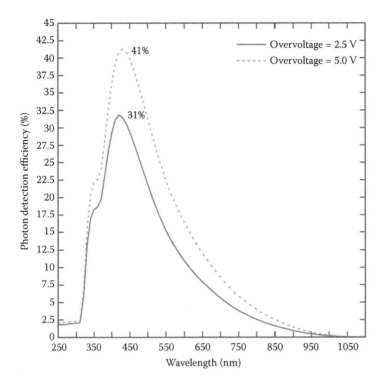

FIGURE 15.8 Photon detection efficiency versus wavelength for a 30035 SiPM at overvoltages of 2.5 and 5.0 V.

The PDE varies with the microcell size according to the fill factor for the different types of SiPM. The PDE does not vary with the active area dimension. Therefore, the fill factor values listed in Table 15.1 can be used to scale the PDE shown in Figure 15.8 to different sensors. To scale the PDE to different sensor types, first divide the PDE of the known SiPM by the fill factor of the known SiPM and then multiply by the fill factor of the unknown SiPM. As an example, to scale the 5.0 overvoltage PDE at 420 nm from the 30035 to the 30050 SiPM the following equation would be used: 41%/0.64 * 0.72 = 46%.

15.4.2 DARK COUNT

The dark count rate is defined as the pulse rate measured in the dark, with a leading-edge trigger at 0.5 times the single photoelectron amplitude. The single photoelectron amplitude is determined by observing when the count rate first decreases as the leading-edge threshold is increased. By increasing the threshold until the dark rate changes, it is possible to determine the maximum pulse height of the single photoelectron counts. A measurement of the dark count rate is not impacted by cross talk but can be influenced by afterpulsing. In Section 15.4.5, afterpulsing will be shown to be very low for the SiPM sensors discussed here and not a significant factor in the measured dark count rate. The dark count rate for 10035, 30035, and

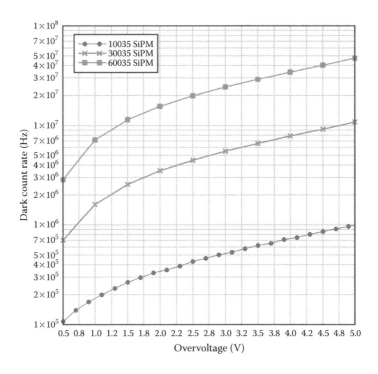

FIGURE 15.9 Measurement of dark count rates for 10035, 30035, and 60035 SiPMs.

60035 SiPM is shown versus overvoltage in Figure 15.9. This demonstrates a very well-controlled dark count rate versus an overvoltage that is suitable for mass-production applications that require SiPM sensor output stability over a wide range of overvoltage conditions and across a wide range of temperatures.

For scintillators with a short decay time, the dark count rate can be largely ignored. For example, Table 15.4 shows that $LaBr_3$ has a decay time constant of 16 ns and typically pulses are integrated on the order of 100 ns to achieve good energy resolution and full signal collection. For a 100 ns signal integration time using a 60035 sensor, and using the dark count values as shown in Figure 15.9, the average number of dark counts integrated can be calculated as $5 \times 10^7 * 100 \times 10^{-9} = 5$. The dark count rate is clearly not a dominant factor for an $LaBr_3$ which emits ~63 photons per kiloelectron volt and therefore would emit more than 40,000 visible photons in response to a single 662 keV gamma ray. For scintillators with longer characteristic decay times such as CsI(Tl), combined with the desire to detect low-energy gamma rays in the tens of kiloelectron volts, the dark count rate becomes a more important factor. The dark count rate can degrade the signal-to-noise ratio. Advances in SiPM production are leading to improved dark count rates. An example of next-generation dark count rates for SiPMs is shown in Figure 15.10 compared with the existing family of SiPMs. This shows the achievement of a dark count rate of less than 100 kHz/mm^2.

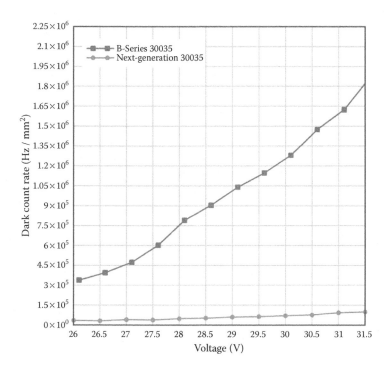

FIGURE 15.10 Dark count rate of next-generation SiPMs. Dark count rates are based on 35 μm-based microcell sensors representing dark count rates per millimeter square for 10035, 30035, and 60035 sensors.

15.4.3 GAIN

The gain of an SiPM is defined as the mean number of charge carriers generated when a single charge initiates an avalanche process in the depletion region of the sensor. This is a unitless parameter and determines the amount of charge flowing through the sensor during detection. The gain can be derived theoretically using the following equation:

$$G = C_{\text{microcell}} * \frac{V_{\text{bias}} - V_{br}}{q}, \tag{15.5}$$

where:

$C_{\text{microcell}}$ = capacitance of a single microcell
V_{bias} = operating voltage
V_{br} = breakdown voltage
q = electronic charge

Knowledge of the gain of the SiPM aids in the conversion between the dark count rate and the measured dark current in the SiPM by the following equation:

$$I_{\text{dark}} = DCR \times G \times q \times (1 + CT) \times (1 + AP), \tag{15.6}$$

where:

   DCR = dark count rate in hertz
   G = gain
   CT = cross-talk probability
   AP = afterpulse probability

It is possible to approximate the SiPM by ignoring the CT and AP terms if they are deemed sufficiently small. This is appropriate for first-order calculations but cross talk and afterpulsing should always be considered for the most accurate estimations.

To measure the gain of an SiPM requires knowledge of the current flowing through the sensor measured while in saturation using a pulsed light source. A standard pulsed light-emitting diode (LED) of 470 nm can be used for the measurement. The basic concept of the measurement is as follows: if all microcells of an SIPM are illuminated simultaneously and with a known repetition rate, it is possible to infer the gain of the sensor by measuring the DC current flowing through the standard anode–cathode. Once this is obtained, it is necessary to subtract the dark current, I_{dark}, from the photocurrent, I_{photo}. The gain is then

$$G = \frac{I_{photo} - I_{dark}}{q * f_{laser} * N},\quad\quad\quad (15.7)$$

where:

   q = electron charge
   f_{laser} = laser pulse frequency
   N = number of microcells

The gain measurement completed in this way does not contain afterpulsing or cross talk and is a true representation of the gain of an SiPM. A plot of the gain for the different SiPM microcell types is shown in Figure 15.11. No difference was noted between samples of different active area type that utilize the same microcell structure.

The gain values for the fast output terminals were designed to be between 2% and 4% of the standard anode–cathode gain and were set so that the gain was not dependent on the microcell size. A measured gain for the fast output was therefore found to be 4.3×10^4 regardless of microcell type at 2.5 V overvoltage. The gain measurement technique was also used to accurately determine the temperature coefficient of the breakdown voltage. The point of the zero crossing gain versus the voltage curve was measured at a range of temperatures. The breakdown voltage was found to vary by 21.5 mV/°C over standard operating temperature ranges.

15.4.4 CROSS TALK

A cross-talk event can be seen in the output signal of an SiPM, as shown in Figure 15.3b. Cross talk is defined as the ratio of the pulse rate measured at 1.5 times the single photoelectron amplitude to 0.5 times the single photoelectron

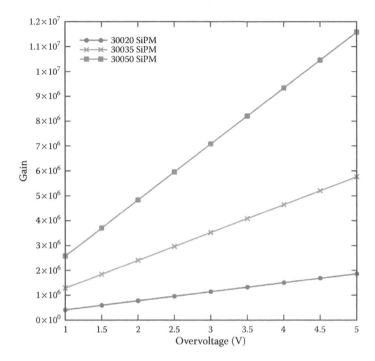

FIGURE 15.11 Gain measurement for the standard anode–cathode output for B-series 30020, 30035, and 30050 SiPMs.

amplitude. Cross talk is an undesirable phenomenon, in that a single-firing microcell fires one or more neighboring microcells almost instantaneously. Cross-talk pulses can therefore appear as multiple output signals in response to a single photon, and are therefore a form of multiplication noise impacting the excess noise factor. Increasing the bias voltage increases cross talk. Cross talk can be caused by electrical charge leakage or it can be optically generated due to the emission of light in SiPM sensors.

To measure the cross talk in SiPM sensors, the rate of dark counts as a function of an increasing leading-edge trigger threshold is measured, and is repeated at various bias voltages. An example of the output from this type of measurement is shown in Figure 15.12. The y-axis indicates the frequency of the dark count measured for the input trigger threshold on the x-axis.

As shown in Figure 15.12, this results in a staircase dependence of rate versus trigger threshold, and progresses to lower rates as the trigger threshold increases. The 1.5 (height) and 0.5 (height) photoelectron rates can then be measured and the ratio computed for each overvoltage. Using this procedure, the cross talk measured for 30020, 30035, and 30050 SiPMs is shown in Figure 15.13.

Cross talk was not seen to depend on the size of the individual die and has been found to only depend on the size of the microcells. To investigate this phenomenon, the cross talk on 10035 and 60035 SiPMs was measured. The cross talk on these SiPMs was within measurement accuracy for the 30035 SiPM and can be considered

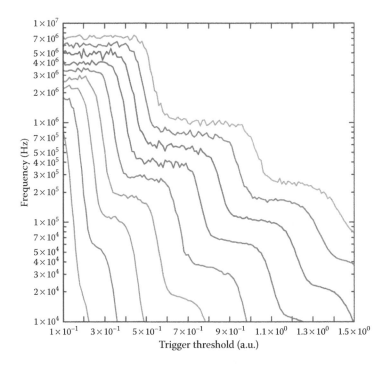

FIGURE 15.12 Cross-talk threshold plot for a 30035 SiPM. Data are taken from 1 to 5 V overvoltage.

identical regardless of the die size. For the 30020 sensors, the cross-talk rates were found to be extremely low and difficult to measure without long measurement cycle times. The measurements below 2 V overvoltage were in fact terminated due to the length of time that it was taking for the measurement. These 10020 SiPM exhibit cross talk of sub 6% at the 5.0 V level. In all cases, the cross talk at typical operating overvoltages for the SiPM reported in this work were found to be extremely low. It is expected that for most applications, the 30035 will provide the best combination of low cross talk and high PDE.

15.4.5 Afterpulsing

Afterpulsing is the phenomenon of an SiPM microcell randomly discharging with higher probability shortly after a previous discharge than at the expected thermal generation rate. Afterpulse events that occur after the recovery time cannot be distinguished from genuine, photon-induced events and therefore deteriorate the photon-counting resolution of an SiPM. Typical timescales for this phenomenon are tens of nanoseconds, comparable with the microcell recovery time. As a result, many afterpulse events are of partial discharge compared with the single photoelectron discharge of a microcell, due to partial recharge of the microcell.

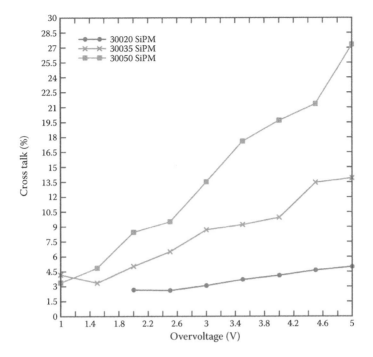

FIGURE 15.13   Measurement of cross talk for 30020, 30035, and 30050 SiPMs.

Afterpulsing is an undesirable effect because it increases the variance of the single cell charge, reduces the dynamic range, and increases cross talk in the SiPM. The degree of afterpulsing depends on the bias voltage, the temperature, the design of the deep doping implant, and the recovery time.

Afterpulse probability measurements are made by binning the time between consecutive pulses and extracting via the formula:

$$P_{ap} = \frac{\int_0^\infty \xi * n_{ap} d\Delta t}{\int_0^\infty \xi * \left(n_{ap} + n_{tp}\right) d\Delta t}, \tag{15.8}$$

where:

$$n_{tp}(\Delta t) = \frac{n_{tp}}{\tau_{tp}} * e^{-(\Delta t / \tau_{tp})}, \tag{15.9}$$

$$n_{ap}(\Delta t) = \frac{N_{apf}}{\tau_{apf}} * e^{-(\Delta t / \tau_{apf})} + \frac{N_{aps}}{\tau_{aps}} * e^{-(\Delta t / \tau_{aps})}, \tag{15.10}$$

where:

n_{tp} = thermal rate probability density

n_{ap} = afterpulse probability rate, consisting of fast and (optionally) slow
        components

ξ = charge fraction of a second pulse time Δt after a given pulse

Afterpulsing was not found to be dependent on the SiPM die size but only on the microcell geometry. The afterpulse rates measured for 30020, 30035, and 30050 SiPMs are shown in Figure 15.14.

The results obtained for 10035 and 60035 SiPMs were within measurement error to the 30035 SiPM. This demonstrates the low level of afterpulsing present in the SiPM, and for the 30020 and 30035, afterpulse rates are less than or equal to 5% up to 5.0 V overvoltage. The afterpulsing measured on the 30050 was found to be similar to the 30035 until 4.5 V. After 4.5 V, the afterpulsing of the 30050 increased more rapidly than the 30035. It was believed that the high gain of the 30050, as shown in Figure 15.11, was the cause for the increase in afterpulsing at overvoltages in excess of 4.5 V. The low level of afterpulsing in the SiPM presented demonstrates the high quality of the start material and processing steps used to fabricate these SiPM sensors.

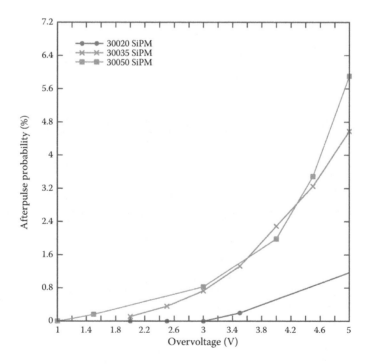

FIGURE 15.14   Afterpulse measurements on 30020, 30035, and 30050 SiPMs.

15.4.6 DYNAMIC RANGE

The dynamic range of the SiPM is determined by the number of microcells over the active area that are available to detect incoming photons. For fast scintillation decay time crystals, such as $LaBr_3$ and LYSO (Table 15.4), the decay time is shorter than the decay time of a typical SiPM (Table 15.1) and to the first order the crystal emission can be considered as a pulse source. For longer decay crystals such as CsI(Tl) this becomes more complicated as microcells can recover and respond multiple times during a scintillation emission. For simplicity, only the pulse dynamic range will be considered here. To the first order, the dynamic range of the SiPM can be considered simply by the following equation:

$$N_{\text{fired}}(M, \lambda) = M\left(1 - \exp^{-(PDE(\lambda)N_{ph}/M)}\right), \tag{15.11}$$

where:

M = number of microcells per die
PDE = photon detection efficiency at a specified lambda
N_{ph} = number of incident photons

A characteristic curve can be plotted for this equation using basic information derived from the datasheet of SiPM products. For Figure 15.15, the dynamic range for a pulsed light is shown from the single photon level up to 10^7 photons. For each die type, the parameters of M and PDE have been obtained from the datasheet for the B-Series SiPM [18].

Taking the photon yield of a scintillator such as LYSO from Figure 15.5 as 32 photons per kiloelectron volt, and assuming an incident gamma-ray photon of 511 keV, approximately 16,352 visible photons should be emitted. Assuming a 50% loss from the light collection and transfer to the SiPM and a PDE of 41% at the peak wavelength of 420 nm from Figure 15.8, there should theoretically be ~3352 photons incident on the sensor. From Figure 15.15 it can be seen that with this number of photons incident on the sensor it would generate a signal equivalent to ~700 microcells firing. Additionally, this signal level is at the top end of the linear range of the sensor. For this reason, the 30035 sensor type with 35 μm microcells works well with LYSO crystals though careful analysis of the linearity should be performed to produce the most accurate energy resolution measurements.

15.5 SiPM UNIFORMITY

The uniformity of the sensor operating parameters is important for using SiPM sensors in high-volume applications. The sensor uniformity minimizes the variation between sensors and reduces or eliminates voltage adjustment between sensors that are used in a system. To demonstrate uniformity, this section will review end-of-line test data obtained during the manufacturing of SiPMs. The sensors described in this section are from SensL's B-Series sensors. These sensors are manufactured on 200 mm wafers in a CMOS foundry, typically in lots of 25 wafers. On

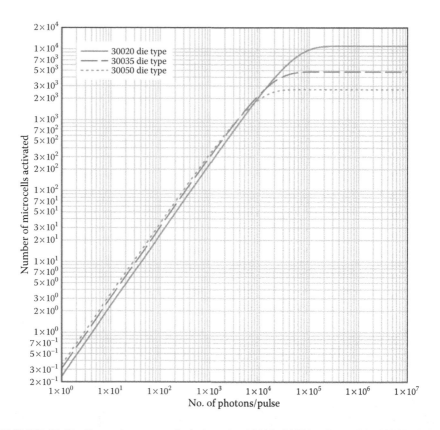

FIGURE 15.15 Dynamic range calculations for 30020 (10998 microcells), 30035 (4774 microcells), and 30050 (2668 microcells) die types.

completion of the fabrication process, each wafer is comprehensively tested, first in a wafer acceptance test (WAT) using a process control monitor (PCM), and then at component probe (CP) using a product test. The WAT verifies that the wafer was processed correctly and that key technological parameters such as the breakdown voltage and diffused and deposited film resistances are on target and within specified test limits. These test limits were established based on technology target values, material and process characteristics such as film thickness and resistivities, and from data collected during the technology development phase from skew or split lots. The measurement data from WAT are monitored wafer to wafer and lot to lot and trends are analyzed to verify process capability and stability. This process is a standard part of SensL's B-series manufacture and is repeated for all commercial SiPM sensors produced.

Following the PCM test, all SiPMs are tested at CP using tester hardware that has been developed for SiPM sensor testing. For B-Series SiPM sensors, the breakdown voltage, dark current, and optical response to short wavelength light are critical parameters that are measured on the SiPM product die. SensL has established a wafer probe flow that measures every die on the wafer under dark and illuminated conditions with appropriate optical filters, so that the response of the SiPM sensor under broadband and short wavelength conditions is measured. Die with failing characteristics are identified on electronic wafer maps, and only those die that pass all of the quality and performance screens are subsequently assembled as packaged sensors. Further details of the key parameters that are measured at CP are provided in the following sections.

15.5.1 BREAKDOWN VOLTAGE

The breakdown voltage is defined as the value of the voltage intercept of a parabolic line fit to the current versus the voltage curve, and in the CP wafer test flow this parameter is calculated from measurements of the dark current at several bias points. Figure 15.16 shows the distribution of the breakdown voltage values as measured from several production lots obtained during the qualification of a single die-size product. For the breakdown voltage variation plot shown here, a total sample size of ~204,000 SiPM sensors of the 30035 type was included. A mean breakdown voltage of 24.69 V was measured. The measured values are seen to be tightly distributed with a standard deviation of 73 mV, and all die have breakdown voltage values within ±0.25 V of the mean.

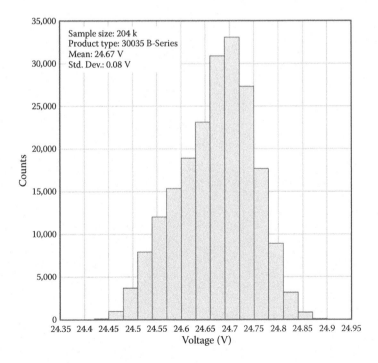

FIGURE 15.16 Histogram of the breakdown voltage distribution for a 30035 SiPM. Data are from ~204,000 commercial sensors.

The uniformity of the breakdown voltage is believed to be due to the tight process control and design of the SensL SiPM. Similar data were obtained for the 60035 SiPM at the wafer level.

15.5.2 Dark Current

Dark current is a key parameter for the SiPM, and is measured in the CP flow at several bias points below and above the breakdown voltage. Die with dark currents that are outside specification are screened out and inked electronically so that they are not assembled into a package. Figure 15.17 shows the tight distribution of dark current values for the 30035 SiPM. For Figure 15.17 all SiPMs were measured at an overvoltage equal to 2.5 V.

Figure 15.17 shows the tight distribution of the SiPM dark current and is within product specification. Additional work is currently under way to reduce the dark count for future SiPM sensors through process improvements.

15.5.3 Optical Current

To ensure the quality of the SiPM, the output current from the sensor die is measured in the CP flow under tightly controlled illumination conditions at a number of overvoltages. For the measurements shown here, all SiPMs were measured at a

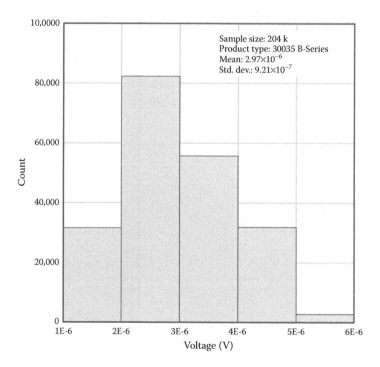

FIGURE 15.17 Histogram of dark current data from a 30035 SiPM. The sample size is ~204,000 and all SiPMs were measured at an overvoltage equal to 2.5 V.

fixed constant voltage of 29.5 V with a uniform and calibrated blue light source. The output current measurement allows an integrated assessment of the sensor gain and PDE. Any variation in gain or PDE has the potential to impact the optical current. By measuring all die on the wafer, it is possible to develop a full understanding of the optical uniformity of the SiPM. Figure 15.18 shows the distribution of output current values as measured from several production lots, with a total sample size of approximately 142,000 for the SiPM sensor of the 30035 type. This measurement procedure was only established during the latter portion of this work and therefore the size of the data set is smaller than the data set that represented the breakdown voltage and dark current measurements.

In a high-volume application it is desirable to be able to operate all of the SiPMs in a system at the same voltage. The use of a fixed voltage in this test, as opposed to a fixed overvoltage, fully demonstrates the uniformity of the SiPMs. By measuring with a fixed voltage, the effect of breakdown voltage variations from die to die, wafer to wafer, and lot to lot was included and variations in parameters such as film thickness, doping level, and gain were integrated into the final result shown. The measured distribution illustrates the tightly controlled uniformity characteristics of the B-Series SiPM manufacturing flow, with uniformity in the range ±10% for optical currents at these bias conditions.

While the uniformity of the SiPM sensors has been shown to be extremely good, it is useful to investigate the major cause of the uniformity variation in commercial SiPM. This can be accomplished by plotting the breakdown voltage of each SiPM

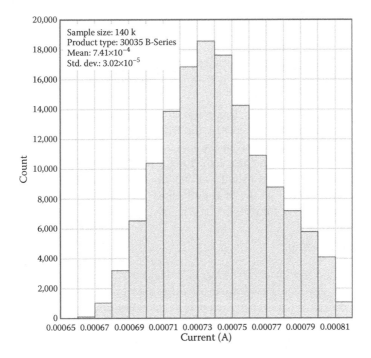

FIGURE 15.18 Optical uniformity at a constant voltage equal to 29.5 V.

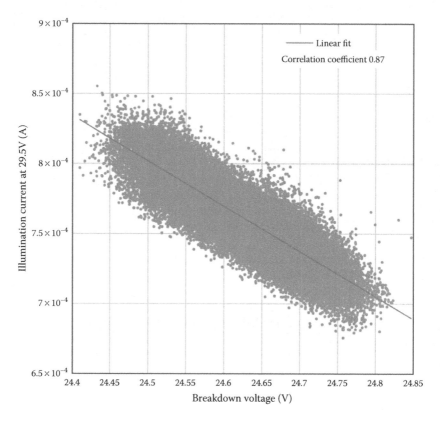

FIGURE 15.19 Correlation plot generated from wafer-level data showing illumination current at 29.5 V against the measured breakdown voltage from over 50k 30035 type SiPMs.

from a wafer lot to the optical uniformity that was measured at 29.5 V. This is shown in Figure 15.19 for over 50 k die from a single wafer lot. The die type is 30035 and all data are plotted in the figure. This shows a correlation coefficient of 0.87 to a linear fit of the data indicating a strong link between the breakdown voltage and the optical uniformity measured. This demonstrates the advantages of tight process control on the final uniformity that can be achieved using SiPMs manufactured in this manner.

15.6 PACKAGING

The packaging of the sensor plays a large role in the overall performance of the system and one that is not often considered in assessing sensor performance. Multiple factors such as wavelength, allowable operating temperature range, soldering conditions, and reliability vary depending on the package type and construction. In this section, the main types of packaging used for SiPMs that are suitable for radiation detectors will be described. A key criterion for many radiation detectors is the need for a large-area optical sensor to read out a large scintillator or a large scintillator array. This necessitates packages that have the ability to be closely tiled together to form large-area sensors with single readouts or sensor arrays with multiple SiPM readouts. The package types discussed in this section are all suitable for use in large-area arrays and can be pixelated or summed as required by the user.

There are three major packaging types that are suitable for use with scintillation crystals in a radiation detector. These package types are:

1. Poured epoxy-coated die in a ceramic or laminate carrier
2. Clear micro leadframe package (MLP)
3. Wafer-scale packaging through silicon via (TSV)

Each of these package types has positive and negative aspects, which should be understood during a product design process. To assist in understanding the variations between packages, the following criteria are critical to understand and can be used to evaluate package suitability for an application. The criteria used to decide on a package type are:

1. Array fill factor: The ratio of the optically active area to the peripheral area. The fill factor is important when creating arrays of SiPMs, which are typically available in 1, 3, and 6 mm^2 sizes. This parameter is not critical if a single sensor can be used alone in a detector. For applications requiring large-area coverage, a high package fill factor is critical to capture as many incident photons coming from the scintillation crystal.
2. Optical transmission: The transmission characteristics of a package vary depending on the package type chosen. Compounding this problem is that the optical transmission can be wavelength dependent. As shown in Section 15.8.1, the peak emission wavelength varies between crystal types. Therefore, it is important to make sure that the package does not limit the light collection from the scintillation material.
3. Operating conditions: The temperature that a package can physically operate in depends on the package construction and material.
4. Reliability: The various package types used for sensors have different failure mechanisms, which lead to different inherent reliability.
5. Service life: The service life or wear out time of the sensor will be determined primarily by the degradation in transmissivity of the package material.
6. Uniformity and reproducibility: The uniformity of the package surface is important to maintain optimum crystal coupling to an SiPM. The reproducibility of the package is important so that consistent results can be obtained in large-area array applications.
7. Cost: The cost of a package must be understood and is important for high-volume applications. Low cost is obviously desirable if the performance and operating-life criteria can be met.

The following subsections will review each of the three main package types with respect to the criteria that have been outlined here.

15.6.1 POURED EPOXY PACKAGES

This category covers a broad range of packaged parts that all use poured epoxy as the covering layer over the SiPM. The epoxy layer provides protection to the SiPM surface and prevents damage to wire bonds. For large-area arrays, the fabrication process entails attaching multiple die to a ceramic or PCB carrier using die attach epoxy. Following die attach and cure, the contact pads on the SiPM are wire bonded to the substrate. For the creation of scalable large-area arrays with uniform performance and wiring track-out, only packages in which the wire bonds are connected directly to the substrate beside the SiPM sensor will be discussed. Stitch bonding between die can be used to create higher-density arrays; however, this leads to variable SiPM track-out lengths on large arrays and is not scalable to large area. Stitch bonding is therefore not discussed here in detail. Following wire bonding, the SiPM and wire bonds are protected by a poured coat of clear epoxy, which is subsequently cured. An example of the completed package is shown in Figure 15.20.

This package has external dimensions of 57.4×57.4 mm^2 and contains 64 6×6 mm^2 SiPMs. The SiPMs are placed with a dead space of 0.80 mm between the die, which allows for wire-bond access. The fill factor that can be achieved with this type of package is shown in Figure 15.21.

Poured epoxy packaging has some advantages. The robust substrate enables manual handling during processing. This and the low capital investment required make it suitable for high-mix, low-volume manufacturing. The fill factor can also be advantageous because the common substrate enables a minimum distance between dies (generally 0.20 mm) for wire bonding. Additionally, this package type can be made without ferromagnetic material, which makes it a candidate for use in magnetic resonance (MR) imaging applications where ferromagnetic materials will degrade MR images.

The most significant disadvantages of this product are related to the epoxy itself. Generally, this is done by mixing two-part epoxy and pouring it onto the die surface. Poured epoxies will cure at room temperature though an elevated temperature is often used to decrease the hardening time. It is difficult to control this process and

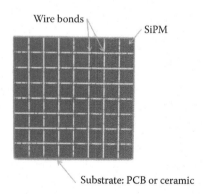

FIGURE 15.20 **(See color insert)** Epoxy-coated ceramic carrier SiPM array. This example is a 8×8 array of 6×6 mm^2 SiPMs mounted on a PCB substrate.

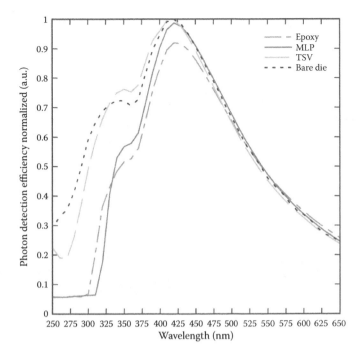

FIGURE 15.21 Comparison of PDE versus wavelength measured on bare die, poured epoxy, clear MLP, and TSV package types.

inconsistent epoxy material properties result, which affect product performance in the following ways:

1. Inconsistent PDE due to variations in optical transmission, particularly at short wavelengths.
2. Degradation in optical transmission during operating life ("yellowing" of the organic epoxy).
3. Nonuniformity of adhesion to the die surface. This will result in delamination during the operating life and consequent loss of transmission or failure of the wire-bond interconnects.
4. Nonuniformity and high values of mismatch between the thermal expansion coefficient of the epoxy and that of the substrate and wire bonds, leading to failure of the wire bonds.
5. Poured epoxy products require a package sidewall to contain the epoxy. Consequently, there is always dead space at the edges of the array. This adds dead space between arrays that are placed beside other arrays and limits the total fill factor.
6. Connection to the array must also be made to the back of the substrate. This increases the PCB traces and connections between the sensor and the subsequent read-out electronics, which can degrade the performance when compared with the other package technologies discussed here.

TABLE 15.2

Summary of Comparison Points for SiPM Array Package Options

Requirement	Poured Epoxy	Clear MLP	TSV
Array fill factor	Near optimum	Limited by peripheral leadframe	Optimum
Optical transmission	Poor	Good	Best
Operating conditions	0°C–40°C	–40°C to 85°C	–40°C to 95°C
Reliability	Manual processing: reduced reliability	Good	Best
Service life	Yellowing of potted epoxy is not well controlled	Good	Best
Uniformity and reproducibility	Poor	Good	Good
Cost	Low cost for laboratory use: not recommended for use in volume arrays	Lowest high-volume sensor cost: low array assembly costs	Highest high-volume sensor cost: low array assembly costs

A summary of the most important points related to the poured epoxy package are detailed in Table 15.2.

15.6.2 Clear Micro Leadframe Package

The clear MLP process entails attaching SiPMs to a large leadframe with die attach epoxy, wire bonding the SiPM contact pads to the leadframe terminals, molding in clear compound, and sawing into individual packages. Since the SiPMs are mounted onto a much larger leadframe, a fully automated process can be used in their manufacture. A drawing of a clear MLP SiPM is shown in Figure 15.22a. This figure details the SiPM placement on the leadframe, the wire-bond placement, and the clear molding compound that encases the SiPM. Examples of three different sizes of clear MLP-packaged SiPM are shown in Figure 15.22b. This figure shows

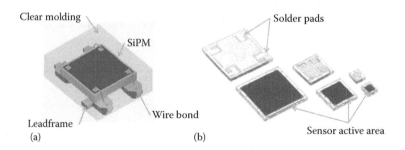

FIGURE 15.22 **(See color insert)** (a) Clear micro leadframe package drawing. (b) Images of the top and back side of 1×1, 3×3, and 6×6 mm^2 SiPM-packaged sensors.

6×6, 3×3, and 1×1 mm^2 SiPM sensors. The back contact of the sensor is shown at the top of Figure 15.22b.

The epoxy in a clear MLP is molded and cured at temperatures between 140°C and 160°C. The material and mold process is more controllable than the mixed and poured epoxy process described in Section 15.6.1. The MLP SiPMs have the following key benefits over poured epoxy SiPMs:

1. Better control of optical properties and thermal expansion/contraction of the material during operating life.
2. A flatter more uniform surface and optical interface allows better coupling to scintillation crystals.
3. Clear MLP SiPM sensors can be surface mounted directly to the read-out PCB and it is possible to minimize the PCB trace length and reduce the number of connectors in a large-area array.
4. A moisture sensitivity level (MSL) of 3 can be obtained which allows 168 h of ambient humidity exposure before the part needs to be soldered to a PCB.
5. Clear MLP sensors can be soldered using standard reflow solder conditions at 260°C.
6. MLP materials and processing are suited to high-volume manufacturing and provide the lowest cost product compared with any other package suitable for arrays.
7. Arrays can be fabricated with minimal dead space at the edge. This allows arrays to be tiled at a constant sensor pitch.

Examples of four different types of arrays, manufactured using clear MLP sensors, are shown in Figure 15.23. In all cases, these arrays were assembled on a PCB with

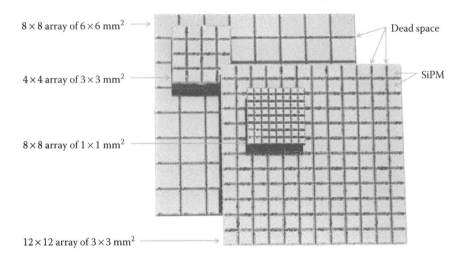

FIGURE 15.23 Example of four different large-area arrays of clear MLP arrays sensors mounted on a PCB. The dead space between the clear MLP sensors is 200 µm.

0.2 mm of dead space between sensors. The largest size array is the 8×8 array of 6×6 mm^2, which is 57.4 mm per edge. For this array, the sensors were placed on a 7.2 mm pitch. The pitch comprises a 6 mm sensor, 1 mm of package dead space, and 0.2 mm of dead space between packages. The 1×1 mm^2 sensor is packaged in a 1.5×1.8 mm^2 package enabling a high fill factor.

In addition to the performance and reliability benefits of the clear MLP, the flexibility to define the array size through the PCB layout and the ability to use high-volume PCB manufacturing, mean that the clear MLP is considered a superior package type to those utilizing poured epoxy.

15.6.3 THROUGH SILICON VIA

Both poured epoxy and clear MLP require packaging of the die after they have been processed. TSV packaging, however, is a wafer-level package and is made while the die are still on the wafer and have not yet been singulated. A true wafer-scale TSV process entails bonding a glass substrate to the top of the silicon wafer and back-grinding the silicon to a thin layer. Vias are etched in the silicon and metal is deposited to make contact with the terminals in the die. Solder bumps are attached to the deposited metal on the back side of the sensor. The wafer is then sawn into individual die and can be placed onto a PCB substrate using a similar process to that used to manufacture clear MLP arrays. An example of a TSV-packaged SiPM is shown in Figure 15.24a. The corresponding cross section of an SiPM TSV sensor is shown in Figure 15.24b.

The performance of a TSV-packaged SiPM brings the PDE closer to the bare-die PDE curve, as shown in Figure 15.21. This is because the TSV glass is more transmissive at short wavelengths than the organic materials used in poured epoxy and clear MLP products. The additional benefits of a TSV-packaged SiPM are as follows:

- Transmission in glass is stable during operating life. The yellowing seen in organic compounds does not occur in glass and TSV products have the longest operating lifetime.
- There are no wire bonds so each TSV component will be the same size as the sensor. This optimizes the fill factor allowing the highest density arrays to be produced. This high fill factor is shown in Figure 15.25.

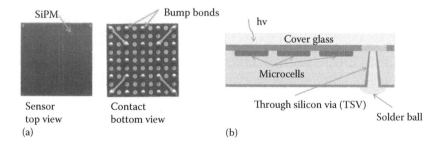

FIGURE 15.24 (a) TSV SiPM example showing top sensor side and bottom contact side. (b) TSV cross-section example.

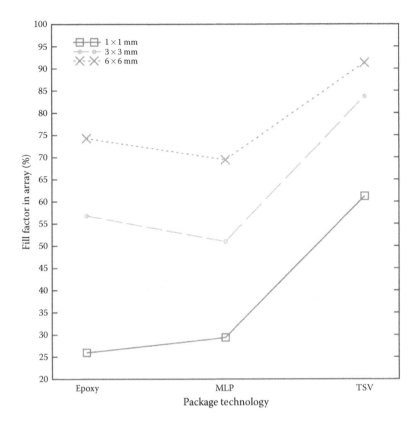

FIGURE 15.25 Comparison of the fill factor between poured epoxy, clear MLP, and TSV package types when placed in an array.

- The failure mechanisms associated with wire bonds and epoxy are absent. Therefore, TSV products will have the highest reliability of any SiPM package.
- TSV packages are MSL 1 compatible, reducing costs at PCB contract manufacturers.
- TSV processing is inherently conducive to tight control and high volume.
- TSV can be placed into arrays with low dead space between sensors. Due to excellent wafer saw tolerances, the dead space between TSV die in arrays can be 0.1 mm. This improves the overall TSV package fill factor compared with clear MLP.

The key disadvantage of the TSV package is a slightly increased cost compared with clear MLP. It is up to the user to decide if the advantages outweigh a sensor cost increase in their product design.

15.6.4 SUMMARY INFORMATION

The following figures and table summarize the various packages discussed in this section. Figure 15.21 compares the PDE measured wavelength for four differently packaged 3×3 mm^2 SiPM sensors. All data have been normalized to allow easy comparison between package types. To provide a baseline of the core performance, the bare-die curve was obtained with an SiPM mounted directly onto the substrate with no poured epoxy over the surface area. It therefore represents the PDE of the sensor. The epoxy curve represents typical PDE measured with poured epoxy products, which is characterized by a reduced peak PDE due to transmission losses in the epoxy. The clear MLP package has better peak PDE than the poured epoxy but does have a short wavelength cut-off at 320 nm due to the clear molding compound. The TSV package curve demonstrates the best performance at the peak wavelength and a greatly improved performance in the ultraviolet (UV) range. The TSV package most closely approximates the bare-die curve and represents the highest performance package for the highest PDE.

A comparison of the fill factors of various package types was performed using the detailed knowledge of each SiPM sensor size; 1×1, 3×3, and 6×6 mm^2 combined with typical design rules for contract manufacturer placement of sensors into arrays. In all cases, the TSV package provides the highest package fill factor possible when creating large-area arrays due to the low dead space. A comparison of the three package technologies is summarized in Table 15.3.

15.7 SiPM RELIABILITY

In this section, the reliability stress assessment completed on SiPMs will be discussed. Additionally, the justification for the adoption of industry standard reliability stress assessment flows from the Joint Electron Device Engineering Council (JEDEC) will be reviewed.

15.7.1 Issues of SiPM Reliability Assessment

SiPM sensor reliability is a key requirement for high-volume applications. At present, there is a lack of industry-accepted standards for reliability assessment of SiPM sensors. This is believed to be due to the varied and complex manufacturing processes required for optical sensors. This is contrasted with the integrated circuits industry where there are strong standards that are widely adopted and used for assessing the reliability of products. As the sensors discussed in this chapter are developed using CMOS, high-volume fabrication facilities, it is applicable and desirable to follow the standards adopted by the integrated circuits industry. A major publisher of standards used by the integrated circuits industry is JEDEC [20], which freely publishes test standards that can be adopted by manufacturers to assess product reliability. For the reliability assessment of SiPMs, the JEDEC standards have been adopted.

15.7.2 Reliability Assessment of SiPMs

All stress and test steps were carried out as per the JEDEC standard conditions. The JEDEC standard for the main reliability test procedure can be obtained from the JEDEC [19]. To assess the silicon reliability, SensL used a combination of the standard commercial packages available, including SensL clear MLP, and metal can

TABLE 15.3
Reliability Stress Program for SiPM

Test	Objective	Required Condition	Lot Size	Duration/ Acceptance	Status
High-temperature operating life	Junction stability	Ambient temperature = 125°C; bias = 30 V	3 lots of 77 units	1000 h/no change in any parameter >10%	100% pass
High-temperature operating life	Junction stability over longer stress time	Ambient temperature = 125°C; bias = 30 V	256 units	2000 h/no change in any parameter >10%	100% pass
High-temperature operating life	Package stress to examine chemical stability (e.g., discoloration of package)	Ambient temperature = 85°C; bias = 27 V	1 lot of 77 units	1000 h/no change in any parameter >10%	100% pass
Unbiased highly accelerated stress	Package stress to examine delamination, transmission loss, and wire bond failure	110°C, 85% relative humidity; passive no bias	3 lots of 25 units	264 h/no change in any parameter >10%; no critical package delamination	100% pass
Temperature cycling	Package stress to examine delamination, transmission loss, and wire bond failure	-40°C to 85°C cycle, 15 s transition, 15 min dwell time; passive no bias	3 lots of 77 units	500 cycles/no change in any parameter >10%; no critical package delamination	100% pass
High-temperature storage test	Package stress to examine chemical stability (e.g., discoloration of package)	504 h at 125°C; passive no bias	3 lots of 25 units	504 h/no change in any parameter >10%	100% pass

(TO5 style) packages with sealed lids containing an optical window. The use of TO5 style packages was a requirement for the high-temperature stress tests designed to test the reliability of the silicon, independent of the package material. For the assessments, packaged die were obtained typically from three separate silicon fabrication lots and were packaged as required. All tests were made in accordance with the relevant JEDEC standards referenced in Table 15.3. Of significant note from Table 15.3 are the high temperature operating-life tests that were completed to 1000 h with an additional higher endurance stress to 2000 h. This represents the longest high-temperature stress to date on a statistically relevant sample set of SiPMs that the authors have seen reported.

All SensL SiPM sensors passed the tests. An example of the PDE measured before and after a high-temperature operating-life stress test is shown in Figure 15.26. No change in performance was observed.

As can be seen, further stress data were taken to assess product reliability, including temperature cycling, moisture and humidity stressing, and high-temperature storage. The excellent results demonstrate that the SiPM manufacturing process is robust and enables the production of reliable, high-quality sensor components.

For high-volume applications, the reliability of the product is a key concern. It is believed that the adoption of the reliability assessment standards used by integrated circuit manufacturers provides the most comprehensive method to assess SiPM product reliability. The standard methods outlined by the JEDEC have been applied to B-Series SiPMs from SensL, with all parts passing standard reliability assessment flows. The adoption of integrated circuit industry standard reliability test flows is considered of vital importance for the adoption of the SiPM in high-volume applications.

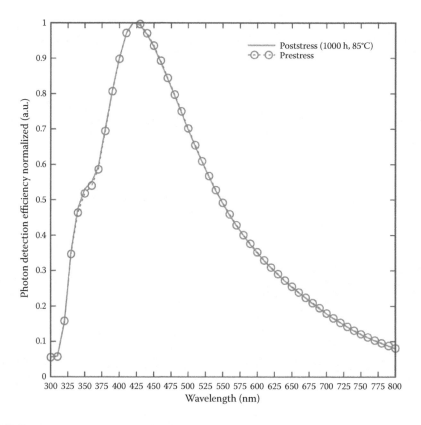

FIGURE 15.26 Measurement of PDE pre- and poststress. The stress condition is 85°C for 1000 h.

15.8 SCINTILLATION CRYSTAL READOUT

The focus of this section is on the use of scintillation crystals for the detection of gamma rays. Such crystals are used for a variety of applications related to medical imaging and radiation detection. Both markets use a combination of a scintillator and an optical sensor to detect gamma rays.

15.8.1 GAMMA-RAY DETECTORS

A gamma-ray detector must detect, amplify, and output an electrical signal that accurately conveys the gamma-ray energy and in some cases the photon arrival time. The most common type of gamma-ray detector relies on a combination of scintillation crystal and optical sensor, as shown in Figure 15.1. A variety of crystals are available that can be used to convert the gamma radiation into optical photons and the work of Dorenbos [21] shows a useful time line of the development and performance limits for various crystal types. Some of the most popular scintillators, such as $LaBr_3$, LYSO, BGO, and CsI(Tl), which are relevant to PET and radiation detection will be considered in this chapter. These crystals have a variety of primary characteristics such as light yield, density, emission wavelength, and decay time, which directly impact the overall performance of the crystal and optical sensor assembly. Typically, to optimize the performance of a gamma-ray detector, the goal is to maximize the number of optical photons that are detected from the scintillator while minimizing the amount of noise coming from the optical sensor.

Figure 15.5 shows the normalized light yield for a variety of standard scintillation crystals that are used for radiation detection. The data for Figure 15.5 were obtained directly from the datasheets available for each crystal type from Saint-Gobain [22] and are in agreement with published data found in the extensive scintillation crystal database made available online by Derenzo [23]. The photon emission has been digitized from the manufacturer's datasheets and normalized according to the peak light yield, which is expressed in photons per kiloelectron volt. A high light yield and an emission spectrum matched to the sensitivity of the SiPM will typically improve detection performance.

Another parameter that plays an important role in the performance of radiation detectors is the emission decay time. Typical decay times for the crystals discussed in this chapter are shown in Table 15.4. The decay time of the crystal determines the amount of time required for a measurement and the amount of noise that will be integrated during the signal acquisition. Longer measurement times lead to increasing amounts of noise included in the recorded signal.

The refractive index of the crystal is also important as this will determine the light coupling between the crystal and the sensor. Mismatched refractive indices lead to Fresnel losses of photons transitioning from the scintillation crystal to the optical sensor. Index-matching coupling material should always be used between an SiPM sensor and a scintillation crystal for optimum light collection.

The scintillation crystal emission and the decay time of the scintillator combined with the sensors' sensitivity and noise play a large role in optimizing the radiation detector performance. Additional scintillation crystal parameters such as density,

TABLE 15.4

Typical Crystal Information from Saint-Gobain and Energy Resolution Obtained in the Literature from Dorenbos

Scintillation Crystal	Emission Peak λ (nm)	Light Yield (photons/ keV)	1/e Decay Time (ns)	Refractive Index at λ (peak)	Energy Resolution (%) at 662 keV
LaBr₃:Ce	380	63	16	1.9	~2.6
LYSO:Ce	420	32	41	1.81	~8
BGO	480	10	300	2.15	~10
CsI(Tl)	550	54	1000	1.79	~5.5

Source: Saint-Gobain, http://www.crystals.saint-gobain.com/; Dorenbos, P., *IEEE Trans. Nucl. Sci.*, 57, 1162, 2010.

thermal expansion coefficients, hygroscopic nature, and cost will also play a role in determining the suitability of a detector for a given application but these are outside the scope of this chapter on SiPM sensors.

15.8.2 ENERGY RESOLUTION AND COINCIDENCE TIMING MEASUREMENTS

Applications involving scintillation materials fall into two main application areas: those that require fast timing and those for spectroscopy requiring optimum energy resolution. These two application areas have slightly diverging requirements in the scintillating crystals used and in the optimal operating voltage settings for the sensor. For fast-timing applications, a high light yield scintillator with a fast rise and short decay time is required. For applications requiring the best (lowest) energy resolution, the nonproportionality of the crystal is a driving factor to obtain the lowest energy resolution. Energy resolution values that have been obtained in the literature are summarized in Table 15.4.

For the best energy resolution measurements, the maximum of the sensor PDE should be matched as closely as possible to the peak wavelength of the scintillator. The energy resolution percentage (full width at half maximum [FWHM]) is related to both the sensor performance as well as the crystal. A good review of this is found in the work by Dorenbos [21] and Grodzicka [24], which determines the energy resolution through the following equation:

$$R^2 = R_{in}^2 + R_{np}^2 + R_{stat}^2, \tag{15.12}$$

where:

R_{in} is related to the crystal inhomogeneity from crystal quality, wrapping variation, and crystal finish differences

R_{np} is due to the nonproportional response of the scintillator

R_{stat} is related to the photon detection statistics and the excess noise of the SiPM

R_{stat} is given by the following equation [25], which has been modified to include the dark count rate in Equation 15.13:

$$R_{\text{stat}} = 2.35 * \frac{\sqrt{\text{ENF} * N_{dp} + N_{dc}}}{N_{dp}}. \tag{15.13}$$

In this equation, N_{dp} is the number of detected photons and N_{dc} is the number of dark counts during the measurement integration time. ENF is the excess noise factor. An ENF of 1 is a perfect detector while an increasing ENF degrades R_{stat}. The ENF for the SiPM is dominated by cross talk and afterpulsing. These unwanted effects increase the variability in the output of the SiPM. When used in a detector, increasing the overvoltage will increase PDE and therefore the number of detected photons, N_{dp}; however, cross talk and afterpulsing will also increase, which can degrade the energy resolution obtained. For typical PET applications looking at signal levels from 511 keV gamma rays, R_{stat} is the predominant factor while R_{dcr} dominates measurements at lower energies.

For coincidence timing measurements the scintillating crystal emission should be short and the PDE and rise time of the sensor should be high. Cross talk, afterpulsing, and dark count rates are considered secondary factors until high overvoltages are reached. This operating point is best determined experimentally as it depends on the scintillator, sensor, and read-out method. Crystals for use in fast coincidence timing measurements emit light in the blue to UV portion of the spectrum [21]. The SiPM sensors shown in this chapter have been optimized to allow the detection of these short wavelength emission crystals.

15.8.3 SiPM Scintillation Crystal Performance

To demonstrate the performance of the SensL SiPM in an application environment, the sensors were coupled with cerium-doped lutetium yttrium orthosilicate (LYSO:Ce) scintillation crystals. This is similar to the system configuration in a PET time-of-flight (ToF) system. When the SiPM sensor is combined with the scintillation crystal, it forms a gamma-ray detector. It is desirable to demonstrate a minimum energy resolution and a minimum coincidence resolving time (CRT) for the pair of 511 keV gamma rays that are produced in electron-positron annihilation. LYSO is well matched to the sensor PDE presented in Figure 15.8 as the output peak at 420 nm matches the photon detection peak of the sensor. The measurement setup for energy resolution and CRT is shown in Figure 15.27. This shows the crystals in "head-on" configuration, which more closely mimics the way that crystals are positioned in a standard PET system. The crystals are directly mounted on the SiPM using coupling grease (BC-630) [22] and amplifiers are used to amplify the output voltage level to an appropriate range for the Wavecatcher USB digitizer [26]. A variety of amplifier combinations can be chosen to suit most crystal testing needs [27].

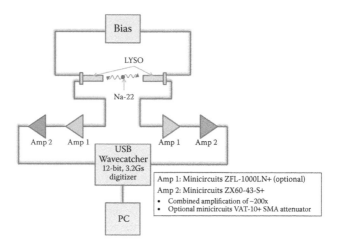

FIGURE 15.27 Example of the energy resolution and coincidence resolving time (CRT) experiment setup used in this work.

15.8.4 COINCIDENCE RESOLVING TIME

CRT is an important PET ToF system measurement parameter. A low CRT allows a more accurate determination of the emission source position through triangulation. For this work, CRT was evaluated by determining the arrival time of coincident 511 keV photon pairs at a corresponding pair of scintillation-based detectors. For this measurement, a ^{22}Na 511 keV source was placed between two facing $3 \times 3 \times 20$ mm^3 LYSO coupled to SiPM sensors, positioned head-on, and the resulting electrical signal from the detectors was amplified and recorded with a high-speed digitizer (USB Wavecatcher 12-bit, 3.2 GS/s). In analyzing the resulting pairs of signal traces, energy filtering was performed at $\pm 9\%$ around the peak value selecting signals that the charge integral matched with the peak in the charge spectrum attributed to 511 keV; time walk correction was performed removing the correlation between the energy difference and the time difference of the scintillator pairs. In addition, the optimal choice of the leading-edge threshold to timestamp each of the electrical signals further optimized the CRT value. Figure 15.28 shows typical CRT histograms measured on the fast output of the 30035 SiPM. Similar results have been shown in the literature [28,29].

Measurements on standard anode–cathode signals were also performed and a value of 306 ± 3 ps was obtained. The fast output provides superior results due to its fast signal rise time, low output capacitance, and low parasitic impedance.

15.8.5 ENERGY RESOLUTION

In energy resolution measurements, the detector principle is that of a proportional counter, in that the number of scintillation light photons are proportional to the energy of the gamma photon penetrating the crystal. Integrating the electric signal from the sensor during each scintillation event and then binning the result yields a spectrum of counts versus an observable proportional to the gamma photon energy.

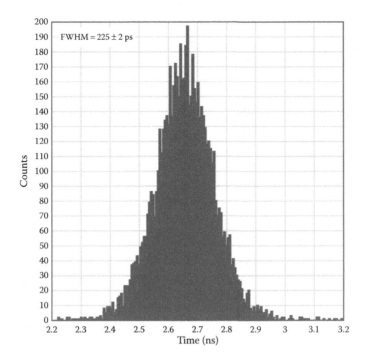

FIGURE 15.28 Coincidence resolving time measured for sample pairs of a 30035 SiPM.

To determine the energy resolution for a typical PET setup, a $3 \times 3 \times 20$ mm^3 LYSO crystal was coupled to a 30035 SiPM sensor using BC-630 silicone optical grease from Saint Gobain. The SiPM was biased at 5 V over the breakdown voltage, and the signal from the SiPM was amplified and traces were recorded using a high-speed digitizer.

The energy resolution spectrum for a 30035 SiPM is shown in Figure 15.29. This measurement was made from the fast output signals and was similar to measurements taken on the standard anode–cathode output. When coupled to a relatively bright and fast crystal such as LYSO, the signal from the fast output is sufficient to accurately represent the signal pulse and an accurate energy resolution plot can be obtained. For this measurement, the energy resolution was limited by the crystal proportionality, statistical uncertainty in photon counting by the sensor, and electrical noise in the sensor and associated electronics.

Additional work on measurements using CsI(Tl) can be found in the work of Becker [30] who details a compact and portable gamma spectrometer measurement system that achieves 5.66% energy resolution.

15.9 SUMMARY

This chapter has demonstrated the operation of SiPMs developed for high-volume CMOS fabrication. The review started from the basic physics of the operation of SiPMs and linked various key operating parameters such as the breakdown voltage and the

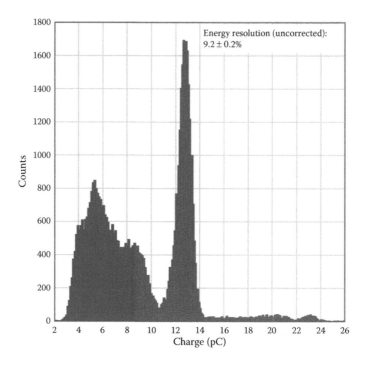

FIGURE 15.29 Charge spectrum full width half maximum measured on fast output for a 30035 SiPM.

wavelength sensitivity to the physical structure of the sensor. The main SiPM parameters were reviewed including PDE, dark count rate, gain, cross talk, afterpulsing, and dynamic range. The key highlights show that a PDE of 41% and low dark count rates of <100 kHz can be obtained with low afterpulsing and cross-talk rates. The manufacturability and high-volume production of SiPMs were demonstrated through wafer-level electrical and optical testing of over 204k SiPM sensors demonstrating excellent breakdown voltage, dark current, and optical uniformity. A review of the packaging requirements and suitable packages for large-area arrays was presented demonstrating the key parameter trade-offs such as PDE, fill factor, and cost. It was shown that clear MLP- and TSV-packaged sensors have clear advantages for high-volume array applications. For the reliability assessment of the SiPM, industry standard integrated circuit testing procedures were adopted. The tests are believed to provide a comprehensive testing regime for SiPMs and should be used for SiPM product qualification. The use of SiPMs for scintillation crystal readout was reviewed theoretically and a clear demonstration of the high performance that can be obtained was shown with LYSO:Ce. CRT of 225 ps and energy resolutions of 9.2% were obtained experimentally.

REFERENCES

1. R. H. Haitz, Model for the electrical behavior of microplasma, *J. Appl. Phys.*, 35(5), 1370–1376, 1964.

2. R. H. Haitz, Microplasma interaction in silicon p-n junctions, *Solid-State Electron*, 7, 439–444, 1964.

3. R. H. Haitz, Mechanisms contributing to the noise pulse rate of avalanche diodes, *J. Appl. Phys.*, 36,(10), 3123–3131, 1965.

4. R. H. Haitz, Studies on optical coupling between silicon p-n junctions, *Solid-State Electron*, 8, 417–425, 1965.

5. W. G. Oldham, et al., Triggering phenomena in avalanche diodes, *IEEE Trans. Electron Devices*, 19(9), 1056–1060, 1972.

6. R. J. McIntyre, Multiplication noise in uniform avalanche diodes, *IEEE Trans. Electron Devices*, ED-13, 164–168, 1966.

7. R. J. McIntyre, On the avalanche initiation probability of avalanche diodes above the breakdown voltage, *IEEE Trans. Electron Devices*, 20(7), 637–641, 1973.

8. S. Cova, et al., Towards picosecond resolution with single-photon avalanche diodes, *Rev. Sci. Instrum.*, 52(3), 408–412, 1981.

9. R. G. W. Brown, et al., Characterization of silicon avalanche photodiodes for photon correlation measurements. 1: Passive quenching, *Appl. Opt.*, 25(22), 4122–4126, 1986.

10. R. G. W. Brown, et al., Characterization of silicon avalanche photodiodes for photon correlation measurements. 2: Active quenching, *Appl. Opt.*, 26(12), 2383–2389, 1987.

11. V. Saveliev, et al., Silicon avalanche photodiodes on the base of metal-resistor-semiconductor (MRS) structures, *Nucl. Instrum. Methods Phys. Res. A*, 442, 223–229, 2000.

12. D. Bisello, et al., Metal-resistive layer-silicon (MRS) avalanche detectors with negative feedback, *Nucl. Instrum. Methods Phys. Res. A*, 360, 83–86, 1995.

13. V. Golovin, et al., Limited Geiger-mode microcell silicon photodiode: New results, *Nucl. Instrum. Methods. Phys. Res. A*, 442, 187–192, 2000.

14. A. B. Kravchenko, et al., A linear array of avalanche MIS photodetectors, *Sov. J. Quantum Electron.*, 8(11), 1399–1400, 1978.

15. A. B. Kravchenko, et al., Feasibility of construction of a pulsed avalanche photodetector based on an MIS structure with stable internal amplification, *Sov. J. Quantum Electron.*, 9(9), 1086–1089, 1978.

16. M. Polyanskiy, Silicon refractive index, Refractive Index Database, http://refractiveindex. info, accessed June 2014.

17. J. T. M. de Haas, et al., Advances in yield calibration of scintillators, *IEEE Trans. Nucl. Sci.*, 55, 1086–1092, 2008.

18. SensL B-Series Datasheet, Fast, Blue-sensitive silicon photomultiplier sensors, http:// www.sensl.com/downloads/ds/DS-MicroBseries.pdf, accessed June 2014.

19. P. Eckert, et al., Characterisation studies of silicon photomultipliers, *Nucl. Instrum. Methods Phys. Res. A*, 620, 217–226, 2010.

20. JEDEC Standards Organization, Global Standards for the Microelectronics Industry. http://www.jedec.org/, accessed June 2014.

21. P. Dorenbos, Fundamental limitations in the performance of Ce^{3+}, Pr^{3+}, and Eu^{2+} activated scintillators, *IEEE Trans. Nucl. Sci.*, 57, 1162–1167, 2010.

22. Saint-Gobain, http://www.crystals.saint-gobain.com/.

23. S. Derenzo, Scintillation Properties, Lawrence Berkeley National Laboratory, http:// scintillator.lbl.gov/, accessed June 2014.

24. M. Grodzicka, et al., MPPC array in the readout of CsI:Tl, LSO:Ce:Ca, LaBr3:Ce, and BGO scintillators, *IEEE Trans. Nucl. Sci.*, 59(6), 3294–3303, 2012.

25. M. Moszynski, et al., A comparative study of silicon drift detectors with photomultipliers, avalanche photodiodes and PIN photodiodes in gamma spectrometry with LaBr3 crystals, *IEEE Trans. Nucl. Sci.*, 2009.

26. Wavecatcher USB Digitizer, http://www.heptech.org/Phocea/file.php?file=Ast/140/fiche_LAL_Wave_Catcher.pdf, accessed June 2014.

27. Mini Circuits website, RF Electronics Engineering, http://www.minicircuits.com, accessed June 2014.

28. S. Dolinsky, et al., Timing resolution performance comparison for fast and standard outputs of SensL SiPM, IEEE NSS/MIC]2013.

29. J. Y. Yeom, et al., Fast timing silicon photomultipliers for scintillation detectors, *IEEE Photon. Technol. Lett.*, 25, 1309–1312, 2013.

30. E. Becker, The MiniSpec: A low-cost, compact, FPGA-based gamma spectrometer for mobile applications, thesis, http://ir.library.oregonstate.edu/xmlui/handle/1957/42365, accessed June 2014.

Index